T0135689

A Model-Independent Security Architecture for Distributed Heterogeneous Systems

Vom Fachbereich Informatik
der Technischen Universität Darmstadt
genehmigte

Dissertation

zur Erlangung des akademischen Grades
Doktor-Ingenieurs (Dr.-Ing.)

von
Stephen D. Wolthusen
aus Bürstadt

Referenten:	Prof. Dr. Johannes Buchmann PD Dr. Werner Stephan
Tag der Einreichung:	26. August 2003
Tag der mündlichen Prüfung:	11. November 2003

Darmstädter Dissertationen 2003
D 17

Darmstädter Dissertationen
D 17

Bibliografische Information Der Deutschen Bibliothek

Die Deutsche Bibliothek verzeichnet diese Publikation in der Deutschen Nationalbibliografie; detaillierte bibliografische Daten sind im Internet über http://dnb.ddb.de abrufbar.

ISBN 3-8325-0416-8

Logos Verlag Berlin
Comeniushof, Gubener Str. 47,
10243 Berlin
Tel.: +49 030 42 85 10 90
Fax: +49 030 42 85 10 92
INTERNET: http://www.logos-verlag.de

CONTENTS

"Fortune," said Don Quixote to his squire, as soon as he had seen them, "is arranging matters for us better than we could have hoped. Look there, friend Sancho Panza, where thirty or more monstrous giants rise up, all of whom I mean to engage in battle and slay, and with whose spoils we shall begin to make our fortunes. For this is righteous warfare, and it is God's good service to sweep so evil a breed from off the face of the earth."

(Miguel de Cervantes Saavedra: *Don Quixote [165]*)

Preface

The threats partially addressed in the following are indeed more real than those faced by the Knight of the Sad Countenance. However, given the history of adoption of security mechanisms from research to general purpose computer systems, the choice of frontispiece seems somewhat appropriate.
This dissertation describes a security architecture that is neutral with regard to the policies and security models to be enforced and sufficient to express any such policy equally enforceable on multiple operating system platforms conforming to an abstract model that is representable by an interpretation of a formal theory. The latter assertion is shown by describing retrofitting elements requisite for implementing the policies on a commercial system without resorting to either source code access or modification.
Since particularly the use of automated reasoning mechanisms can yield results that are not immediately obvious and, moreover, the assurance of the security architecture and hence the ability to have the architecture evaluated against a CCITSE EAL 5 assurance level (assuming a suitable security policy instantiation for the purposes of defining a security target) was considered of paramount importance, mature formal methods were used in the specification, modeling, and implementation of the security architecture and its implementation; excerpts of this model are provided in this dissertation.
The implementation mechanisms described here focus firstly on operating systems and secondly on a single platform of which, again, excerpts are discussed.

Acknowledgments

De nihilo nihil. Assembling this dissertation has been an enterprise that has drawn — directly or indirectly, and for several years — on the time, talents, and goodwill of many individuals and institutions. Most of all I am indebted beyond measure to my parents for incessant encouragement and support.

I would like to thank my advisors, Prof. Dr. Johannes Buchmann, PD Dr. Werner Stephan, and Prof. Dr. mult. José Luis Encarnação for their time, patience, and advice in bringing this to fruition. I would also like to thank the members of the reading committee, Prof. Dr. Sorin Huss and Prof. Dr. Tsuyoshi Takagi.

I am deeply indebted to the talent and dedication of the researchers and staff on whose support I was able to draw in turning concepts to operational systems. These were — in alphabetical order — Silviu Burtescu, Hartmut Chodura, Paul Cook, Thomas Corell, Henning Daum, Keith Dunwoody, Martin Einhoff, Thilo Frotscher, Wolfgang Funk, Frank Graf, Ralf Hayn, Ingo Jankowski, Urs Krafzig, Lucie Langer, Thomas Leyer, Zhi Liu, Elena Maronova, Emanuel Mandache, Timm Müller, Collin Mulliner, Michael Nagel, Alexander Nouak, Jan Peters, Alex Pleiner, Claudia Press née Justin, Ero Rademer, Nelly Salehi, Eduard Singer, Andreas Truffel, Carsten Wald, Christian Weyer, and Armin Zeidler.

For material support that enabled me to pursue part of my research, I am grateful to the Mitsubishi Corporation of Tokyo, Japan — particularly to Mr. Teruyuki Nakazawa, Mr. Osamu Kanzaki, and Mr. Makoto Saito.

Other people have contributed in countless ways to the rich and invigorating working environment of the Security Technology department at Fraunhofer-IGD, particularly the department head Dr. Christoph Busch who as well as his predecessor in the department, Dr. Eckhard Koch.

Abstract

This dissertation shows that information assurance properties can be both specified within a formal model of systems to which the properties are applied using the same formal theory for modeling, specification, and reasoning and enforced in such a way that enforcement is performed consistently across multiple, heterogeneous nodes and organizational domains while retaining the semantics of the formal model.

The properties, referred to as security policies, can be specifically enforced at the operating system level and are constructed in such a way that automated reasoning mechanisms derive lower abstraction layer properties from higher semantic levels specified by administrative personnel based on the formal abstract model and interpretations thereof. Moreover, operations to be performed are permitted based on proofs obtained within the formal model while required operations are also derived within the model.

To permit the consistent enforcement of an arbitrarily large set of security policies and scalability across large organizations and networks, externally controlled reference monitors and external reference monitors are introduced which control layered enforcement mechanisms that can be implemented both in systems constructed ab initio and as an add-on to existing, particularly commercially available operating systems even if no source code is available for modification to ease the transition to secure systems while permitting mission fulfillment based on legacy systems. These aspects are demonstrated using the Microsoft Windows 2000 operating system as an example.

Enforcement mechanisms are described using the reference interpretation including modification and augmentation of file system and network protocol stack behavior along with the implicit benefits derived from the use of these enforcement mechanisms. Specifically, the implementation of dynamic distributed network firewalling and intrusion detection as well as multilevel security capabilities under the control of consistent policies are discussed. For this purpose the suitability of the framework for modeling multisensor data fusion as applied to intrusion detection is discussed.
To demonstrate the capabilities of the layered enforcement system for application-specific domains, the use of visible and invisible labeling mechanisms for hard copy output is furthermore discussed.

CHAPTER

ONE

Introduction

The assertion that information systems of non-trivial extent are likely to consist of multiple systems or nodes – each with heterogeneous operating systems, application programs, and other pertinent characteristics – that are increasingly interconnected, again possibly by multiple heterogeneous networking or other connection mechanisms, appears readily justifiable.
Moreover, defense and intelligence systems require that operations of such information systems meet certain positive and negative constraints [518], a requirement that is even more pronounced in civilian application areas as requirements can derive from multiple sources such as regulatory compliance (from multiple domains and jurisdictions), corporate policy, or contractual obligations.
In addition, coalition environments or similar cooperation among entities in civilian environments may require that multiple sets of such requirements are fulfilled simultaneously, protecting assets and information as well as ensuring that each coalition element can fulfill its own mission without being jeopardized or compromised by other elements.

It is imperative that such requirements are enforced rigorously and at all times by the information system itself to ensure adequate levels of assurance. The requirements themselves, typically called (security) policies or security models, are frequently formulated in natural language. This leads to several problems addressed in this dissertation. First, there may exist contradictions within a single security policy or multiple, concurrently applicable policies that are not readily apparent given a natural language formulation. Similar issues exist with regard to the completeness of a given policy.

Second, there may exist a potentially large semantic gap between the natural language formulation and a technical realization of said formulation.
This not only implies that the verification and validation of the assertion that a given technical realization indeed conforms to the natural language formulation is difficult if not impossible, the formalism chosen may also be of inadequate expressiveness and therefore either cannot capture all requirements and falls short of the natural language formulation, or it may be overly restrictive and therefore impose additional, unnecessary burdens on the operation of the information system.

Third, policies may not be enforced in such a way as to yield equivalent results regardless of the technical realization (e.g. operating system, database, or application program) within an information system these are applied to.

Therefore, in this dissertation a security architecture for distributed heterogeneous systems is described, which permits both the specification and enforcement of confidentiality, authentication, integrity, and non-repudiation aspects of information assurance (cf. section 1.1) within an abstract model.
These are enforced in such a way as to be semantically equivalent regardless of the individual information system component.
The abstract model itself is to provide a basis for formulating policies at a sufficiently high semantic level that the verificiation and validation of the correspondence between a natural language formulation and the formalism provided by the model can be performed manually while automated reasoning mechanisms can derive valid conclusions as to the requirement or permissibility of operations at lower abstraction levels.

1.1 Information Assurance

The term Information Assurance (IA) has been used increasingly since the mid-1990s, primarily in defense-related areas to indicate a focus extending beyond the confidentiality aspect traditionally associated with the term security or more specifically information security. The working definition of the IEEE Task Force on Information Assurance (derived from the National Information Systems Security Glossary [502], this definition is also reflected in other publications [738, 519]) is:

Definition 1.1 ***Information Assurance*** *(IA) are Information Operations (IO) that protect and defend information and information systems by ensuring their*

- *confidentiality,*
- *authentication,*
- *integrity,*
- *availability, and*
- *non-repudiation.*

This includes providing for restoration of information systems by incorporating

- *protection,*
- *detection, and*
- *reaction capabilities.*

This requires two supporting definitions, namely

Definition 1.2 ***Information Operations*** *(IO) are actions taken to affect adversary information and Information Systems while defending one's own information and Information Systems.*

and

Definition 1.3 ***Information Systems*** *(IS) are the entire infrastructure, organization, personnel, and components for the collection, processing, storage, transmission, display, dissemination, and disposition of information.*

With the exception of offensive information operations that do not apply to the civilian sector, this definition is generally applicable and certainly of a daunting scope since it subsumes already problematic areas such as software assurance. The main issue to consider in the ramifications of IA are the frequently ill-defined, often globally dispersed components of the infrastructure and their interdependencies.

1.2 Securing Systems

There exists a certain dichotomy between the need for security frequently stated by users of information systems ranging from individual consumers to sensitive defense areas and the decisions taken when information assurance interferes with various other objectives. As is made evident by the definitions of section 1.1, at least part of the problem can be attributed to the very fact that the aspects requiring consideration for ensuring the assurance of an information systems are covering such a broad spectrum, making even the definitions regarding what precisely constitutes a secure or high assurance system subject to debate.

Some of the differences between the information assurance objectives stated and actually pursued may also be explained by observing that the prerequisite definitions of an organization's mission and objectives for information systems are not necessarily in place or are not updated with sufficiently small delays so as to reflect the actual mission (as is e.g. recommended by the relevant ISO standard [329]). Most likely the cause for this is that information assurance is considered as an afterthought to an information infrastructure and operations that have been created and extended ad hoc over a period of time.

As a corollary to the inexistence of such mission and objective definitions, the creation of risk analyses is severely hampered. Without the ability to clearly evaluate objectives against risks that are preferably expressed in quantitative form, the issue of justifying any imposition of restrictions on performing operations using information systems or even disallowing operations based on the determination that the risk resulting from such an operation is not commensurate with the benefit derived from it regardless of ameliorating mechanisms used becomes difficult at best.

Similar outcomes in conflicts between objectives can also be observed in both design and realization of existing systems, particularly systems which are deployed widely.

Given that even apparently innocuous systems may constitute fault causes for critical systems and even national infrastructure [445], the call for improvements in information assurance appears uncontroversial, is reflected widely [749] and is certainly not a recent development:

> *Looking into the future, we can foresee that computer utilities are likely to play an increasingly large part in human affairs. Communities will design systems to perform various functions—intellectual, economic, and social—and the systems in turn undoubtedly will have profound effects in shaping the patterns of human life. The coupling between such a utility and the community it serves is so strong that the community is actually a part of the system itself. Together the computer systems and the human users will create new services, new institutions, a new environment, and new problems... To what ends will the system be devoted,* and what safeguards can be designed to prevent its misuse? *It is easy to see that the progress of this new technology will raise many social questions as well as technical ones.*
>
> — [198] (emphasis added)

More than three decades after this observation there are few clichés in the field of information assurance as well worn as "there is no perfect security" and given the abovementioned conflicts, particularly in conjunction with the need to employ COTS components or systems, that can only be affirmed [749] even if perfection is merely a distant goal and has not been supplanted by fatalistic complacency.

What has emerged in recent years as a multiplier of individual threats is the ubiquitous internetworking of devices ranging from cellular phones to mainframe systems.
In analogy to Metcalfe's Law[1], which holds that for every n users in a network, the value of the network increases by n^2 for some arbitrary metric, one might assume that this observation apparently also models the utility derived from internetworking devices, systems, and ultimately networks.
The same can, however, also be assumed for the risks emanating from — among others — interdependencies of these nodes as well as the dangers to confidentiality and integrity due to transitive vulnerabilities emerging only in internetworked settings, which hardly requires emphasis or additional analysis arguing for their existence.

At least part of the risks stem from the fact that the system constituted by the rapidly changing and reconstituting components apparently can not be modeled adequately; this aspect of complexity in safety critical systems and the human inability to successfully cope with the resulting effects or at least comprehend the risks associated with it has been documented amply and is not limited to information systems [593, 184, 546, 130] (see also section 12.2.1).
However, it is not merely the internetworking aspects of systems that are to be considered in evaluating the challenges to information assurance. The complexity particularly of widely deployed commercial operating systems is also a contributing factor to an increase in risks.
Some of this can be attributed to design practice or lack thereof, some to a placement of emphasis in the development on factors other than information assurance ranging from ease of use to cost of development.
One of the conclusions supported both by experience from long--time practitioners [631], statistical analysis [353], and mathematical modeling [105, 381] is that even at the complexity levels of currently deployed systems the development processes of such COTS systems results in systems with very large numbers of potentially critical

[1]actually invented by George Glider and later attributed to Robert Metcalfe

implementation defects not removable by testing (regardless of formalized procedures or the more typical reaction to published reports of defects) in addition to design and architectural flaws.
As will be discussed in chapter 3, even if one were able to provide components designed and implemented specifically for the purpose of meeting high assurance criteria [458, 459, 460, 770] a number of desirable properties do not compose.
Intuitively, this results in the observation that security properties are inherently "brittle"; ostensibly minor changes to a secure — even provably secure — system result in the violation of information assurance properties.

All of the above observations do not lend themselves as a foundation for an optimistic outlook on the assurance for the vast majority of systems. On the contrary, as custom high assurance components are increasingly replaced with COTS components [179, 514, 513] it may be argued that the converse is true.
This is in spite of a large body of research that has resulted in knowledge on the construction of secure systems to the extent that this knowledge has been available in textbooks for a number of years [240, 171] as well as a number of high assurance systems primarily used in defense applications.

If one resorts to intuition and visualizes the conflicts of interest between some of the various objectives to be met by an information technology system as a cone segment as shown in figure 1.1 from which an embedded rectangular object must be chosen, an apparently valid observation is that while systems have made great strides in terms of the speed achievable as well as in terms of the functionality offered, they are still not using mechanisms initially researched and deployed more than three decades ago.

One of the objectives of this dissertation is an attempt at describing an architecture for enforcing information assurance properties within distributed, heterogeneous systems.
Given the observations discussed above a pragmatic aspect of the architecture is the possibility of being retrofitted to existing widely

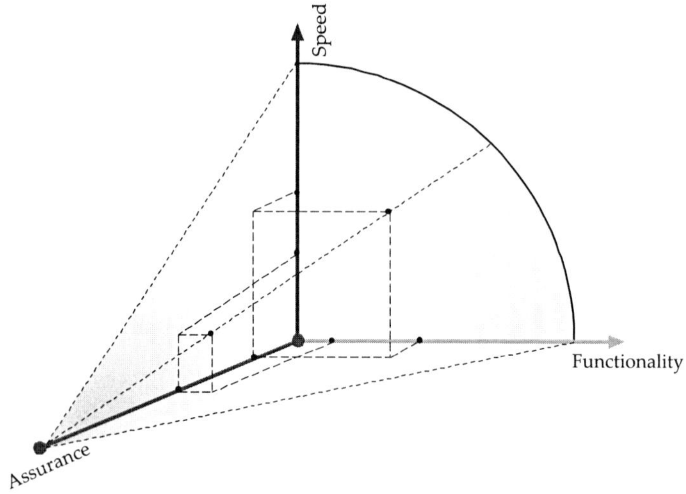

Figure 1.1: The assurance dilemma

deployed systems, particularly to COTS systems even in cases where source code is unavailable for modification.

By defining appropriate security policies, the enforcement of such policies can occur without modifications to applications or legitimate user behavior.

This is pragmatic in that it permits the use of COTS operating systems as a foundation as well as that of application programs which may not be available for modification or revision to higher assurance standards.

However, the fundamental problems of low assurance particularly of the protection mechanisms inherent in the base operating system into which the architecture's components are embedded remain and one can certainly argue that the very fact of augmenting the base operating system leads to additional risks due to unpredictable interactions.

It is in full awareness of this apparently irreconcilable contradiction (the fact that a multilevel secure system could not be created using the *"penetrate and patch"* approach and was unlikely to yield sufficiently high assurance systems was already made evident by experiments and studies conducted during the early 1970s [359]) that the exemplary or reference interpretation and implementation mechanism described in this dissertation is mainly based on one of the most controversial platforms in terms of the code base that must be considered trustworthy[2] — Microsoft Windows 2000 — which, due to decisions mainly based on a determination to remain compatible with previous ill-designed operating systems, is as such ill suited for retrofitting any kind of assurance mechanism.

The Microsoft Windows NT family of operating systems can also be considered one of the most complex operating systems in terms of the amount of function points implemented at maximum privilege levels, largely circumventing existing hardware as well as internal protection mechanisms. No claims are made to the overall assurance of the resulting retrofitted system; rather, it is demonstrated that the security architecture can be adapted to one of the most prevalent — and also arguably ill-designed from the assurance perspective[3] — systems deployed.

The main focus of this dissertation, however, is the thesis that the properties relevant for describing and reasoning over information assurance of information systems, particularly but not limited to operating systems, can be modeled abstractly based on formal logic and that security policies can thus be formulated as well-formed formulae over the model. Moreover, not only can policies be specified within this model, but the enforcement of these policies can be performed by formulating the intent or requirement to execute operations as hypotheses and using automated reasoning mechanisms to prove such hypotheses directly or to obtain more general results that subsume the given hypothesis.

[2]In the sense that a "trusted" component is one that can violate the security properties of a system.

[3]As a corollary to the bon mot of C.A.R. Hoare *"The unavoidable price of reliability is simplicity."*

1.3 Structure of this Dissertation

This dissertation is divided into twelve chapters. The following provides a brief overview of the structure of the dissertation and the topics addressed in the individual chapters.

Chapter 1 defines the terminology for Information Assurance and Information Operations and provides a brief motivation for the relevance of the subject of the dissertation. The actual motivation and hypotheses are formulated in chapter 4.

Chapter 2 provides background information on basic formal logic, automated deduction, and lattices solely as a convenience for readers not familiar with the subject.

Chapter 3 briefly discusses prior research in several areas related to the subject of this dissertation, particularly in the theory of secure systems, policy mechanisms, automated deduction, and operating system and network security mechanisms. Additional references to related work are given throughout the remaining chapters where appropriate.

Chapter 4 contains a concise statement of the motivation, underlying assumptions, and hypotheses discussed throughout the dissertation; the start of the chapter provides a brief perspective on the terminology used while the remaining sections discuss underlying assumptions and hypotheses backed by references to related work. In addition, core elements of the basic model underlying the architecture and realization are formulated in the Z notation.

Chapter 5 introduces the basic syntax and the semantics for the formulation of security policies over operating systems within the formal model as well as exemplary predicates for modeling.

Chapter 6 provides several instantiations of general termini introduced in chapter 5 required for an actual instance of an implementation and furthermore contains the outline of an interpretation of the basic model given in the preceding chapter.

Chapter 7 briefly discusses several security models and their properties and shows the formalization within formal logic for an information flow control model as well as for a role-based access control model; additional models for information flow based explicitly on information theory, integrity and availability are also discussed but not formalized.

Chapter 8 begins with a review of the reference monitor as introduced by the Anderson panel and introduces the logical and physical separation mechanisms leading to the externally controlled reference monitor model as well as ancillary considerations such as the embedding within tamper-resistant secure coprocessors.

Chapter 9 describes the requisite capabilities of the enforcement modules that are required to represent the various abstraction layers and predicates of the security policy and provides an example of the actual implementation of the file system enforcement module under Microsoft Windows NT/2000.

Chapter 10 discusses the application of the previously described model and mechanisms to network security; a brief threat model is followed by the analysis of the effects of the model on access and flow control mechanisms and intrusion detection and reaction, particularly with regard to multisensor data fusion capabilities.

Chapter 11 provides an example of the flexibility and extensibility of the overall security architecture by providing an application-specific capability (labeling as required by MLS policies) in the form of an additional enforcement module.

Chapter 12 summarizes the results obtained and provides an outlook on additional research subjects as well as a critical view of the assurance obtainable for civilian general-purpose systems.

CHAPTER

TWO

Background Information

Readers familiar with elementary formal logic, abstract algebra, and set theory can safely omit this chapter; to ensure that the terminology and symbols used are both well-defined and immediately accesible, it has been attempted to provide index entries wherever possible.

Section 2.1 briefly discusses the relation of specifying security policies and models in terms of formal, primarily first order, logic with specification, verification, and proof of algorithms and concurrent systems.
Section 2.2 covers the formal logic necessary for describing policies; as the terminology is already well established in the other remaining areas required as tools for describing policies throughout this dissertation, namely set theory and abstract algebra, the reader is referred to the literature, e.g. [74, 320, 191] for the latter.
Section 2.3 briefly describes an elementary and well-established mechanism (resolution theorem proving) that can be used for unifying a hypothesis with the set of predicates and functions that make up the security policy which, while not the most the most efficient mechanism, has the virtues of simplicity and universality.
Section 2.4 introduces the requisite algebraic (lattice) definitions and an observation required in section 5.1 which introduces the elements of the formal theory used for describing reasoning on security policies.

2.1 Related Uses of Formal Mechanisms

The use of formal logic for the specification of security models and policies as outlined in chapters 5 and 6 is closely related to the use of such methods for the specification and proof of correctness of computational systems in that information assurance properties may be considered part of the specification to which a program must conform.

It may be argued that the concepts for using formal methods for specification and proof was already in place quite early indicated by the works of von Neumann and Turing [721, 260] (after significant results obtained during the early 1930s by Turing, Gödel, Herbrand, and Gentzen [252, 253, 254, 242, 243, 720, 298]).

However, this approach was not pursued to a significant extent until a seminal paper by Hoare [302] who was building on earlier work by Floyd [213], apparently unaware of the earlier work by von Neumann and Turing, and expanded later leading to formal proof methods [192, 215, 303, 304, 137, 305, 306, 36, 307, 308] that have been studied intensively [438, 164, 32, 33, 420].

However, while formal specification and proof, particularly the automation of parts thereof, has since been pursued by research [377, 151], this has not readily translated to the use of such techniques in deployed systems to the extent originally envisioned in [302].

The actual applications of formal methods are mainly in the verification of hardware [529, 268, 300, 370] while software verification and generation are limited to highly critical systems such as avionics [763, 585, 154].

Instead, implementation defects in software have apparently come to be accepted as well as ill-defined specifications [373] although a limited approach for using formal mechanisms to ensure security properties within programs itself was introduced by Necula in the form of proof-carrying code [507, 142, 506], intended primarily for mobile code [505]; in the general case such proofs cannot be generated automatically although for limited properties this process can be automated [508, 486, 641]. Bernard and Lee also introduced tem-

poral logic within the conceptual framework of proof-carrying code [68].
The mechanisms developed by Hoare *et al.* also share the common trait that they are primarily focused on systems based on transformational computation such as algorithms; the systems considered in this discussion, however, are generally reactive in nature as per definition 2.1.

Definition 2.1 *A **reactive system** is modeled by*

1. *A formal language I, called the input language*
2. *A formal language O, called the output language*
3. *A formal language S, called the state-space*
4. *A function $f : I \times S \to O \times S$, called the transformation.*

Such systems commonly encountered in safety-critical systems present certain unique challenges for specification and proof; consequently, the use of formal methods for the specification and analysis of reactive systems has been a research topic for more than two decades [389, 391, 553, 555, 556, 5]. As discussed in section 3.5.1, this has also been applied to complex reactive systems in the form of operating systems [543], typically with particular emphasis on confidentiality properties. For this as well as the use of formal proof mechanisms in other larger systems to occur, automated tools were instrumental [202, 580, 658, 18, 263, 262, 12, 655, 529, 319].

2.1.1 Temporal Logic

The term temporal logic is used to describe approaches to the representation of temporal information within a logical framework; with regard to this discussion it refers more specifically to the modal-logic type of approach introduced around 1960 by Prior [565] under the name of tense logic which was subsequently developed further by logicians and computer scientists [552] and remains an active research area. The temporal operators used in section 5.2.5 was introduced by Kamp [352]. The applicability to reactive systems as discussed here was demonstrated in [554]; while most of the research has concentrated on specification, verification, and proofs, particularly of concurrent systems [439, 190, 133], temporal logic has also been applied to the specification and verification of security-related problems in particular. A formalism based on temporal logic, TLA, and a refinement thereof, TLA+ were introduced by Lamport [392, 393, 394] and is used for specification and reasoning of concurrent systems [211]. See also [230, 232, 231] for a rigorous treatment.

2.2 Formal Logic

As the literature tends not to use fully standardized terminology, let alone symbology, it is necessary to define the required terms and symbols here. In the interest of brevity, proofs have generally been omitted; the reader is referred to [9, 162]; the terminology and symbology of the following section follows [162] wherever possible. Similarly, in the interest of readability some of the definitions are somewhat less rigorous than what might be expected; applying the necessary rigor would result in a more cumbersome and lengthy notation and arguably does not contribute to the goals pursued here.

Definition 2.2 *A **formal theory** $\mathcal{T}$ consists of the following:*

1. *A countable set of symbols (a finite sequence of symbols of $\mathcal{T}$ is called an **expression** of $\mathcal{T}$).*

2. *A subset of the expressions, called the **well-formed formulae** of $\mathcal{T}$. The well-formed formulae are the legal expressions of the theory.*

3. *A subset of the well-formed formulae called the **axioms** of $\mathcal{T}$.*

4. *A finite set of relations $R_1, \dots, R_n$ on well-formed formulae called **rules of inference**. For each R_i there is a unique positive integer j such that for every set of j well-formed formulae and each well-formed*

formula A one can effectively decide whether the given j well-formed formulae are in the relation R_i to A. If so, A is called a ***direct consequence*** *of the given well-formed formulae by virtue of R_i.*

Definition 2.3 *A* ***schema*** *is a statement form. It provides a template showing the form of a well-formed formula while leaving some elements unspecified. These unspecified elements are indicated using metavariables (i.e. variables that are not symbols in the theory and are substituted by well-formed formulae in the theory).*

Metavariables are represented typographically like this: A.

Definition 2.4 *An* ***instance*** *of a statement form or schema is a well-formed formula obtained from the statement form by substitution where all occurrences of the same metavariable are replaced by the same well-formed formula.*

Definition 2.5 *Let $\tilde{S}$ be a set of well-formed formulae and let P be a well--formed formula in the formal theory $\mathcal{T}$. P is called* ***deducible from $\tilde{S}$ in $\mathcal{T}$*** *(denoted as $\tilde{S} \vdash_{\mathcal{T}} P$) if there exists a finite sequence of well-formed formulae $P_1, \ldots, P_n$ such that $P_n = P$ and for $1 \leq i \leq n$, P_i is either an axiom, a formula in $\tilde{S}$ (called a* ***hypothesis****), or a direct consequence of previous P_i by applying one of the rules of inference.*

Lemma 2.1 $\vdash A$ *for any axiom A.*

Definition 2.6 *The finite sequence of well-formed formulae $P_1, \ldots, P_n$ such that $P_n = P$ and $1 \leq i \leq n$, P_i is called a* ***derivation*** *of P from $\tilde{S}$, or a* ***proof*** *of P from $\tilde{S}$.*

Definition 2.7 *If P is deducible from the empty set, this is denoted as $\vdash_{\mathcal{T}} P$ and referred to P as a* ***theorem****; alternatively, P is called* ***provable*** *in $\mathcal{T}$.*

It should generally not be necessary to distinguish between theories as we will generally use only one theory at a time in this discussion, so as a shorthand deducibility will be denoted as $\vdash$.

Observation 2.1 *The following properties of deducibility are consequences of definition 2.5.*
Let $\tilde{S}_1$ and $\tilde{S}_2$ be sets of well-formed formulae and A be a well-formed formula, then the following statements hold:

1. *If $\tilde{S}_1$ is a subset of $\tilde{S}_2$ and $\tilde{S}_1 \vdash A$, then $\tilde{S}_2 \vdash A$.*

   This property is also referred to as ***monotonicity****.*

2. *$\tilde{S}_1 \vdash A$ is true if and only if there is a finite subset $\tilde{S}_2 \subseteq \tilde{S}_1$ such that $\tilde{S}_2 \vdash A$. The requirement for the subset being finite is due to definition 2.5.*

   This property is also referred to as ***compactness****.*

3. *If $\tilde{S}_2 \vdash A$ and for each well-formed formula $B \in \tilde{S}_2$ we can prove $\tilde{S}_1 \vdash B$, then $\tilde{S}_1 \vdash A$ holds.*

A system of logic is called **non-monotonic** if the addition of a new axiom can invalidate a proof that was possible prior to the addition of the axiom or axioms.
Such systems of logic can be useful in simplifying some models since it is possible to state general rules as axioms and add exceptions to these general rules as additional axioms.
However, since this would add to the complexity of the deduction mechanisms this discussion is limited to monotonic systems of logic.
It should be noted that the term *proof* is constructed narrowly, i.e. as a sequence of well-formed formulae obeying certain restrictions and that this term refers to the last statement of such a sequence. However, by considering only the *n*th statement of the sequence, one can trivially obtain a proof of that statement.
The following is a preliminary and intuitive definition of the term interpretation; for first order languages this is formalized later in definition 2.22.

Definition 2.8 *An* ***interpretation*** *imbues a meaning into each symbol of a formal theory such that any well-formed formula can be understood as a statement that is either true or false in the interpretation.*

Definition 2.9 *An interpretation is a* **model** *for a set of well-formed formulae $\tilde{S}$ if every well-formed formula in $\tilde{S}$ is true in the interpretation. It can be stated that an interpretation provides a model for a formal theory $\mathcal{T}$ if it provides a model for the set of theorems of $\mathcal{T}$.*

Definition 2.10 *A theory is* **complete** *if every statement that is true in all interpretations of a formal theory is provable in the theory.*

Definition 2.11 *A theory is* **sound** *if every provable statement of the formal theory is true in all interpretations.*

Definition 2.12 *A formal theory is* **decidable** *if there exists an effective procedure that will determine for each statement of the theory whether or not that statement is provable in the theory.*

Definition 2.13 *A formal theory is* **consistent** *if it contains no well-formed formula such that both the well-formed formula and its negation are provable.*

Definition 2.14 *The countable set of* **first order languages** *is defined as follows:*

- *Logical connectives* $\neg$, $\rightarrow$, *and* $\forall$
- *Auxiliary symbols:*), (, *and* ,
- *A countable number of variables* $\mathsf{x}, \mathsf{x}_1, \mathsf{x}_2, \ldots$
- *A countable, possibly empty, set of constant symbols* $\mathsf{a}, \mathsf{a}_1, \mathsf{a}_2, \ldots$
- *A countable, possibly empty, set of function symbols* f^n_k *for all positive integers k and n*
- *A countable, non-empty, set of predicate symbols* A^n_k *for all positive integers k and n*

Definition 2.15 *The well-formed formulae (abbreviated as* wff *here) of a first order language are defined as follows:*

term	::=	variable | constant | function_symbol '(' termlist ')'
termlist	::=	term | term ',' termlist
atomic_formula	::=	predicate_symbol '(' termlist ')'
wff	::=	atomic_formula
	|	'(¬' wff ')'
	|	'(' wff ' → ' wff ')'
	|	'((∀' variable ')' wff ')'

Definition 2.16 *∀ is the* ***universal quantifier*** *and may be read as "for all". ∃ is the* ***existential quantifier*** *and may be read as "there exists". One can use the existential in formulae as a shorthand; it is not defined as a primitive in definition 2.15 since it can be defined in terms of the primitives of that definition:* $((\exists x)A)$ *can be substituted by* $\neg((\forall x)\neg A)$.

Definition 2.17 *The* ***scope*** *of a quantifier is the well-formed formula to which it applies.*

Definition 2.18 *The following shorthands are introduced for use as connectives:*

$$\begin{array}{lll} A \vee B & \textit{for} & (\neg(A \rightarrow (\neg B))) \\ A \wedge B & \textit{for} & ((\neg A) \rightarrow B) \\ A \equiv B & \textit{for} & ((A \rightarrow B) \wedge (B \rightarrow A)) \end{array}$$

As an example for definition 2.17 consider the expression $((\forall x)\neg A \vee B)$. Here, $\neg A \vee B$ is the scope of the quantifier $(\forall x)$.
Besides the fully parenthesized form one can also eliminate some parentheses in informal descriptions. The outer pair of parentheses may be omitted, i.e. $(A \rightarrow B)$ may be written as $A \rightarrow B$, $(\neg A)$ as $\neg A$. For informal descriptions terms are defined to be left-associative, thus being able to write $(((A \rightarrow B) \rightarrow A) \rightarrow C)$ as $A \rightarrow B \rightarrow A \rightarrow C$. Finally, a precedence on the connectives from highest to lowest is defined as follows: $\neg, \wedge, \vee, \exists, \forall, \rightarrow, \equiv$.

Definition 2.19 *An **occurrence** of a variable x is **bound** in a well-formed formula if and only if it is either a variable of a quantifier* $(\forall x)$ *or* $(\exists x)$ *in the well-formed formula or it is the same variable that is quantified and is within the scope of the quantifier.*
*Otherwise the occurrence is said to be **free** in the well-formed formula.*

Definition 2.20 *A variable is said to be free (bound) in a well-formed formula if and only if it has a free (bound) occurrence in the well-formed formula.*

Definition 2.21 *If A is a well-formed formula and t is a term, t is said to be **free for*** x_i ***in** A if and only if no free occurrences of* x_i *in A lie within the scope of any quantifier* $(\forall x_j)$ *where* x_j *is a variable in t.*

Definition 2.22 *An **interpretation** of a well-formed formula or a set of well-formed formulae of a first order language consists of*

1. *A nonempty set* $\mathcal{D}$*, called the **domain***
2. *An assignment to each n-ary predicate symbol* A_k^n *of an n-ary relation in* $\mathcal{D}$
3. *An assignment to each n-ary function symbol* f_k^n *of an n-ary operation closed over* $\mathcal{D}$ *(i.e.* $\mathcal{D}^n \mapsto \mathcal{D}$*)*
4. *An assignment to each individual constant* a_i *of some fixed element of* $\mathcal{D}$*.*

Variables range over $\mathcal{D}$ and the logical connectives $\neg$ and $\rightarrow$ (and hence also $\equiv$, $\vee$, and $\wedge$) are given the usual meaning.
$((\forall \mathsf{y})A)$ means that for all values of y taken from the domain, property A holds. If A does not contain the variable y, $((\forall \mathsf{y})A)$ means the same as A.
$((\exists \mathsf{y})A)$ means that there exists some value for y taken from the domain such that A holds.

Definition 2.23 *A **closed** well-formed formula is a well-formed formula containing no free variables.*

Definition 2.24 *The closed well-formed formula obtained from A by prefixing A with universal quantifiers that appear free in A is called the* ***closure*** *of A.*

For a given interpretation, a closed well-formed formula represents a statement that is true or false. A well-formed formula with free variables represents a relation on the domain that may be true for some values of the free variables and false for others.
Definition 2.25 shows what is meant by a sequence $\mathbf{s} = (\mathbf{b}_1, \mathbf{b}_2, \ldots)$ in the set of denumerable sequences of elements of $\mathcal{D}$, called $\aleph$.

Definition 2.25 *Given an interpretation with domain $\mathcal{D}$, let $\aleph$ be the set of denumerable sequences of elements of $\mathcal{D}$. $\mathbf{s}^*(t)$ is defined by cases. A term can be:*

1. *A variable; in this case, if t is $\mathsf{x_i}$ then $\mathbf{s}^*(t) = \mathbf{b_i}$.*

2. *A constant; in this case, if t is an individual constant, then $\mathbf{s}^*(t)$ is the fixed element of $\mathcal{D}$ assigned to that constant by the interpretation.*

3. *A function symbol; in this case if $\mathsf{f_j}$ is a function symbol, g is the corresponding operation in $\mathcal{D}$ assigned by the interpretation and $t_1, \ldots, t_n$ are terms then $\mathbf{s}^*(\mathsf{f_j}(t_1, \ldots, t_n)) = g(\mathbf{s}^*(t_1), \ldots, \mathbf{s}^*(t_n))$.*

With $\mathbf{s}^*$ one has a function which maps a syntactic expression into its values under the interpretation. For a sequence $\mathbf{s} = (\mathbf{b}_1, \mathbf{b}_2, \ldots)$ and a given term t, $\mathbf{s}^*(t)$ is the element of $\mathcal{D}$ obtained by substituting $\mathbf{b_i}$ for all occurrences of x_i in t for each i, then applying the operations of the interpretation corresponding to the function symbols in t.

Definition 2.26 *Satisfiability is defined by cases. A well-formed formula A can be:*

1. *An atomic well-formed formula $\mathsf{A_j}(t_1, \ldots, t_n)$. Let $\mathbf{A_j}$ be the corresponding relation of the interpretation; the sequence* **s** ***satisfies*** *A if and only if $\mathbf{A_j}(\mathbf{s}^*(t_1), \ldots, \mathbf{s}^*(t_n))$ is in the relation $\mathbf{A_j}$.*

2. *A negated well-formed formula. In this case* **s** *satisfies* $\neg A$ *if and only if* **s** *does not satisfy A.*

3. *An implication. In this case* **s** *satisfies* $A \rightarrow B$ *if and only if either* **s** *either does not satisfy A or* **s** *satisfies B.*

4. *A quantified well-formed formula. In this case* **s** *satisfies* $(\forall x_i) A$ *if and only if every sequence of* $\aleph$ *that differs from* **s** *in at most the i-th component satisfies A.*

Definition 2.27 *A well-formed formula A is* **true** *for a given interpretation if and only if every sequence in* $\aleph$ *satisfies A. A is* **false** *for a given interpretation if and only if no sequence in* $\aleph$ *satisfies A.*

Note that definition 2.27 permits a well-formed formula to be neither true nor false.
One can now state several conclusions from the above definitions.

Observation 2.2 *A is false for a given interpretation if and only if* $\neg A$ *is true for that interpretation.*
Conversely, A is true if and only if $\neg A$ *is false.*

Observation 2.3 *If A and* $A \rightarrow B$ *are true for a given interpretation, then so is B.*

Observation 2.4 *For a given interpretation,* $A \rightarrow B$ *is false if and only if A is true and B is false.*

Observation 2.5 *For the satisfiability of connectives the following holds:*

- **s** *satisfies* $A \wedge B$ *if and only if* **s** *satisfies A and* **s** *satisfies B.*
- **s** *satisfies* $A \vee B$ *if and only if* **s** *satisfies A or* **s** *satisfies B.*
- **s** *satisfies* $A \equiv B$ *if and only if* **s** *satisfies A and* **s** *satisfies B.*
- **s** *satisfies* $(\exists x) A$ *if and only if there exists a sequence* **s**$'$ *differing from* **s** *in at most the i-th place such that* **s**$'$ *satisfies A.*

Observation 2.6 *$A(x)$ is true if and only if its closure $(\forall x)\,A(x)$ is true.*

Observation 2.7 *Every instance of a tautology is true for any interpretation.*

Observation 2.8 *If the free variables of a well-formed formula A occur in the list $x_1, \ldots x_k$ and if the sequences* **s** *and* **s′** *have the same components in the 1st,. . .,kth positions, then* **s** *satisfies A if and only if* **s′** *satisfies A.*

Observation 2.9 *If A does not contain x_i as a free variable then $(\forall x_i)(A \rightarrow B) \rightarrow (A \rightarrow (\forall x_i)B)$ is true for all interpretations.*

Lemma 2.2 *If t and u are terms,* **s** *is a sequence in $\aleph$, and t' results from t by substitution of u for all occurrences of x_i in t, and* **s′** *results from* **s** *by substituting $\mathbf{s}^*(u)$ for the i-th component of* **s***, then $\mathbf{s}^*(t') = (\mathbf{s}')^*(t)$.*

Therefore Lemma 2.2 states that it is permitted to use a symbol table to simulate substitution.

Corollary 2.1 *If $(\forall x_i)A(x_i)$ is satisfied by* **s***, then so is $A(t)$. Hence $(\forall x_i)A(x_i) \rightarrow A(t)$ is true for all interpretations.*

Definition 2.28 *A well-formed formula is said to be* ***logically valid*** *if and only if A is true for every interpretation.*

Definition 2.29 *A well-formed formula is said to be* ***satisfiable*** *if and only if there is an interpretation for which A is satisfied by at least one sequence in $\aleph$.*

Definition 2.30 *A well-formed formula A is said to be* ***contradictory*** *(or* ***unsatisfiable****) if and only if $\neg A$ is logically valid, i.e. if and only if A is false for every interpretation.*

Definition 2.31 *A well-formed formula A is said to* ***logically imply*** *B if and only if in every interpretation any sequence satisfying A also satisfies B.*
B is a ***logical consequence*** *of a set $\hat{S}$ of well-formed formulae if and only if in every interpretation every sequence that satisfies every well-formed formula in $\hat{S}$ also satisfies B.*

Definition 2.32 *The well-formed formulae A and B are said to be* ***logically equivalent*** *if and only if they logically imply each other.*

Definition 2.33 *Any statement of a formal language that is an instance of a logically valid well-formed formula is called* ***logically true****.*
An instance of a contradictory well-formed formula is said to be ***logically false****.*

Again, some observations based on the above definitions are stated:

Observation 2.10 *A logically implies B if and only if $A \to B$ is logically valid.*

Observation 2.11 *A and B are logically equivalent if and only if $A \equiv B$ is logically valid.*

Observation 2.12 *If A logically implies B and A is true in a given interpretation, then so is B.*

Observation 2.13 *If B is a logical consequence of a set $\hat{S}$ of well-formed formulae and all well-formed formulae in $\hat{S}$ are true in a given interpretation, then so is B.*

Observation 2.14 *Every instance of a tautology is logically valid.*

Observation 2.15 *If A does not contain x as a free variable, then $(\forall x)(A \to B) \to (A \to (\forall x)B)$ is logically valid.*

Observation 2.16 *If t is free for x in A then $(\forall x)A(x) \to A(t)$ is logically valid.*

So far only the symbols and well-formed formulae of a first order theory have been defined. In order to complete the definition of a first order theory one needs to define the axioms and rules of inference.

Definition 2.34 *The symbols and well-formed formulae of a **first order theory** are those of a first order language (cf. definition 2.14). The **axioms** of a first order theory are split into two classes, **logical axioms** and **proper axioms**.*

The logical axioms of any first order theory are those given below; proper axioms may vary from one first order theory to another.
*If a first order theory does not contain proper axioms it is called a **first order predicate calculus**.*

Axioms *If A, B, and C are well-formed formulae then the following are axioms:*

A1 $A \rightarrow (B \rightarrow A)$

A2 $(A \rightarrow (B \rightarrow C)) \rightarrow ((A \rightarrow B) \rightarrow (A \rightarrow C))$

A3 $(\neg B \rightarrow \neg A) \rightarrow ((\neg B \rightarrow A) \rightarrow B)$

A4 $(\forall x)A(x) \rightarrow A(t)$ *if $A(x)$ is a well-formed formula and t is a term free for x in $A(x)$.*

A5 $(\forall x)(A \rightarrow B) \rightarrow (A \rightarrow (\forall x)B)$ *if A is a well-formed formula containing no free occurrences of x.*

Rules of Inference *There are two rules of inference:*

Modus Ponens (MP) $A, A \rightarrow B \vdash B$

Generalization (GE) $A \vdash (\forall x)A$

A first order predicate calculus is denoted as $\mathcal{PC}$.

Lemma 2.3 *Every well-formed formula A which is an instance of a tautology is a theorem of $\mathcal{PC}$ and may be proved using only axioms **A1–A3** and the rule of inference **MP**.*

Lemma 2.4 $A \to A$.

Proof

1. According to **A2** one can obtain
   $(A \to ((A \to A) \to A)) \to ((A \to (A \to A)) \to (A \to A))$
2. According to **A1** one obtains
   $A \to ((A \to A) \to A)$
3. By applying **MP** on steps 1 and 2 one obtains
   $(A \to (A \to A)) \to (A \to A)$
4. Using **A1** one obtains
   $A \to (A \to A)$
5. Finally, applying **MP** on steps 3 and 4 results in the desideratum.
   $A \to A$

□

Definition 2.35 *For each well-formed formula A let* $\mathbf{h}(A)$ *be the expression obtained by deleting all quantifiers and terms in A.*

Lemma 2.5 $\mathbf{h}(A)$ *is a tautology whenever A is a theorem of* $\mathcal{PC}$.

Proof To show that $\mathbf{h}(A)$ is a tautology whenever A is a theorem, it must be shown that $\mathbf{h}(A)$ is a tautology for each axiom and that the rules of inference lead from well-formed formulae whose images under **h** are tautologies to well-formed formulae whose images under **h** are also tautologies.
By definition, $\mathbf{h}(\neg A) = \neg(\mathbf{h}(A))$ and $\mathbf{h}(A \to B) = \mathbf{h}(A) \to \mathbf{h}(B)$. For every axiom A given by **A1–A5**, $\mathbf{h}(A)$ is a tautology.
For **A1–A3** this is obvious since they are already free of quantifiers and terms.

For **A4** it can be seen immediately that any instance of $(\forall x)A(x) \to A(t)$ is transformed under **h** into a tautology of the form $A \to A$.
For **A5**, any instance of the axiom $(\forall x)(A \to B) \to (A \to (\forall x)B)$ becomes $(A \to B) \to (A \to B)$ under **h**.
If $\mathbf{h}(A)$ and $\mathbf{h}(A \to B)$ are tautologies, then by the definition of $\to$, so is $\mathbf{h}(B)$; and if $\mathbf{h}(A)$ is a tautology, then so is $\mathbf{h}((\forall x)A)$, which is transformed **h** into $\mathbf{h}(A)$. □

Theorem 2.1 *$\mathcal{PC}$ is consistent.*

Proof If there were a well-formed formula B such that $\vdash B$ and $\vdash \neg B$, then, by Lemma 2.5, both $\mathbf{h(B)}$ and $\mathbf{h}(\neg B)$ (or $\neg\mathbf{h}(B)$) would be tautologies, which is impossible. □

Definition 2.36 *Let A be a well-formed formula in a set $\hat{S}$ of well-formed formulae. Assuming that a deduction $B_1, \ldots, B_n$ from $\hat{S}$ together with justification of each step of the deduction is given, one can now state that B_i **depends upon** A in the proof if and only if*

1. *B_i is A and the justification for B_i is that it belongs to $\hat{S}$*
2. *B_i is justified as a direct consequence by **MP** or **GE** of some preceding well-formed formulae of the sequence where at least one of those well-formed formulae depends upon A.*

Lemma 2.6 *If B does not depend upon A in a deduction $\hat{S}, A \vdash B$, then $\hat{S} \vdash B$.*

Proof Let $B_1, \ldots, B_n = B$ be a deduction of B from $\hat{S}$ and A, in which B does not depend upon A. Proof that $\hat{S} \vdash B_n$ is true follows by induction.

Basis Case $n = 1$. $B_1 = B$, which does not depend upon A, therefore $B \in \hat{S}$, or B is an axiom, thus $\hat{S} \vdash B$.

Induction Step Assuming $\hat{S} \vdash B_i$ for $i < n$, it is shown that $\hat{S} \vdash B$:

If B belongs to $\hat{S}$ or is an axiom, then $\hat{S} \vdash B$. If B is a direct consequence of one or two preceding well-formed formulae, then, since B does not depend upon A, neither do these preceding well-formed formulae. By the inductive hypothesis, these preceding well-formed formulae are deducible from $\hat{S}$ alone, therefore so is B.

□

Theorem 2.2 *Assuming there exists a proof* $\hat{S}, A \vdash B$, *where in the deduction, no application of* **GE** *to a well-formed formula that depends upon* A *has as its quantified variable a free variable of* A, *then* $\hat{S} \vdash A \rightarrow B$.

Proof Let $B_1, \ldots, B_n = B$ be a deduction of B from $\hat{S}, A$ that satisfies the assumption of the theorem. Proof follows by induction on i showing that $\hat{S} \vdash A \rightarrow B_i$ for each $i \leq n$.

Basis Case $i = 1$.

1. B_i is an axiom or belongs to $\hat{S}$.

$\hat{S} \vdash B_i$	Definition 2.5
$B_i \rightarrow (A \rightarrow B_i)$	**A1**
$\hat{S} \vdash A \rightarrow B_i$	**MP**

2. B_i is A

$\vdash A \rightarrow A$	Lemma 2.4
$\hat{S} \vdash A \rightarrow B_i$	Definition 2.5

Induction Step Assume $\hat{S} \vdash A \rightarrow B_I$ for $i < k$; show that $\hat{S} \vdash A \rightarrow B_k$:

1. If B_k is an axiom or an element of $\hat{S}$, the result follows as in case 1 of the Basis Case.
2. If B_k is A then the result follows as in case 2 of the Basis Case.

3. If B_k follows by **MP** from B_i and $B_j = B_i \to B_j$ where $i, j < k$, then

$$\begin{array}{ll} \hat{S} \vdash A \to B_i & \text{induction hypothesis} \\ \hat{S} \vdash A \to (B_i \to B_k) & \text{induction hypothesis} \\ \vdash (A \to (B_i \to B_k)) \to & \\ \quad ((A \to B_i) \to (A \to B_k)) & \textbf{A2} \\ \hat{S} \vdash ((A \to B_i) \to (A \to B_k)) & \textbf{MP} \\ \hat{S} \vdash A \to B_k & \textbf{MP} \end{array}$$

4. If B_k follow by **GE** from B_k, where $j < k$ (i.e. $B_k = (\forall x)B_j$), then

$$\hat{S} \vdash A \to B_j \qquad \text{induction hypothesis}$$

   (a) If B_j does not depend upon A in the deduction of B from $\hat{S}$ and A then

$$\begin{array}{ll} \hat{S} \vdash B_j & \text{Lemma 2.6} \\ \hat{S} \vdash (\forall x)B_j & \textbf{GE} \\ \vdash B_k \to (A \to B_k) & \textbf{A1} \\ \hat{S} \vdash A \to B_k & \textbf{MP} \end{array}$$

   (b) If x is not a free variable of A then

$$\begin{array}{ll} \vdash (\forall x)(A \to B_j) \to & \\ \quad (A \to (\forall x)B_j) & \textbf{A5} \\ \hat{S} \vdash A \to B_j & \text{induction hypothesis} \\ \hat{S} \vdash (\forall x)(A \to B_j) & \textbf{GE} \\ \hat{S} \vdash A \to (\forall x)B_j & \textbf{MP} \\ \hat{S} \vdash A \to B_k & \end{array}$$

The theorem follows as the case where $i = n$. □

Corollary 2.2 *If a deduction $\hat{S}, A \vdash B$ involves no applications of* **GE** *in which the quantified variable is free in A, then $\hat{S} \vdash A \to B$.*

Corollary 2.3 *If A is closed and $\hat{S}, A \vdash B$, then $\hat{S} \vdash A \to B$.*

2.3 A Resolution Mechanism

Definition 2.37 *A **ground** expression, term, or clause is one that does not contain any variables.*

Modifying the grammar from definition 2.15 somewhat, a grammar for first order predicate calculus can be obtained.

Definition 2.38 *The well-formed formulae of $\mathcal{PC}$ are expressed as follows (with ∪ denoting the set union):*

term	::=	variable | constant | function_symbol '(' termlist ')'
termlist	::=	term | term ',' termlist
atomic_formula	::=	predicate_symbol '(' termlist ')'
literal	::=	atomic_formula | '¬' atomic_formula
clause	::=	'□' | {literal} | {literal} ∪ clause
sentence	::=	{clause} | {clause} ∪ sentence

Definition 2.39 *A well-formed formula of $\mathcal{PC}$ is in **clausal form** if it is a sentence in the form of definition 2.38.*

Intuitively, a clause is a disjunction of its elements (literals) and a sentence is a conjunction of its elements (clauses). Compared to definition 2.15, quantifiers have been omitted. The reason for this is that a well-formed formula of $\mathcal{PC}$ is provable if and only if its closure is provable (by **A4** and **GE**) and that a well-formed formula is logically valid if and only if its closure is logically valid. Every well-formed formula in clausal form is implicitly closed and one can consider each variable to be universally quantified over the clause in which it appears.
Resolution depends on the fundamental operation of substituting terms for variables in well-formed formulae. A substitution provides an assignment of terms to variables. Each such assignment is denoted by an ordered tuple (term, variable).

Definition 2.40 *A **substitution** ς is a set of substitution components*

$$\varsigma = [(t_1, v_1), \ldots, (t_n, v_n)]$$

The first element of each substitution is a term and the second element is a variable such that

1. $v_i \neq v_j$ *if* $i \neq j$

2. v_i *does not occur in* t_j *for any* i *and* j.

Definition 2.41 *If a variable v is replaced at each of its occurrences in the well-formed formula C by the term t, the resulting well-formed formula is called a **substitution instance** of C, and is denoted $C[(t,v)]$, read "C with t for v".*
Similarly, the result of simultaneously replacing all occurrences of different variables $v_1, \ldots, v_n$ in C by terms $t_1, \ldots, t_n$ respectively, is denoted by $C[(t_1, v_1), \ldots, (t_n, v_n)]$.

The independence of the components of a substitution is ensured by the two conditions of definition 2.40 and guarantees that two substitutions will be set-theoretically equal if and only if they have the same effect on all expressions.
In applying a substitution to an expression, the ordering of the components in the substitution therefore does not affect the result. $C\varsigma$, the result of applying ς to C is the substitution instance $C[(t_1, v_1), \ldots, (t_n, v_n)]$.
Because of condition (2) in definition 2.40 no new occurrences of any variables affected by the substitution can be introduced through any of the substitution components.
To allow the definition of an algorithm for resolution an ordering on the expressions of $\mathcal{PC}$ is required.

Definition 2.42 *Well-formed formulae and sequences of well-formed formulae can be lexically ordered as follows:*

1. *The primitive symbols are ordered according to type (i.e. variables, constants, function symbols, predicate symbols, connectives), then within each type according to the number of arguments (i.e. superscripts for function and predicate symbols), then subscripts (assuming that symbols without subscripts are sorted in before subscripted symbols), and finally alphabetically. The connectives are ordered according to their scope with the least in scope coming first (i.e.* $\neg, \wedge, \vee, \rightarrow, \equiv$*).*

2. *Well-formed formulae and terms are first ordered by length, then two well-formed formulae of equal length are placed in the order of the symbols in the position in which they first differ.*

   Similarly, sequences of well-formed formulae are ordered by length and then by the order of the first members at which they differ.

The well-formed formulae are required to be in **Skolem free variable form**, defined as follows:

Definition 2.43 *A well-formed formula is in* ***prenex form*** *(also called* ***Skolem free variable form****) if every variable is quantified and all quantifiers (the* ***prefix*** *of the well-formed formula) precede a quantifier-free sentence (the* ***matrix*** *of the well-formed formula).*

Every well-formed formula of $\mathcal{PC}$ is equivalent to a well-formed formula in prenex form. To transform a well-formed formula into prenex form, one eliminates any occurrences of $\equiv$ and $\rightarrow$ by applying the following rewrite rules:

1. $A \equiv B$ is transformed into $(A \rightarrow B) \wedge (B \rightarrow A)$

2. $A \rightarrow B$ is transformed into $\neg A \vee B$.

Subsequently, all quantifiers are moved to the left of all connectives. At this point the well-formed formula can only contain the connectives $\neg$, $\wedge$, and $\vee$. It is therefore sufficient to show how to move each quantifier across each type of connective.

1. $\neg\exists xA$ is transformed into $\forall x\neg A$
2. $\neg\forall xA$ is transformed into $\exists x\neg A$.

One can now proceed by renaming all bound variables so all distinct variables have distinct names, then

1. $(\wedge \ldots \forall xA \ldots)$ is transformed into $\forall x(\wedge \ldots A \ldots)$
2. $(\wedge \ldots \exists xA \ldots)$ is transformed into $\exists x(\wedge \ldots A \ldots)$
3. $(\vee \ldots \forall xA \ldots)$ is transformed into $\forall x(\vee \ldots A \ldots)$
4. $(\vee \ldots \exists xA \ldots)$ is transformed into $\exists x(\vee \ldots A \ldots)$.

The following special cases can be used to simplify the prefix.

1. $(\wedge \forall x_1 A_1(x_1) \ldots \forall x_n A_n(x_n))$ is transformed into
$$\left(\bigwedge \forall xA_1(x) \ldots \forall xA_n(x)\right)$$
which in turn is transformed into $\forall x(\wedge A_1(x) \ldots A_n(x))$
2. $(\vee \exists x_1 A_1(x_1) \ldots \exists x_n A_n(x_n))$ is transformed into
$$\left(\bigvee \exists xA_1(x) \ldots \exists xA_n(x)\right)$$
which in turn is transformed into $\exists x(\vee A_1(x) \ldots A_n(x))$.

The closure of the well-formed formula is then formed by universally quantifying all free variables (by observation 2.6).
The prenex form thus obtained is not necessarily unique but is guaranteed to be equivalent to the original formula.

Definition 2.44 *The* ***Herbrand domain*** *of A' is the set of the terms $\mathcal{H}$ obtained from A' using the following rules:*

1. *If a is a constant (or a 0-ary function symbol) in A', then $a \in \mathcal{H}$. If A' has no constants, an arbitrary constant symbol is included in $\mathcal{H}$*

2. *If $t_1, \ldots, t_n$ are terms in $\mathcal{H}$, and f is an n-ary function symbol in A', then $f(t_1, \ldots, t_n) \in \mathcal{H}$.*

The notation $P((\hat{S}))$ denotes the set of instances obtained by applying all possible substitutions with terms belonging to a set P of terms to a set $\hat{S}$ of of well-formed formulae.

Definition 2.45 *The set of all substitution instances of A' obtainable by replacing each variable in A' by a term from $\mathcal{H}$ is called the* ***Herbrand expansion*** *of A' and denoted by $\mathcal{H}((A'))$.*

At certain intervals the conjunction of the instances so far generated can be tested for inconsistency since the basis is a refutation hypothesis. If the conjunction is consistent, further instances can be generated and the larger conjunction can be tested etc. until an inconsistent set of instances is found.
That this will happen precisely in the case where A' is unsatisfiable (and hence A is logically valid) follows from Herbrand's theorem:

Theorem 2.3 *If A is a finite conjunction of clauses and $\mathcal{H}$ is its Herbrand domain, then A is unsatisfiable if and only if some finite subset of $\mathcal{H}((A))$ is inconsistent.*

For a proof of theorem 2.3 refer to [116].

The problem with the approach for showing that A is unsatisfiable outlined above is not the time taken to test a subset of $\mathcal{H}((A))$ but that if $\mathcal{H}((A))$ is generated according to the increasing complexity of the ground terms in the instances, a prohibitively large number of instances may have to be generated before an inconsistent conjunction arises. Often a small inconsistent subconjugation can be found. Therefore, the problem is to avoid generating consistent instances. One way to achieve this is to predict which instances of two clauses will contain complementary literals (i.e. will "clash") using the information at hand about the way in which terms of $\mathcal{H}$ are constructed. The goal is therefore to apply a succession of substitutions of terms (not necessarily ground terms) for variables, each substitution being

chosen so that the result of applying it to two clauses is that some of the literals in the first clause become equal (or "collapse") to a literal L and some of the literals of the second clause collapse to $\neg L$.
In order to formalize these notions, the concept of a most general unifier must be introduced.

Definition 2.46 *Let $\hat{L} = \{L_1, \ldots, L_n\}$ be a set of expressions. $\hat{L}$ is* ***unified*** *by ς if $L_1\varsigma = L_2\varsigma = \cdots L_n\varsigma$.*
L_1 is the ***common instance*** *of the elements of $\hat{L}$ determined by ς.*

Definition 2.47 *ς is a* ***most general unifying substitution*** *(or* ***most general unifier****) of the set of expressions $\hat{L}$ if for any other unifier θ there is a substitution χ such that $\hat{L}\theta = \hat{L}\varsigma\chi$.*

Intuitively this means that a most general unifier is one that makes the fewest and most general (in the sense of leaving variables wherever possible) substitutions necessary to unify the given set of expressions.
In a number of cases there will be no most general unifier. However, if a set of literals $\hat{L}$ is unifiable, then there is a most general substitution unifying $\hat{L}$ and, moreover, an algorithm can be given for finding such a substitution.
This algorithm, due to Robinson [577], starts with the empty unifier and builds up, step by step, a most general ς_0 that unifies the set $\hat{L}$ of literals. If, at the kth step, the substitution obtained so far is ς_k, and the literals $L_1\varsigma_k, \ldots, L_n\varsigma_k$ in $\hat{L}\varsigma_k$ are not all equal, the procedure changes ς_k on the basis of a disagreement set containing the first well-formed expression in each $L_i\varsigma_k$ which needs to be changed.

Definition 2.48 *Let each $L_i\varsigma_k$ be a sequence of symbols. The* ***disagreement set*** *of $\hat{L}\varsigma_k$ is the set of all well-formed subexpressions of the literals in $\hat{L}\varsigma_k$ that begin at the first symbol position at which not all of the literals have the same symbol.*

The disagreement set is either $\hat{L}$ itself (in which case $\hat{L}$ is not unifiable) or it contains a term or subterm from each literal in $\hat{L}\varsigma_k$.

Unifying substitutions can be constructed component by component. The substitution component (t, x) can be combined with the substitution $\varsigma = [(t_1, v_1), \ldots, (t_n, v_n)]$ provided that $x \neq v_i$ and v_i does not occur in t (cf. definition 2.40) for all i, $1 \leq i \leq n$ as follows:

$$\varsigma[(t, x)] = [(t_1[(t, x)], v_1), \ldots, (t_n[(t, x)], v_n), (t, x)]$$

Algorithm 2.1 ***Robinson's unification algorithm*** *[577]:*

1. *Set $\varsigma_1 = []$ (the empty substitution), set $k = 0$, and proceed with step 2.*

2. *Set $k = k + 1$. If the elements of $\hat{L}\varsigma_k$ are all equal, set $\varsigma_0 = \varsigma_k$ and stop; otherwise go to step 3.*

3. *Let s_k, t_k be the two earliest expressions in the lexical ordering of definition 2.42 of the disagreement set of $\hat{L}\varsigma_k$; if s_k is a variable and does not occur in t_k, set $\varsigma_{k+1} = \varsigma_k[(t_k, s_k)]$ and go to step 2, otherwise stop (there is no unifier).*

Since $\hat{L}\varsigma_{k+1}$ contains one less variable than $\hat{L}\varsigma_k$, the procedure must stop in at most m steps if $\hat{L}$ has m variables.
Also, it is obvious that if the procedure stops in step 2, ς_0 is uniquely determined and its terms contain only function symbols occurring in $\hat{L}$. It is also true that if $\hat{L}$ is unifiable then ς_0 is defined and is a most general unifying substitution. For a proof of the last statement see [400].

Definition 2.49 *A **resolvent** of two clauses C_1 and C_2 is a third clause C_3 obtained as follows:*

1. *If $v_1, \ldots, v_m$ are the variables of C_2 and the highest variable of C_1 in lexical order is u_k, let $\theta = [(u_{k+1}, v_1), \ldots, (u_{k+m}, v_m)]$. Thus none of the variables in $C_2\theta$ occurs in C_1. This is called **standardizing apart** the clauses C_1 and C_2.*

2. *If there exists a pair of sets of literals* $\hat{L}$ *and* $\hat{M}$ *such that* $\hat{L} = \{L_1, \ldots, L_k\} \subseteq C_1$ *and* $\hat{M} = \{M_1, \ldots, M_n\} \subseteq C_2$*, and the set* $\{L_1, \ldots, L_k, \neg M_1, \ldots, \neg M_n\}$ *is unifiable, let* ς_0 *be the most general unifying substitution according to algorithm 2.1 so that* $\hat{L}\varsigma_0$ *and* $\hat{M}\theta\varsigma_0$ *are complementary literals (i.e. they are singleton sets whose elements are complementary literals). Then* C_3 *is the set of literals*

$$(C_1 \setminus \hat{L})\varsigma_0 \ \cup \ (C_2 \setminus \hat{M})\theta\varsigma_0$$

Any pair of clauses has at most a finite number of resolvents, because there are only finitely many pairs of literals $\hat{L}$ and $\hat{M}$, and for each pair at most one substitution ς_0. The resolvents of a pair of ground clauses are ground resolvents. A resolution deduction is defined as follows:

Definition 2.50 *Given an initial set of clauses* $\hat{S}$*,* $\mathfrak{R}(\hat{S})$ *denotes the union of* $\hat{S}$ *with the set of all possible resolvents of clauses in* $\hat{S}$*.* $\mathfrak{R}^n$ *is defined as follows:*

$$\begin{aligned}\mathfrak{R}^0(\hat{S}) &= \hat{S} \\ \mathfrak{R}^{n+1}(\hat{S}) &= \mathfrak{R}(\mathfrak{R}^{n+1}(\hat{S}))\end{aligned}$$

Definition 2.51 *There is a* ***refutation*** *of a set of clauses* $\hat{S}$ *(that is,* $\hat{S} \xrightarrow{*} \Box$*) if and only if* $\Box \in \mathfrak{R}^n(\hat{S})$ *for some* n*.*

There is an infinite number of possible instances of a literal any time the Herbrand domain contains a single constant and a single function letter. Herbrands theorem ensures that, given an inconsistent set $\hat{S}$, some finite subset of $\mathcal{H}(\hat{S})$ is inconsistent, thus $\Box \in \mathfrak{R}^n(\hat{S})$ for some n.

However, it is not possible to put an upper bound on n. That is, there is no decision procedure for determining whether or not $\Box \in \mathfrak{R}^n(\hat{S})$ for any n for an arbitrary set of clauses $\hat{S}$.

The unification principle outlined here has been applied in most computational proof calculi, albeit in much more sophisticated form; for details refer e.g. to [405, 578], see also section 3.3.

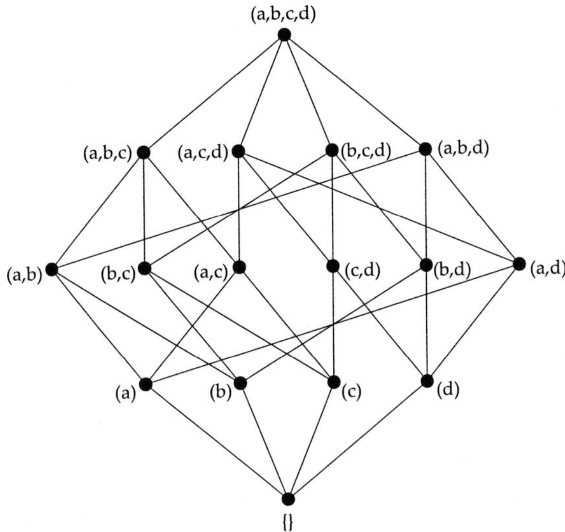

Figure 2.1: Hasse diagram for the lattice $\mathcal{P}(\{a, b, c, d\})$ ordered by set inclusion

2.4 Lattices

Lattice theory supplies a well-understood mechanism for mathematical modeling [78, 273]. Based on lattice algebra, Birkhoff also introduced the tool of formal concept analysis based on lattice theory [236].

Definition 2.52 *Let R be a relation on the set X. R is said to be*

- ***Reflexive*** *if xRx for all $x \in X$*
- ***Symmetric*** *if xRy implies yRx for all $x, y \in X$*
- ***Antisymmetric*** *if xRy and yRx imply $x = y$ for all $x, y \in X$*
- ***Transitive*** *if xRy and yRz imply xRz for all $x, y, z \in X$*

If R is reflexive, symmetric, and transitive, then R is called an ***equivalence relation*** *on X. If R is reflexive, antisymmetric, and transitive, then R is called a* ***partial order*** *on X.*

Definition 2.53 *Let $(X, \leq)$ be a set where $\leq$ imposes a partial order*[1] *on X and let $x, y \in X$. If $x \leq y$, then x is said to be* ***contained*** *in y. If $x \leq y$ and $x \neq y$, then x is said to be* ***properly contained*** *in y which is denoted as $x < y$. If $x < y$ and there is no element $a \in X$ such that $x < a < y$, then y is said to* ***cover*** *x.*

Definition 2.54 *Let $(X, \leq)$ be a set where $\leq$ imposes a partial order on X. Let S be a subset of X.*
An ***upper bound*** *of S is an element $b \in X$ such that $x \leq b$ for all $x \in S$. A* ***least upper bound*** *or* ***lub*** *of S is an element $m \in X$ such that $x \leq m$ for all $x \in S$ and if $x \leq m'$ for all $x \in S$, then $m \leq m'$.*
A ***lower bound*** *of S is an element $b \in X$ such that $b \leq x$ for all $x \in S$. A* ***greatest lower bound*** *or* ***glb*** *of S is an element $m \in X$ such that $m \leq x$ for all $x \in S$ and if $m' \leq x$ for all $x \in S$, then $m' \leq m$.*

Definition 2.55 *A partially ordered set $\mathcal{L}$ is called a* ***lattice*** *if every pair of elements in $\mathcal{L}$ has a least upper bound and a greatest lower bound.*
The least upper bound of x, y is denoted as $x \triangledown y$ and called ***join****, the greatest lower bound is written as $x \vartriangle y$ and called* ***meet****.*

Observation 2.17 *For any set S the power set $\mathcal{P}(S)$, ordered by set inclusion, is a lattice.*

Figure 2.1 on page 39 shows the Hasse diagram for such a lattice.

[1] Relation symbols are chosen for ease of reading with their intuitive meaning on real numbers, no restriction on X is implied by this notation

CHAPTER

THREE

Related Work and State of the Art

This chapter is intended to provide an overview of both directly related work and of basic results that determine the functional bounds of the architecture discussed here.

Additional references to and discussion of related research is also provided in subsequent chapters, the decision for inclusion of material in each case being somewhat arbitrary.

Section 3.1 recapitulates basic results imposing boundaries on the expressiveness (provided that deterministic mechanisms are used for decision) of security policy mechanisms and the ability to prove security and safety properties for arbitrary policies.

In addition, section 3.2 identifies strands of research and development towards expressive, flexible multipolicy systems.

Section 3.3 then briefly provides references to material on automated reasoning with the primary focus on automated deduction as an example of a sufficient mechanism for the requirements formulated in chapter 5.

Subsequently, section 3.4 discusses several approaches to network security directly related to the mechanisms described in chapters 9 and 10.

Finally, section 3.5 discusses operating system security mechanisms related to the enforcement mechanisms described later.

3.1 Some Results in the Theory of Secure Systems

Given that comprehensive formal specification and proof appears not feasible (cf. section 12.2.1), research has concentrated on mechanisms which enforce certain properties of programs.
The feasibility of proving properties about high-level abstract models was first studied by Harrison, Ruzzo, and Ullman [293] based on a model of protection systems initially proposed by Lampson [396] for operating systems; Conway, Maxwell, and Morgan simultaneously and independently developed an equivalent model for database systems [145]. The model defines a system as a triple (S, O, A) where

1. S is a set of subjects, the active entities of the model. Subjects are also considered objects in this model: $S \subseteq O$.

2. O is a set of objects representing the protected entities of the model. Each object is uniquely identified by a name.

3. A is an access matrix with rows corresponding to subjects and columns to objects. An entry $A[s, o]$ contains a list of privileges $\{r\}$ of subject s for object o.

Harrison, Ruzzo, and Ullman identified six primitive operations used for characterizing access control (enter r into $A[s, o]$, delete r from $A[s, o]$, create subject s, create object o, destroy subject s, and destroy object o) and defined their effect on the access matrix (cf. section 7.1); the effect on the system is derived by defining state transitions from a state Q to a state $Q' = (S', O', A')$ under the conditions of the access matrix and for a given operation op, denoted as $Q \vdash_{op} Q'$. The effect of an operation on the state of protection (i.e. security) of the system is that $Q' = Q$ if the condition for the operation is not satisfied. The most fundamental result from [293] is theorem 3.1.

Theorem 3.1 *It is undecidable whether a given state of a given protection system is secure for a given generic right.*

Proof (sketch). An arbitrary Turing machine can be encoded in a protection system such that the granting of a right corresponds to the Turing machine entering a halting state. Therefore, if security is decidable, then so is the halting state (for a detailed proof see [293] or [171]). □

Theorem 3.2 *There exists an algorithm that decides whether a given system which for each operation* op_i *performs a single primitive operation (called* ***mono-operational****) and initial state* Q_0 *is secure for a given generic right r.*

For a proof of theorem 3.2 refer to [293, 171]. Even for such restricted systems which are not readily applicable to prevalent operating system design, Harrison, Ruzzo, and Ullman showed that the problem is $\mathcal{NP}$-complete, although algorithms polynomial in the size of the initial access matrix exist [726, 239, 537].
Theorem 3.1 implies that the set of secure protection systems is not recursive. Since one can enumerate all protection systems and all sequences of operations causing a violation of the security property one obtains corollary 3.1.

Corollary 3.1 *The set of insecure systems is recursively enumerable.*

Based on these results Denning, Denning, Garland, Harrison, and Ruzzo showed the corollaries 3.2 and, from theorems 3.1 and 3.1 obtained corollary 3.3 [173].

Corollary 3.2 *Any theory* $\mathcal{T}$ *adequate for proving security must be undecidable.*

Corollary 3.3 *There is no recursively axiomatizable theory* $\mathcal{T}$ *for proving security.*

Finally, theorem 3.3 establishes that systems for proving security are of necessity incomplete.

Theorem 3.3 *Given any recursively axiomatizable theory $\mathcal{T}$ and any representation of security in $\mathcal{T}$, one can construct a protection system p for which $\mathcal{T} \vdash t_p$ if and only if p is insecure. Furthermore, if $\mathcal{T}$ is sound, then p must be secure, but its security is not provable in $\mathcal{T}$.*

For a proof of theorem 3.3 refer to [173]. Even if only protection systems of bounded size in the number of objects and domains of access are considered, there exists no efficiently computable verification mechanism; the argumentation for information flow controls for the above theorem follows by analogy.
For concurrent systems, a distinction of correctness criteria into safety and liveness properties was proposed by Lamport [389].
Safety is intuitively defined as ensuring that "nothing bad happens", liveness as "something good happens eventually". Lamport defined safety properties more precisely in [391]. For the following, let ϵ denote a finite or infinite execution having i or more steps, τ' a finite execution, $\epsilon[i]$ denote the prefix of the first i steps of ϵ, $\tau'\epsilon$ the execution of τ' followed by that of ϵ, Ψ the set of all finite and infinite sequences, and Π^- the set of all finite prefixes in the set Π of finite and infinite sequences.
Properties are defined in the literature on linear-time concurrent program verification as a set of executions for which set membership is determined by each element alone and not by other members of the set, leading to definition 3.1.

Definition 3.1 *A property Γ is a safety property if and only if for any finite or infinite execution ϵ equation 3.1 holds.*

$$\epsilon \notin \Gamma \Rightarrow (\exists i : (\forall \tau \in \Psi : \epsilon[i]\tau \notin \Gamma)) \tag{3.1}$$

This definition is due to Alpern and Schneider [15] which extended Lamport's formulation by representing a finite prefix of an execution as the set of all possible continuations from that point on; similar proposals were also made by Lichtenstein, Pnueli, and Zuck [410] as well as Manna and Pnueli [440].

Alpern and Schneider also showed that safety properties are closed under conjugation and finite disjunction, resulting in a topology where safety properties are closed sets and liveness properties dense sets in the Cantor topology [155, 769, 99, 100, 702]; this correspondence can be used to show that every property can be represented as a conjunction of a safety and a liveness property (the decomposition theorem) although topological arguments are not necessary for proving the theorem [642].
The theorem can also be viewed as a transformation of Büchi automata [113, 114, 115, 702] corresponding to the properties [16]. The abovementioned characterizations of properties agree on the description of safety properties via closed sets in the Cantor topology; there are differences in the treatment of liveness properties in the descriptions [410, 440]. The latter papers establish correspondences between the Borel hierarchy level G_δ within the framework of temporal logic. However, this discussion is concerned only with safety properties and does not consider liveness and fairness [538, 564, 280, 47].
Schneider characterizes a class of security policies P in the form

$$P(\Pi) : (\forall \epsilon \in \Pi : \hat{P}(\epsilon)) \tag{3.2}$$

called Execution Monitoring (EM) policies which monitor execution steps of a program (target) and terminate the execution of the target if it violates a security policy; policies in this terminology are properties.
EM policies must be prefix closed

$$(\forall \tau' \in \Psi^- : \neg\hat{P}(\tau') \Rightarrow (\forall \epsilon \in \Psi : \neg\hat{P}(\tau'))) \tag{3.3}$$

and must terminate executions violating a policy after a finite number of executions:

$$(\forall \epsilon \in \Psi : \neg\hat{P}(\epsilon) \Rightarrow (\exists i : \neg\hat{P}(\epsilon[i]))) \tag{3.4}$$

Schneider also proposed the modeling of said policies and their enforcement via a subclass of Büchi automata [188, 189, 702]. Since the prefixes to be considered can be arbitrarily large, this would imply that the state space of the Büchi automata is also unbounded, which must be remedied by way of an arbitrary upper boundary [643]. Regardless of this observation, the above leads to observation 3.1:

Observation 3.1 *If a reference monitor does not use traces of executions in the determination of security policy compliance and does not modify the target it satisfies equation 3.2 and is hence limited to enforcing the EM class of security policies.*

Trace-based security models and equivalent formalisms such as process algebrae, temporal (modal) logic or CSP approaches are briefly covered in section 7.2.2. Generally, research has focused on static models of the systems under consideration in conjunction with the specification of security properties over this model and reasoning over the model thus generated.
Modal logic in general has been considered by several researchers (see also section 2.1); Glasgow and MacEwen discussed such mechanisms for both confidentiality and integrity security models [247, 246] (cf. section 7.3) and proposed the use of an S5 modal logic [317] and possible worlds semantics [379] in unified branching time [440] to obtain a logic for reasoning about security [248].
However, this mechanism is limited in its efficiency and, since it is based on a propositional calculus, does not permit the axiomatic specification of policy instances and proof of hypotheses within the theory as proposed by Glasgow and MacEwen. Another approach using modal logic for reasoning about authentication was proposed by Abadi *et al.* [1].
Similarly, other types of modal logic have been investigated for specification and reasoning about security models; deontic logic [437, 25, 299] as introduced by Mally and formalized by Anderson was also proposed by Bieber [76, 77]. Syverson as well as Biber [76, 691] proposed the modeling particularly of information flows within epistemic logic [232, 301, 741]; this is an approach commonly used in protocol verification in the form of e.g. BAN (Burrows, Abadi, Needham) logics [117, 588].

3.2 Flexible Policy Mechanisms

Systems supporting a single security policy or model (see chapter 7) may not be adequate for an application area or provide the assurances (e.g. integrity or confidentiality) required [277, 778, 48], and even within a given application area policy elements can change, in some instances dynamically [684]. Restriction to a single, possibly mismatching policy such as the multilevel security policy described in the TCSEC [733] is therefore inadequate (TCSEC itself also did not take into account all requirements set forth in DoD 5200.1-R [736, 735] even though the Originator Controlled Policy prescribed in DCID 1/7 predates the TCSEC [182]).
Hosmer provided a survey of the concept of the multipolicy paradigm for supporting the enforcement of different protection policies as well as a summary of earlier work [312, 313, 316]. This approach is based on the concept of metapolicies, i.e. originating in work on database access policies; the term itself was introduced by Hosmer [311]. Early work in this area is summarized in [315, 167]; a model of the multipolicy machine [314] has been provided by Bell [61].
The mechanisms for representing multiple policies (albeit not at the same time) by means of separate decision modules enforcing access control rules were developed at MITRE in the Unified Access Control project under the leadership of Abrams and LaPadula [7, 403] based on earlier work on rule bases by Page *et al.* [532] in a in feasibility study performed for the SDIO. The resulting framework model is also known as the GFAC.

Moffett *et al.* [482] view policy (not restricted to security alone [483]) as composed of two components, an imperative policy that controls initiating or inhibiting actions from the secondary policy, the authority policy which permits or prohibits the actions. In this framework, the functions that implement the policy components are separated into a Situation Monitor and Reference Monitor, respectively and are modeled as multiple objects that can themselves be manipulated and grouped into hierarchies to resolve conflicts between policies. This work was later extended by Lupu and Sloman for metapolicies for the specification of logical predicates over multiple policies and conflict resolution; the conflict resolution is to occur prior to the actual deployment of the policies [432, 431, 433].

The DGSA[1] [517] specified, beginning in 1993 security principles and target security capabilities that were intended as guidance for system security architects in creating specific security architectures based on earlier this and other earlier work such as the separation of policy and enforcement first proposed for the HYDRA system by Levin *et al.* [777, 408, 140]. Among the DGSA security principles were those organizing information in a manner consistent with mission objectives and allowing unique security policies to be applied to distinct sets of subjects and objects.
For this the DGSA defined common terminology including the Security Policy Decision Function (SPDF) as the function within a reference monitor deciding policies and the Security Policy Enforcement Function (SPEF), both of which are extensions of the ISO Access Control Framework [327] which specified access control mechanisms only; the SPDF has also been referred to[2] as a Custodian [384].

The influence of the DGSA furthermore is evident in the policy flexibility introduced in DTOS (see section 3.5.1) and later Synergy--sponsored as well as other research [386, 385, 41]. Fraser introduced an object-oriented model-based framework for the specification and implementation of multiple concurrent discretionary, mandatory, and device-specific SPDF mechanisms [227].
In addition to GFAC, several later research systems have concentrated on using language-based security policy specification. Woo and Lam proposed a mechanism for authorization using multiple policies. In their model, the SPDF is a model which decides on authorization requests formulated as $\mathsf{req}(r, s, o)$ for a type of access $r \in R$ to an object $o \in O$ by the subject $s \in S$ with either $\mathsf{grant}(r, s, o)$ (a successful positive authorization), $\mathsf{deny}(r, s, o)$ (a successful negative authorization), or the failure to reach a decision $\mathsf{fail}(r, s, p)$. A

[1]Since superseded by the C^4ISR framework [518], but substantially unchanged in the goals pursued.

[2]While referencing the DGSA directly

policy consists of a 4-tuple $\langle P^+, P^-, N^+, N^- \rangle$ where each component is a subset of $\{(r,s,o) \mid r \in R, s \in S, o \in O\}$ and soundness constraints $P^+ \cap N^+ = \emptyset$; policies for which the constraints hold satisfy $\mathsf{grant}(r,s,o) \Leftrightarrow (r,s,o) \in P^+$, $\mathsf{deny}(r,s,o) \Leftrightarrow (r,s,o) \in N^+$, and $\mathsf{fail}(r,s,o) \Leftrightarrow (r,s,o) \notin P^+ \cup N^+$.

The policy language is restricted to a fixed number of variables and constants for which the semantics are defined by a fixed number of rules of inference which include precedences of certain rules over others as well as the requirement for an explicit ordering in the evaluation of the rule objects, resulting in nonmonotonic semantics [771, 572, 429].

A similar approach was derived based on earlier work by Bertino *et al.* on multipolicy database access control [70, 73, 71, 72] that were generalized to multiple access control policies for general systems by Jajodia et al. [340, 339]. The policy defined arbitrary types of objects and fixed types of subjects as well as groups of subjects and roles for subjects; the scheme defines hierarchical set inclusion relations on both roles and groups.

The policy language is a mapping $\langle o, u, R, a \rangle \rightarrow \{\texttt{authorized}, \texttt{denied}\}$ where o is an object, u a user, R a role set, and a an action; in addition, a number of rules of inference and predicates are defined that can express both the granting and denial of authorization via the predicate $\mathsf{grant}(o, u, R, \mathsf{op}(a)) \leftarrow P_1 \wedge \ldots \wedge P_k$ where op is either granting (+) or refusal (−) of authorization and P_i are the predefined predicates cando specifying a positive or negative authorization, dercando specifying a positive authorization for subjects within a hierarchy, done providing axioms over authorizations, resolution predicates do specifying a selection of preferred rules in case of inconsistent rules and several binary predicates: dirin specifying set membership in a superset and in specifying set membership[3], and typeof specifying set comprehension for objects. The predicate error permits the explicit specification of inconsistent states.

As in the case of the proposal by Woo and Lam, the semantics are nonmonotonic while multivalued logic and extraneous rules of in-

---

[3]redundantly

ference increase the complexity and limit expressiveness at the same time. A related formulation, also derived from relational database concepts, was proposed by Bonatti *et al.* which used an algebraic formulation for specifying policies and the derivation of overall policies from multiple component policies.

In this approach, the transformation of algebraic specifications into logic programs as the actual implementation. The expressiveness of the policy is, however, limited to a proper subset of first order logic in that it is restricted to monadic operations with at most unary predicates [94, 95].

Another approach, similar to some extent to the work by Bonatti *et al.*, for supporting multiple policies was proposed by Wijesekera and Jajodia; this approach is based on representing policies as purely syntactical entities over which simple algebraic structures are defined [768].

This limits the expressiveness of the model to the somewhat crude tools provided by the algebraic structures representing policies as atomic entities, although it provides an improvement over an earlier model by Jajodia *et al.* in that it permits the formulation of separation of duty policies [338].

Bai and Varadarajan proposed a logic-based mechanism for changes in policies based on a policy model consisting of a tuple (C, M) where C is a finite set of first order formulae over a language consisting of ground atoms and M is a Herbrand interpretation thereof; C is viewed as an immutable component of the policy while M is changeable.

Transformations on the policies are defined as pre- and postconditions over state transitions within the model, permitting multiple model states satisfying the invariants [46].

While providing limited added expressiveness over the mechanisms described earlier by Sandhu *et al.* [602, 601], the explicit formulation and transformation of policies in model-based semantics is noteworthy.

A variation on the mechanisms proposed by Necula (cf. section 2.1) was proposed by Felten *et al.* for authentication in the form of a higher-order logic mechanism [31, 54] particularly across domains in case of WWW services; this has the obvious disadvantage of being undecidable.

3.3 Automated Reasoning

While the flexibility and expressiveness of the architecture discussed here derive to a considerable extent from the use of techniques in automated deduction, only the existence of such mechanisms, the subject of a mature field, is postulated.

The saturation-based mechanism described in section 2.3 was mainly selected for being the first efficient automated proof mechanism and must therefore be seen as only an example of such a mechanism that can be described with the brevity called for here.

A perspective on seminal work in the field is provided by [656, 657]; this is expanded upon in [161]; the relation between traditional and mechanized proof is discussed in [435, 436].

Given the limited requirements imposed on the proof mechanisms in this dissertation, efficient mechanisms may also, besides resolution, include term rewriting systems, Knuth-Bendix completion, tableaux, term graph rewriting, and term ordering [98, 42, 551, 177, 281].

The collection [578, 579] provides an overview of principles and methods in automated reasoning; the CADE, TABLEAUX, and FTP conferences[4] conferences also provide an overview of current activities in the field [372, 237, 453, 752, 166, 489, 58, 123, 269].

3.4 Network Security Mechanisms

Network security mechanisms are instrumental in realizing a distributed security architecture; this section discusses mainly the presumed initial appearance of relevant related concepts and does not

[4]These have been merged into a multiconference "International Joint Conference on Automated Reasoning" (IJCAR) held jointly in 2001 and to be held in 2003.

represent a comprehensive overview of research and standardization activities in each area.

3.4.1 Guard Mechanisms

Guard mechanisms are employed in multilevel secure systems, i.e. enforcing classification and document handling schemes [515]; this can be accomplished in the internetworking of multilevel secure, partitioned (or compartmented), system high, or dedicated systems in the DoD MLS model (see definition 7.2).
MLS guards control the flow of information across security boundaries; several types of guard mechanisms can be distinguished by taxonomic features, namely the requirement for human review of the data flow, support for unidirectional data flow, and support for bidirectional data flow.
The term itself is derived from the system by said name (the Advanced Command and Control Architecture or ACCAT Guard) developed by Woodward *et al.* at MITRE and Logicon [20, 774, 531] under Navy sponsorship [266, 697] which provided monitoring and sanitization of bidirectional queries and responses between database systems operating at different security levels with human review and was based on the KSOS system for trusted processing (see page 195); although a version based on a security kernel enforcing enforces the axioms of the Bell-LaPadula model was created, the requirement for explicit downgrading and the inability to determine the semantics of sanitization limited the enforceability of the $*$-property.
Even unidirectional guard mechanisms must take into consideration that malicious code such as Trojan horses are transmitted from low to high security levels; Denning [172] describes the Trojan horse problem in an automated security guard.
Deployed guards include the Standard WWMCCS Guard providing the means for DoD organizations to extract Secret and less classified data from the Top Secret WWMCCS, which operates in the system-high mode, and to make that data available automatically to users on Secret command and control systems [714, 715].

Other examples of guard systems include the MMS developed by Landwehr, Heitmeyer, McLean *et al.* [401, 402, 297, 296, 148] and the Standard Mail Guard whose implementation was derived from the LOCK prototype (see section 3.5.1) [667].

The NRL pump by Kang and Moskowitz [354] which was later extended to networked systems [356, 355] represents a guard mechanism explicitly intended for limiting covert channels [397, 417, 362, 363].
Davida *et al.* introduced the systematic use of cryptographic mechanisms for the control of information flow between dedicated units in a multilevel environment [160].
Epstein and Monteith, moreover, proposed the use of probabilistic mechanisms for flow control based on information flow signatures [193, 485]; this enables a probabilistic automated downgrading mechanism.

The labeling of network traffic in support of guard mechanisms was part of the basic Internet Protocol version 4 specification [561] which stipulated compliance to the markings set forth in DIAM65-19 [727] in the form of IPSO.
This was later amended to RIPSO [681, 366] in that additional security information could be specified on a per-datagram basis (if necessary in multiple instances per datagram) but did not see actual deployment or the further specification required and has since been deprecated.
Instead, an attempt was made in parallel by the TSIG to provide a common labeling architecture for civilian and defense applications; for this purpose an IETF working group (CIPSO) was chartered to which a specification was submitted [65, 653].
This draft was also never acted upon although the TSIX led to several implementations including Sun Solaris, SGI IRIX, and Digital MLS+ [144, 717] and also included APIs for setting and reading MLS labels via orthogonal extensions to the BSD socket API.

Security attributes in TSIX could be sent with each packet or exchanged via an out-of-band protocols in the form of tokens to be

mapped later onto actual attributes, SATMP and SAMP, which permitted performance enhancement and the use of multiple labeling schemes identified by a DOI[5].
Even during the development of CIPSO, the DoD promulgated separate labeling standards [730, 731, 734]; MIL-STD 2045-48501 is part of the standards mandated by the JTA [516].

3.4.2 BLACKER

The BLACKER proof-of-concept was initiated by the DoD in the late 1970s by the contractors SDC (responsible for software) and Burroughs (responsible for hardware). It provided a mechanism for centralized key management and access control at the level of granularity of individual systems [762].
This was accomplished by attaching a device referred to as the BLACKER front end (BFE) to the node in the place of the X.25 data communications equipment (DCE) endpoint connecting to the Defense Data Network (DDN) operated by the Defense Communications Agency (DCA) [204].
Nodes using attached BFE devices were able to communicate with each other over insecure packet-switched networks using data paths secured by the encryption services of the BFEs. The nodes in this virtual network were referred to as node of the "Red" network. The packet-switched network that carrying both the data secured by BFEs and any other unsecured data was referred to as the "Black" network.
BFE devices received authorization and address translation services from an Access Control Center (ACC) residing on the Black network. The ACC performed access control decisions determining which hosts were allowed to communicate with each other; for this purpose the individual nodes attached to BFEs were considered subjects within the Bell-LaPadula security model [64].
Communication between the hosts was considered to be the sole object; each of these objects was at a single security level while

[5]This concept was later re-used in the definition of the IPv6 IPSec key exchange mechanisms.

subjects could be operating at either single or multilevel security (MLS).

A Key Distribution Center (KDC) residing on the Black network provided encryption keys and key management services which were used to secure inter-BFE communication; in case of the production system the BFE used a custom-built real-time kernel while both the ACC and the KDC used an off-the-shelf kernel (GEMSOS) developed by Gemini Computers [633].

BLACKER introduced the concepts of cryptographically secured virtual private networks [632] as well as the use of cryptographic co-processing devices for this purpose; its impact on other research was limited, however, due to the classified nature of the work. It also constituted the first distributed system that was evaluated against the TCSEC criteria (the Gemini Trusted Network Processor which shared the GEMSOS kernel with BLACKER was evaluated according to the TNI as compliant to the A1 class which did not exist at the time of evaluation for BLACKER; see section 4.4) for the A1 class including all of its components which was successfully completed in 1991 [498].

3.4.3 Distributed Firewalling

Firewalls for the restriction of network communication in packet--switched networks were initially based on the topological separation of disjoint networks [484, 129, 636]. Based on observations on the use of mobile devices, remote access mechanisms, and the performance requirements for choke point firewalls resulting from increasing network performance, Bellovin proposed the migration of firewall enforcement to the nodes to be protected while retaining a central policy definition for network access control [66].

Instead of relying on topological information for obtaining statements on the identity of an entity, Bellovin proposed to use the mechanisms for the use of public key infrastructures inherent in IPSec [701] which also addressed the issue of support for virtual private networks found in traditional firewalls [371], although this was already done by Chitturi [132] in 1998 within the Fluke project context (see section 3.5.1). This permitted the use of a general policy language for trust management such as KeyNote [88, 87, 39, 86]. A prototype based on this proposal was developed at the University of Pennsylvania [330, 131]; it embedded the enforcement mechanism within the BSD socket API system calls `connect(2)` and `accept(2)` of the OpenBSD operating system while the actual policy was determined by a user level process, the policy daemon.

This exposed the policy daemon to manipulation and introduced considerable latency due to frequent changes between supervisor and unprivileged operation modes. In addition, the restriction to the BSD socket API implied that no control over lower-level operations commonly associated with the term "firewall" were exerted as well as a limitation with regard to other types of APIs that permitted network communication.

Another approach to distributed firewalling, also derived from concepts introduced by Bellovin was pursued by Payne and Markham at SCC on behalf of the Space and Naval Warfare Systems Center. To counter the threat of Trojan horses and tampering by users, Payne and Markham realized the embedding of the firewalling mechanisms (EFW) on a COTS network interface card with cryptographic and limited processing capabilities[6] as initially proposed by Bellovin. In this model, the EFW is assumed to be the sole means of communication to the outside, tamper-resistant to attacks from the host and tamper-resistant to the network. To achieve this, the NIC is programmed in such a way that changes in the rules governing network traffic (simple packet filtering due to limited processing capabilities) may be initiated only via data received from the network interface authenticated by a shared secret[7], nominally a policy server. The policy server specifies the individual packet filtering rules which are requested by each EFW on initial startup from the policy server database after relying on a restricted set of initial rules; the server

[6]Developed in DARPA-sponsored research, the 3Com Corporation *Typhoon* series of network interface cards (3CR990) and later models provide these capabilities.

[7]The 3Com Typhoon NICs do not have asymmetric cryptography capabilities.

also receives audit messages from the EFWs [541, 444]. A similar mechanism for EFWs was also developed by Friedman and Nagle at Carnegie Mellon University [234, 229].

3.5 Operating System Security Mechanisms

A number of security-centric, mainly historical operating system developments are discussed in section 4.2; this section discusses additional related seminal research activities.

3.5.1 The Synergy Program and its Antecedents

During the 1992–1993 timeframe the United States National Security Agency initiated a research program aimed at providing an "open architecture" as well as prototypes of secure distributed systems which were to be based on security policy flexible microkernels. The requirement for such a system was determined by NSA in collaboration with DISA and NIST. The program, outlined in [605] was intended from the outset to be a potential basis for both governmental and commercial systems and therefore was intended to be policy-flexible.

The objectives for the operating system integrated existing research, namely research on a microkernel-based operating system as well as in secure operating systems.

A core concept – capabilities – of one of these components can be traced back to the Codeword facility of the Rice University Computer (R1) designed by Iliffe and its SPIREL [272] monitor. Codewords denoted a 2-tuple (x, r) where x specified the unique logical address of an object and r a set of access rights to x [324] and were generalized for R2[8] [323].

The term capability itself is due to Dennis and VanHorn [176] who modeled the protection domain of a process in a capability list containing said tuples; in some architectures this was provided in soft-

[8]Never completed.

ware only while others implemented the indirection via capabilities to the actual memory location in hardware [195, 492]. To ensure that capabilities were not manipulated by non-system processes, Redell and Fabry proposed the use indirect references to the capability vector (itself protected by hardware mechanisms); this mechanism also permitted the central storage and deletion of capabilities [571].

Capabilities proved influential, other examples of capability-based systems include UCLA Data Secure Unix (see chapter 8), iMAX [351], StarOS [348], HYDRA [777, 140], SWARD [490, 491], CAP [509], KeyKOS [223, 289, 567, 222, 399, 93] and its successor EROS [649, 651, 652, 650] as well as Amoeba [693, 487, 694, 695, 692].

The direct antecedent of the Synergy architecture, however, was PSOS which was designed under the direction of Feiertag and Neumann [203, 510].

It was designed as a strictly hierarchically layered system using a design methodology developed at SRI and implemented as layered abstract machines using a semiformal technique also developed at SRI [539, 102, 103, 580, 658, 409, 128] and based on the object orientation implied by a capability architecture.

Although PSOS itself was not implemented, the concepts developed herein were reinstantiated in the Honeywell Secure Ada Target (SAT).

The SAT design was based on a capability hardware architecture (the Honeywell Level 6) and was designed using a hierarchical methodology similar to that of PSOS using the Gypsy Verification Environment [18, 263, 463, 262, 12] and provided an access matrix security model implementing multilevel security relying heavily on hardware architecture to realize protection mechanisms [358, 91, 92, 784].

SAT also introduced a new terminology; it referred to groups of objects in common access control terminology as types (in analogy to object-oriented design and analysis) and execution domains, or domains for sets of processes [90].

While SAT did provide an implementation, this system also was not deployed mainly due to its dependency on a specific hardware architecture. However, the NCSC initiated the LOCK (Logical Co-processing Kernel) program in 1987 and awarded the contract to Honeywell Corporation, effectively continuing the efforts related to SAT, although based on a Motorola MC68020 CPU.

As with SAT, LOCK was a standalone system and did not incorporate networking support although it was based on the TCSEC A1 class criteria.

LOCK introduced a minimized TCB containing only elements critical to security such as process management and a single level (flat) memory-mapped file system.

Application programming interfaces were provided by a personality layer providing POSIX compatible services, LOCK/ix which emulated Unix file system semantics on the storage services of LOCK in separate instances for each process and provided a simulation of the Unix superuser mechanism in nonprivileged mode, subordinating superuser privileges to the system privileges.

LOCK also integrated cryptographic mechanisms (e.g. for external storage encryption) in the TCB and incorporated an external hardware coprocessor, SIDEARM (System Independent Domain Enforcing Assuring Reference Monitor), for performing all security decision and audit functions.

SIDEARM was contacted on each access to an object and contained the only authoritative access control information [603, 629, 604, 699, 668]. LOCK was later equipped with networking support and was deployed in a limited role (see section 3.4.1).

The other major line of development preceding the Synergy architecture was based on the message-passing-based research operating system Mach developed at Carnegie Mellon University which was itself derived from two earlier CMU systems, Accent and RIG [51, 569, 8, 568, 82].

While early versions of Mach were developed within the framework of 4.2 and later 4.3 BSD kernel [404], Mach 3 reduced the kernel to process handling (tasks and threads in Mach terminology) and scheduling, message-based IPC (ports, which were controlled by a capability mechanism), and virtual memory management while providing the BSD personality as user-mode services outside of the microkernel [261]. The basic abstractions of Mach permitted the development of distributed computing environments by means of message passing between nodes of a distributed system, although it had initially been designed as a standalone operating system based on the BSD Unix model.

While Mach has been the basis for a number of developments both internal and external to CMU, only Trusted Mach (TMach), developed at Trusted Information Systems is of relevance to the discussion here.

TMach was designed as a multilevel secure system intended for evaluation against the TCSEC B3 class or ITSEC F-B3 functionality class, high minimum strength in conjunction with the E5 evaluation level [143].

As in CMU Mach 3.0, the microkernel TCB was segregated from (non-TCB) operating system personalities; additional TCB elements were provided for the administration of the system. Security functionality was encapsulated in a reference monitor access mediation mechanism called the Root Name Server (RNS) which implemented a multilevel security policy based on access control lists for discretionary access control and mandatory access control in the form of the Bell-LaPadula model [62].

The RNS identified objects according to names which were themselves grouped into types, analogous to the mechanism introduced by SAT [106, 707, 706, 705, 711, 710, 708, 709].

The LOCK and TMach stands were joined in 1993 in the Distributed Trusted Mach (DTMach) study by the Secure Computing Corporation[9] (SCC) on behalf of the RADC [606, 212, 607, 609, 608].

DTMach was to follow the distribution approach for Mach developed at CMU [138] and to implement a reference monitor mechanism — albeit referred to as domain and type enforcement; type enforcement is based on two sets of attributes, one associated with the current domain of execution of a subject, the other associated with an object. These attributes define the type of object on which

[9]Secure Computing Corporation had been divested by Honeywell Corporation and retained the staff that had worked on SAT and LOCK.

the access control mechanism restricts operations based on the domain of execution of the subject attempting the operation — in the form of security servers replacing the dedicated SIDEARM server of LOCK which were accessed via messages analogous to other function servers within Mach.

Besides permitting a modular design with the kernel itself enforcing the security policy but deferring to the security server for actual decisions, one of the objectives for this was to permit changes in the security policy to be implemented.

The security servers implemented the domain and type enforcement mechanism by defining relations (tables) representing access rights on each tuple of domains and types, thus permitting policies other than MLS to be implemented.

Work at Carnegie-Mellon University by Jones [347] showed that type enforcement can be used for various aspects of (non-lattice based) memoryless subsystem problems (i.e. variants of the confinement problem [397]) while Jones and Wulf showed that type enforcement can support non-discretionary access controls that can be represented as a lattice [350].

Cohen and Jefferson subsequently showed that type enforcement can support a variety of non-discretionary policies that cannot be represented as a lattice [140] while Boebert and Kain discussed type enforcement as an alternative to hierarchical integrity policies described by Biba [75].

The Distributed Trusted Operating System (DTOS) was the first system to be funded under the Synergy program starting in 1993. One of the primary objectives for DTOS was the design and implementation of a high assurance equivalent to the Mach microkernel using semiformal and formal techniques [679].

Other objectives were carried over from the DTMach study such as policy independence including the ability to implement MLS policies and the related design decision to separate enforcement from policy decisions in the form of security servers.

A number of security models (policies in the terminology of DTOS) were shown to be implementable based on the DTOS prototype.

The limitation to a single subject and object for consideration by the security server imposed limitations as well as the inability to parametrize policies (i.e. only static permissions could be queried). Another limitation (primarily due to caching and the resulting issue of cache coherence) was the retraction of permissions; the behavior for retractions was not entirely deterministic [480, 524, 610, 611, 615, 613, 612, 619, 616, 614, 617, 621, 620, 623, 622, 618]. DTOS has been the basis for several other research systems such as the STAR MLS database [680] and DX, a secure X Windows server [81].

However, the Mach code base used in both DTMach and DTOS was found to be unsatisfactory for high assurance targets. As a result, the Flux nanokernel [217, 219, 218] initially developed at the University of Utah by Ford, Back, and Lepreau between 1996 and 1997 was selected as the basis for additional research beginning in 1997 in a program by the NSA entitled "Assurance in the Fluke[10] Microkernel".

The resulting architecture was called Flask (Fluke Advanced Security Kernel) [624, 625, 626, 627, 678, 540].

This work was performed by SCC together with members of the NSA Information Assurance Research Group led by Loscocco. Flask continued the general design set forth in DTOS with a central security server and decentralized enforcement based on the recursive virtual machine mechanism and associated protection mechanisms in Fluke. As in DTOS, a performance enhancement to this model consisted of each object manager being capable of caching pertinent policy information locally.

Major enhancements over DTOS consisted in the ability to revoke permissions and the ability to model dynamic policies. While, as with DTOS, several security models were implemented, one of the goals that was not reached during the program was the ability to support a hierarchy of policies.

Given that Flask was, like its antecedents, also limited to a single workstation, [624] identified the extension to distributed systems as a problem for additional research.

[10] The Flux μ-kernel Environment

3.5.1.1 Security-Enhanced Linux

Partially in acknowledgment of the limited deployment of the research during the Synergy program, the NSA Information Assurance Research Office in cooperation with SCC and NAI Labs decided to disseminate several results from this research to the general public. To facilitate the use of security policies including mandatory access control in general purpose systems, the Linux operating system was selected for modification in a prototype (as in other systems, e.g. Perseus by Pfitzmann *et al.* [547], although based on the L4 / Fiasco microkernels by Liedtke *et al.* [411]).
The goal of this development was the migration of several aspects particularly of the policy architecture developed up to the Flask prototype and embedding these mechanisms largely transparently within the Linux operating system.
This was accomplished by carrying over the separation of enforcement from policy decisions (in the form of a security server; in SELinux the security server was modeled as a kernel module instead of the user-level service found in its Mach-based antecedents). Similarly, the object managers of Mach-based systems were mapped onto kernel subsystems such as process management, file systems, and System V IPC, albeit with a centralized caching mechanism. To support multiple policies, variable length labels for subjects and objects were managed by the object managers and mapped accordingly with the actual content of the label being determined by the security server.
Process management in SELinux consists of coarse-grained permissions, although capabilities equivalent to those provided by DTOS were considered feasible but not required; file system controls were realized through the use of persistent labels stored within the file system structure representation on physical storage media; this limits the use of labeling (and hence policy enforcement) to file systems local to the SELinux node and, moreover, to a specific revision of the file system. IPC mechanisms were labeled; however, this also limited policy-controlled communication to communication local to the SELinux node [422, 423].

The modifications in SELinux were provided in the form of patches, which made the implementation difficult to maintain. This, along with similar efforts such as the systems developed by Fraser *et al.* [226, 224, 225], Cowan *et al.* [152], Edwards *et al.* [187], Hallyn and Kearns [286], and Ott [528] likewise facing problems led to the initiation of a project for the Linux Security Module (LSM) in 2001 [776]. As a result, several of the mechanisms, particularly SELinux have been adapted to the LSM framework [665, 664]. This framework, however, is – as noted in [776] – restricted in its design objectives to providing access control while being minimally invasive of the overall kernel architecture. As a result, the check mechanism is emplaced as "hooks", typically *after* performing error checks and discretionary access control verification. This significantly limits the information available to and hence the design space for security policy decisions.

3.5.2 Cryptographic Mechanisms for File Systems

Providing security at the file system level is part of the basic security functionality provided by virtually all general-purpose operating systems; however, the limitations of file system protection only via data structures maintained by the operating system have long been obvious. One problem is that such protection mechanisms require the operating system providing the service to be active. This assumption can be violated in non-networked configurations in at least two cases.

One case occurs when storage media are accessed at the local node, but with an operating system other than the one enforcing the security mechanisms (possibly just another instance of the same operating system, only configured differently), the other occurs when the storage media are exposed; typically another system will not honor protections set for removable media. The latter issue is of particular relevance for network-based storage solutions; the provision of standardized cryptographic security mechanisms for storage devices is the subject of the Security in Storage Working Group of the IEEE Task Force on Information Assurance under the leadership of Hughes formed in June 2002.

A solution to the former issue that does not require the modification of existing application programs (CFS, the Cryptographic File System) was proposed by Blaze for the Unix operating system family. In this approach, users individually associated key material with certain directories. The files in these directories including the entries representing the pathname components were transparently encrypted and decrypted with the specified key without further user intervention after setting the key in such a way that plaintext is never stored on a disk or sent to a remote (NFS) file server.

The implementation relied on the redirection of file systems through the mechanism used for NFS external to the operating system kernel and user space processes for communicating key and directory information to this subsystem from individual users [85].

This was later migrated and integrated into the Linux operating system by Catteano and Persiano at the University of Salerno in the TCFS system [449]. This approach has the advantage of not being limited to specific devices attached to a given node as well as being able to discern among multiple users of the same system, problems that are immanent to simple volume encryption systems.

The Secure File System (SFS developed at the University of Minnesota in cooperation with StorageTek by Hughes and Swinarski [318] represents another mechanism for providing cryptographic security to Unix-based systems at the user level based on earlier work by Alexandrov *et al.* on the UFO user-space file system extension mechanisms [13].

Unlike the systems described earlier, SFS permitted the coordinated use of multiple client systems where individual users were authenticated using smart cards and access control was accomplished using a central server, the Group Server which maintained access control lists.

Another approach, based on the observation that the processing capabilities of storage devices are reaching levels previously associated with general purpose computer systems [688, 234, 233, 235] was pur-

sued in the NASD project[11] at the Carnegie Mellon University under the leadership of Gibson which embeds an operating system within storage devices communicating via RPC or translations of RPC to NFS and includes survivability and intrusion detection mechanisms in the form of journaling and version retention for operations.
The approaches for adding security functionality to the Unix operating system pursued by CFS, TCFS, and SFS suffer from severe performance penalties due to the cost of changing between privileged and unprivileged modes of operation. An alternative to this approach for Unix-based systems was proposed by Skinner [663] in the form of an interface at the VFS/VNode level but was not incorporated into mainstream systems; however, a direct manipulation of the modular file system is feasible for most Unix variants as shown by Zadok at Columbia University [789, 790, 787, 791, 788] and Ludwig at Chemnitz University of Technology [428], both of which employ user and key management similar to CFS.
Cryptographic authentication mechanisms for file systems (which, however, may also require the use of semantics different from existing application program interfaces) are part of several research distributed systems including the LegionFS developed as part of the Legion system of the University Virginia [278, 764], Truffles developed as part of the Ficus distributed file system at the University of California at Los Angeles [533, 146], and OceanStore developed at the University of California at Berkeley [383, 382].

[11] An abbreviation of the project name assigned by DARPA ITO, "Secure Distributed and Parallel File Systems Based on Network-Attached Autonomous Disk Drives".

CHAPTER

FOUR

Securing Heterogeneous Distributed Systems

The observation made by Gasser in 1988 [240] that "*many interesting techniques have been implemented in experimental systems but have never seen production use. Some sophisticated features appear in research systems that are used daily at universities, proving that the concepts are viable, but for various reasons (not the fault of the researchers) the systems remain one-of-a-kind*" does not seem to require any significant adjustments more than a decade later; indeed both the recognition of the issues to be solved and a number of viable solutions or at least mechanisms for ameliorating the problems date back two decades before the statement by Gasser.

The basic problems for information systems in computer security (i.e. those related primarily to confidentiality) had already been formulated between 1967 and 1972. ARPA had been requested to form a task force in June of 1967 for studying hardware and software mechanisms for protecting classified information in timesharing systems under the authority of the Defense Science Board. The resulting recommendations of the task force were formulated as a RAND report in 1970 and may be considered the foundation for much of the work in the ensuing decade sponsored by the United States government [756].

One of the most important problems was that of trap doors, i.e. programs that would behave benignly or simulate benign behavior but exhibit malignant behavior under certain circumstances, which was a concern even before its application to timesharing systems in ballistic missile launch control during the early to mid-1960s [562]. The terminology used more commonly today to refer to the general class of programs exhibiting this behavior as Trojan horses[1] was introduced by Edwards in [26]. While the problem description originated from the need to operate multilevel secure time-sharing systems for defense applications, its significance is no less for other areas, particularly given increased use of various mechanisms for mobile executable code.

However, more than three decades after the first multilevel secure operating systems became operational (see section 4.2), the vast majority of information systems is still vulnerable — by their very architecture — to this particular class of attacks that has become particularly relevant due to increased internetworking as discussed in section 4.4. It is this observation, along with others that led to the development of the architecture discussed here.

The remainder of this chapter is divided into two parts. In the first part, the basic premises behind the design and realization of the security architecture are exposed and formulated as a set of hypotheses. The term "ὑπόθεσις" is used deliberately, both from its etymological origin and its common definition as a "supposition for further investigation" [542].

This is mainly because proof is not feasible or in some cases impossible in both the mathematical and strict Popperian sense [560] for these statements since the hypotheses are formulated over insufficiently characterized entities; the validity of the statements can be partially verified only in concrete instantiations and interpretations which are presumably not covered to an sufficient extent in this dissertation.

As the architecture is built on all of the following hypotheses, this necessarily introduces a certain complexity; the individual sections attempt to address this issue by providing a certain amount of supporting evidence that is elaborated in later chapters although in some cases (notably for hypotheses 4.1 and 4.2, existing research pro-

[1] Technically this is incorrect, the wooden horse being of Achaean origin [216]

vides sound argumentation; these hypotheses are identified mainly as premises for latter ones). The second part of the chapter beginning with section 4.8 provides an excerpt of the formal model underlying the architecture and implementation, most which are to be discussed more informally in the following chapters.

4.1 Security at the Operating System Level

This dissertation is primarily concerned with security mechanisms for operating systems; while the general mechanisms discussed are certainly applicable to other areas, this represents both an already large field in itself and at the same time is the very foundation upon which, of necessity, other security mechanisms such as those for virtual machines and mobile code must be built.

This position has been argued forcefully by Loscocco et al. [424] as well as by earlier researchers [594, 414, 406] in response to an increased emphasis on other protection mechanisms espoused in recent research.

The thus criticized deprecation of the protection of local operating systems due to the development of a virtual machine model or variation upon this model to mediate interactions between nodes in a distributed system is certainly intellectually appealing due to the apparent simplicity of the model [583, 357, 361]. It is, however, a problematic proposition for most application areas.

Unless there exists a convincing argumentation or proof that the system composed of the virtual machine mechanism in conjunction with the underlying host operating system does not contain architectural, design, or implementation flaws that permit the violation of information assurance policies imposed at the level of the virtual machine model, the assurance will be no more than what can be provided by the underlying operating system.

The virtual machine system can merely add a layer of indirection and obfuscation to the overall problem, particularly since most systems will not exclusively rely on processes executing within the virtual machine framework.

An additional problem faced by such virtual machine architectures is the granularity of individual virtual machines as the confinement mechanism for individual entities. Not only does the creation of a virtual machine for an entity involve a certain overhead, the use of this mechanism raises new issues as to how information flows or other operations across machine boundaries are to be addressed as most security policies will require operation beyond simple confinement [223].

Similarly, while providing security mechanisms at the level of unprivileged application programs can permit highly adaptive solutions, this assumes implicitly that the application program, or at the very least all application programs handling sensitive operations, are equally capable of enforcing the security properties.

A single Trojan horse program is then capable of subverting an entire system [257, 38, 492]. This, along with the argumentation deriving from [756] and [26] leads to the formulation of hypothesis 1.

Hypothesis 1 *A general information assurance architecture must provide audit, control, and confinement mechanisms at the operating system level controlling non-virtualized resources.*

4.2 Positioning of Security Mechanisms

Very few operating systems were designed with security at least as one of the primary design criteria from the outset. The field of computer security emerged in the mid-60s, primarily driven by the needs for reliable timesharing of mainframe systems among users [291, 310] although the concept of time sharing had emerged almost a decade earlier in a proposal by Strachey [687].

Even the Multics operating system — justly renowned as one of the most security-conscious general purpose operating systems — did not include sophisticated security mechanisms from the outset [147, 753, 159] but rather was predicated on ensuring timesharing operation isolating simultaneous users from one another.

There are, however, some notable exceptions to this general observation. One of the earliest examples of the integration of pervasive security mechanisms is the ADEPT-50 system which used a rigorous semi-formal security model as a guiding principle in design and implementation. This system, developed between 1967 and 1969 under the leadership of Linde and Weissman [413, 761] was based on IBM S/360 hardware and offered multilevel security; the model itself was formulated in 1967 by Weissman and Schaefer [628].
ADEPT-50 is also notable (besides the fact that it was actually used in operational capacity) for the fact that it represents the first foray of the United States government into computer security, being developed in parallel to the ARPA task force.
Other (non-exhaustive) examples for systems with security as the main focus for development, particularly implementing the reference monitor and security kernel concepts, are mentioned in chapter 8 (pp. 195 ff.) as well as the KeyKOS operating system initially developed on behalf of TYMSHARE [223, 289, 567, 222, 399, 93] (see also section 3.5.1) or the Secure Ada Target (later renamed to LOCK) [91, 92, 784, 629, 604] which was itself an outgrowth of the PSOS (Provably Secure Operating System) [581, 510, 220] study that itself commenced in 1973. More recently, the developments outlined in section 3.5.1 have continued the tradition begun with PSOS.

However, security for general purpose systems has mostly been an afterthought and hence has been retrofitted onto existing designs or provided as add-ons for clients with heightened security requirements.
Examples of this approach include the SE/VMS (Secure VMS) which provides mandatory access control mechanisms for the VMS operating system including user registration and login, device and volume management, file creation and access, and the production of labeled printed output [89], IBM's RACF facility for MVS (later renamed first to Security Server in OS/390 and then becoming part of z/OS) [325], the VM/370 effort [258], Argus Pit Bull, and specialized derivative versions of most commercial Unix variants including Sun Trusted Solaris, HP HP-UX BLS, SGI Trusted IRIX, and IBM CMW for AIX

along with earlier proponents of modifications to the Unix environment discussed below.
Also, object-based discretionary access control mechanisms were retrofitted onto the Microsoft Windows NT family of operating systems [690].

This approach of retroactively enhancing and extending information assurance through mechanisms both at the architectural and implementation levels, "*penetrate and patch*", has been discredited early on [359, 412] as noted before; with few exceptions, however, it must still be regarded as the state of the art when considering the majority of systems developed and deployed for which general purpose application programs exist.
This approach, particularly as far as the retroactive improvement at the implementation levels is concerned, will not yield adequate assurance systems [105] and, moreover, even in situations where high assurance systems are not required, this approach is inadequate due to the speed differential at which new attack mechanisms for recently discovered vulnerabilities in systems and application programs propagate compared to the detection of such attacks, their remediation through engineering change orders (ECO) or patches, the distribution of such ECOs, and the diligence with which the majority of operators will apply these. The rigorous approach is certainly not only required for high assurance systems but also more appealing than retrofitting and has been e.g. pursued during most of the Synergy program discussed in section 3.5.1. Yet it has largely failed to translate into operational systems.
The main reason for this is presumably the lack of application programs that are available for such secure systems; commercial vendors may not see the benefit of offering a substantially revised program for what may be considered a limited market by such a vendor. Moreover, even in case of custom programs the creation of revised and adapted versions is likely to face cost and manpower considerations, even if the system is sufficiently documented to permit understanding and revisions of it without unduly affecting operational capabilities (see section 12.2.1).

As a result, the requirements for operational systems may outweigh the risks immanent in the retrofitting approach; this is e.g. reflected in the CMW criteria of the DIA for intelligence processing systems within the DODIIS community [156, 729, 728, 732] which emphasized *"... maximizing use of COTS products, and promoting software reuse ..."* consistent with the TAFIM in place at the time [517][2] which resulted in multilevel secure systems based on COTS standard APIs [194] being deployed widely throughout the defense intelligence analysis community.

Other early examples of the retrofit approach include Linus IV (based on 4.1 BSD Unix) [378] and Secure Xenix [249], followed the same path as the original CMW prototype which was based on Sun Microsystems' BSD-derived operating system (this later evolved into Trusted Solaris 1.1, unrelated to later versions of Trusted Solaris).
The criteria for CMW (CMWEC) assumed that the main threat to the information operations consisted in dissemination of classified material to unauthorized parties; the underlying assumption behind accepting a lower assurance threshold compared to the TCSEC B3 class offering the required MAC and auditing mechanisms — CMWREQ state assurance requirements approximately at the TCSEC B1 class — was that the users of the workstations were assumed to be trustworthy (since access to a workstation requires a Top Secret SCI clearance with a comprehensive SBI (see definition 7.2) this assumption is at least partially justified) and had other means of removing such information surreptitiously so that not only the high assurance criteria but also the covert channel requirements could be relaxed. Instead, the main focus in the CMW requirements was the prevention of attacks by Trojan horses.
The MITRE CMW prototype [156] provided early confirmation that it was possible based on COTS operating system instrumentation.
This earlier research in conjunction with the argumentation outlined above results in hypothesis 2:

[2] Since superseded by the C^4ISR framework [518], but substantially unchanged in the goals pursued.

Hypothesis 2 *The limited assurance that can be provided by retroactive modification of existing systems to insert instrumentation for information collection and enforcement is justifiable under the assumption that*

- *the retrofitted system is a transitory stage in the migration to a high assurance environment, or*
- *that external circumstances mitigate the risk of deliberate information assurance policy violations by legitimate subjects.*

4.3 Policy Neutrality

Secure system development has largely been focused on implementing the mandatory access control mechanisms requisite for conforming to the multilevel security requirements for information processing laid out originally by the Defense Science Board report [756]. However, not only are there a number of well-known models for confidentiality and integrity [63, 255, 256, 402, 90, 135, 472, 344, 470] (see also chapter 7), but organizations will typically require a certain amount of flexibility in defining a policy even if it is based on a general model instead of being a purely ad hoc creation. This was recognized even for defense applications in the feasibility study performed for SDIO and RADC [532]. Moreover, parts of an organization may require the application of different (e.g. more stringent) policies simultaneously.

Since there is no obvious choice in security model or a security model that is sufficiently generic to support an adequate spectrum of policies (this would also contradict with the simplicity desideratum for security models) it seems imprudent to restrict the policies and models an information assurance system is to support in advance, hence hypothesis 3.

Hypothesis 3 *The instrumentation mechanisms for information assurance policies must permit the individual collection of information on subjects, objects, and operations as well as control over operations within a system as determined by one or more policy rules. No presumption as to* *the activation of instrumentation in groups or in response to specific events may be made.*

4.4 Policy Enforcement in Distributed Systems

Most non-trivial information systems encompass multiple platforms and operating systems for such platforms and require the interaction of such heterogeneous nodes within a distributed system for the execution of information operations.

An organization's security policy, however, is rarely restricted to individual nodes; instead such policies will typically formulate requirements in terms of abstract objects such as documents and information flows and hence necessarily encompass multiple nodes within a distributed system. This ultimately requires the consistent and simultaneous enforcement of a given security policy across all nodes, which may be considered self-evident. This viewpoint is also supported by earlier work on assuring networked systems resulting in the TNI. While the TCSEC [733] were originally limited conceptually to the evaluation of isolated computer systems; this was remedied with the introduction of the TNI in 1987 [496, 495] which permitted the evaluation of networked systems. Since there has been no comparable effort within the framework of the Common Criteria [328] to define Protection Profiles, the TNI represents the only document containing guidance to the proper structuring of trusted distributed systems. In addition, one of the first systems evaluated under the TNI was the Boeing MLS LAN [640, 686] which was evaluated against the A1 class according to the TNI and was found in compliance in 1994 [497], providing anecdotal evidence that it is indeed possible to provide high assurance within a networked environment.

An implicit corollary of consistent distributed enforcement is the view of the distributed system under the control of a policy as a single system (as opposed to a federation of individual trusted systems), albeit with a partitioned trusted code base with the source of the security policies to be enforced accessible to each node at all times and each node capable of autonomous operation (provided access to the policy source). Every component that is trusted must enforce applicable security policies at the node level which may be an arbitrary subset of the overall policies. This requires that transitions of subjects and objects or operations spanning multiple nodes within the distributed system can identify the entity or operation in such a way that an applicable security policy can be identified and applied to these. Assuming hypothesis 5, each entity can be labeled in such a way that it is uniquely identified within the distributed system and irrespective of the platform on which an entity resides. Aspects of flexible labeling within a networked environment were e.g. investigated by Loscocco [421]; however, by way of an indirection step via entity identification it becomes not only possible to realize variations on the multilevel security model but arbitrary policies that rely on semantically equivalent entities and operations for the specification. In summary, hypothesis 4 is obtained.

Hypothesis 4 *A given non-empty set of information assurance policies must be enforced on all nodes of a distributed system in such a way that each node enforces the policies with equivalent semantics whereas entities and operations over which those policies are defined can be identified uniquely across a distributed system and using the policies current at the time of an operation.*
If transitions between the distributed system enforcing policies and nodes not enforcing a given policy exist, the nodes must enforce all policy elements regarding composition of such nodes.

4.5 Permissibility of Abstraction Across Operating Systems

As noted by Brinch Hansen [109] there are surprisingly few core concepts that constitute the foundations of virtually all operating systems in existence (exceptions are mainly in the areas of specialized systems and distributed operating systems such as Plan 9 [563, 60], Mach-US [685], or Amoeba [695] which either do not provide all services expected in a general purpose operating system such as process isolation or distribute certain abstractions across a network). This is also reflected in the provision of fundamental resource abstractions such as processes, files and the hierarchical representation of directories, devices, memory, as well as network and interprocess connection mechanisms. Examples of systems that provide such abstractions include the various Unix derivatives [576, 416, 404, 464, 44, 712, 265, 450], OpenVMS [364, 457, 365, 163, 259], the Microsoft Windows NT family of operating systems [157, 158, 675, 503, 676], IBM z/OS and its predecessors and to some extent z/VM and its predecessors [40, 322], hence hypothesis 5.

Hypothesis 5 *There exists sufficient similarity between recent non-experimental operating systems to permit the specification of entities and operations at abstraction levels describing basic primitives as well as of composites of such primitives.*

4.6 Use of Formal Logic for Definition and Enforcement of Policies

Most published security models and policies (see chapter 7) can be described as the specification of a subset of permissible operations of all operations within an information system (either directly or as the inverse of said set) regardless of the mechanisms used for specifying this set.

While most non-trivial protection systems satisfy the requirement for expressiveness set forth in hypothesis 3 in that a procedure for modeling Turing machines in them exists (see e.g. section 3.1), aesthetic and pragmatic considerations preclude the obfuscatory use of such systems to represent general policies.

The use of formal logic (as will be discussed in chapter 5 the discussion in this dissertation is limited to $\mathcal{PC}$ mainly in the interest of

brevity and clarity) to represent such sets of permissible operations permits a relatively elegant formulation of such sets.
Any entity within a computer system[3] can be identified by a unique constant (the set is recursively enumerable, compare observation 6.1); the same argument also applies to unique predicates identifying operations on such entities. Knowledge representation embedded in formal logic is a well-understood approach underlying much of classical artificial intelligence [677, 241, 104, 574, 586, 407, 332].
A specific instance of a permitted operation can thus be modeled as a closed well-formed formula that is logically valid given suitable definitions for constants and predicates. Moreover, predicates can be used to specify arbitrary sets of both entities and operations in a compact way by means of set comprehension.

Assuming hypothesis 5 as a way of addressing the epistemological problem in this instance [455], set comprehension can also be used to define sets representing equivalent entities and operations which must be mapped to each of the operating systems for which the policy is to be applied.
The verification of permissibility under a security policy in an operating system or similar reactive system can be represented by the formulation of a hypothesis describing the operation to be performed. Given a policy formulated as described above, the deducibility of the hypothesis results in the permissibility of the operation. In active systems, the validity of a well-formed formula can induce operations; this can be seen as an equivalent to production systems.
Moreover, by defining relations on the sets of entities, particularly relations establishing hierarchies of abstraction layers and well-formed formulae establishing the derivation within such relations, it becomes possible to define security models at higher abstraction levels along with derivation rules and having an automated deduction system find permissible operations at lower abstraction levels that are consistent with the policy defined at a high abstraction level.

[3]Due to the inclusion of entities that are capable of performing computations outside the scope of the Church-Turing hypothesis, the term *information system* as defined in definition 1.3 cannot be used.

In addition, by permitting deduction at several nodes within a distributed system structured according to hypothesis 7, the necessity for creating and transmitting hypotheses between nodes can be reduced by having an external reference monitor (cf. chapter 8) relating the decision as a set of well-formed formulae from which a closed well-formed formula can then be deduced at the node formulating the hypothesis as well as possibly deducing the validity of other hypotheses.

The decisions regarding hypotheses must occur under real-time constraints; depending on policy and operations to be performed (e.g. involving human interaction), the time constraints imposed are on the order of several to several hundred milliseconds.

Given the limitations of automated deduction this implies that the number of facts that can be considered in the deduction of a hypothesis (or its refutation) must not exceed a certain threshold x, which is on the order of 10^2.

Applicable facts can, however, be identified by efficient algorithms in advance by the labels of entities according to hypothesis 4.4, limiting the number of facts to be used for automated deduction. Moreover, when observing existing security models and policies, the number of applicable rules (or well-formed formulae) is often sharply limited although there exist models (particularly role-based models) where this is not the case.

Given that the set of valid inferences in $\mathcal{PC}$ is recursively enumerable but not recursive and that upper bounds may be imposed on the effort expended at deducing the validity of any given hypothesis, under some policies (e.g. by not imposing restrictions on decidability) there may be logically valid well-formed formulae formulated as hypotheses for which a decision will be made that indicates the contrary.

Such a decision is related to the node formulating the hypothesis outside of the framework of $\mathcal{PC}$ and hence does not introduce contradictory formulae. Policies may, however, accept such a situation since only negative decisions regarding permissibility are ever reached in this way and adequately defined safety properties hence not vio-

lated; such policies must also accept the violation of the idempotence property for hypotheses.
The formulation of policies in formal logic, particularly in the monotonic logic of $\mathcal{PC}$ requires that the well-formed formulae of a policy are consistent both internally and when used as the basis for deduction in conjunction with other policies. This implies that appropriate constraints must be placed on the entities over which policies are formulated and also limits efficiency in the formulation of discretionary access policies.
In summary, hypothesis 6 can therefore be formulated and will be expounded in the following chapters.

Hypothesis 6 *The modeling of information assurance policies and models in formal logic results in compact a compact representation which can be used in such a way that permissible and required operations can also be deduced automatically from abstractions of entities and operations.*
The number of facts to be considered for a decision on any given policy decision is small enough to permit efficient deduction.

4.7 Externalization of the Reference Monitor Concept

The reference monitor mechanism fulfills the requirements of hypotheses 1 and 3. However, to address those levied by hypothesis 4, an extension must be made in that the source of policies to be enforced is equivalent to a single consistent instance (cf. the NTCB concept of the TNI). Separation of enforcement mechanism and decision instance as well as the safety and idempotence properties of a policy formulation according to hypothesis 6 can fulfill the requirements of hypothesis 4.
The decision mechanism of the external reference monitor mechanism can be replicated provided that the decisions reached are idempotent for each instance of the mechanism queried. For this narrowly defined mechanism, the Isolation Property (cf. property 1 on page

194) can be ensured by enclosing the mechanism in a tamper-resistant device such as a cryptographic coprocessor. In summary, this yields hypothesis 7.

Hypothesis 7 *The reference monitor mechanism can be extended by separating the enforcement mechanisms physically as well as logically from the definition and distribution of the decision regarding the operations for which the reference monitor is invoked.*
The Isolation Property can be maintained in distributed systems by embedding the decision mechanism in a tamper-resistant device.

4.8 Basic Model

While this section provides the framework for the security architecture, it is not a self-contained discussion. Instead, it draws on discussions found in later chapters of this dissertation. The exhibition of the external reference monitor mechanism as well as definitions pertaining to structural elements within this mechanism is found in chapter 8 while a discussion on the layering of abstraction layers and the interactions between the layers can be found in chapter 9 with additional discussion on interaction side effects in chapter 10.
The elements formulated in the Z notation [679, 772, 333, 52] are not self-contained and can only cover basic aspects of the overall model; the full formal model of the architecture is beyond the scope of this discussion and can be found in [118] of which the following contains selected excerpts.

The overall architecture can be articulated first at the level of nodes in a distributed system; for the purposes of the discussion here, policy definition and enforcement is assumed to always occur at the level of distinct nodes.
Definition 4.1 deliberately provides a broad scope for entities that constitute nodes; the disjointness characteristic is required to ensure that nodes are not in fact operating as simulated Turing machines under the control of another Turing machine. While the latter cannot be computationally distinguished from disjoint nodes, this would

negate the ability for independent characterization as constraints imposed on the simulating machine would otherwise transitively be imposed on the simulated machine. However, the requirement for disjointness does not imply a need for physical disjointness.

Definition 4.1 *A **node** is an entity capable of performing computations equivalent to a Turing machine that is disjoint from other nodes.*

Within the model, entities can be subject to security policies. To distinguish such entities from other entities, definition 4.2 introduces the concept of the domain of control (cf. definition 6.9 for the analogous security policy domain concept).

Definition 4.2 *The **security architecture domain** is the domain of control over entities through technical or administrative means.*

As security policies require the specification of interactions with entities external to the security architecture domain, definition 4.3 introduces such entities within the scope of the model.

Definition 4.3 *A **non-policy-enforcing entity** (NOPE) is a node which is outside the security architecture domain, but which nonetheless interacts in some way, shape or form with nodes and entities within the security architecture domain.*

Within the security architecture domain, security policies must be injected and deleted. To ensure that injection and deletion occur only under well-defined circumstances, certain nodes must be designated as authoritative sources of security policies, as is the case with definition 4.4.

Definition 4.4 *A **policy controller node** (PCN) is a node which acts as a repository for security policies and which can, for security policies within the security architecture domain*

- *answer a policy request authoritatively and correctly, provide a correct answer to a policy request based on cached information, or provide a correct reply to a policy request by querying an authoritative PCN*

- *issue policy commands to other nodes based on regulated operations.*

Nodes within the security architecture domain that are not PCN are thus required to act on the commands issued by PCN nodes as described in definition 4.5.

Definition 4.5 *A* ***policy-enforcing node*** *(PEN) is a node in which any operation on subjects and objects within the security architecture domain not induced by a PCN must be completed successfully if and only if*

- *based on cached information the PEN can ascertain the conformance to all applicable security policies, or*
- *the operation is formulated as a policy request and positively answered by a PCN.*

Furthermore, a PEN must perform operations if a PCN within the security architecture domain issues a command for this operation.

The basic node types identified in definitions 4.1, 4.3, 4.4, and 4.5 are represented in Z notation as given sets; nodes in general are represented by a disjoint union free type.

Definition 4.6

$$[\,PEN, PCN, NOPE\,]$$

$$\begin{aligned} NODE ::= {} & policyEnforcingNode\langle\langle PEN\rangle\rangle \\ \mid {} & policyControllerNode\langle\langle PCN\rangle\rangle \\ \mid {} & nonPolicyEnforcingNode\langle\langle NOPE\rangle\rangle \end{aligned}$$

Furthermore, entities within the model are named; the name space is, however, not restricted to the basic entities.

Definition 4.7

$$NAME ::= name\langle\langle \mathbb{N}\rangle\rangle$$

A number of naming functions are introduced, which map names to the various node types, namely the functions *nameNOPE*, *namePEN*, and *namePCN*, which are all partial injections.

Definition 4.8

$$
\begin{array}{|l}
nameNOPE : NAME \rightarrowtail\!\!\!\!\!\!+\; NOPE \\
namePEN : NAME \rightarrowtail\!\!\!\!\!\!+\; PEN \\
namePCN : NAME \rightarrowtail\!\!\!\!\!\!+\; PCN \\
topLevelEntityNaming : \mathbb{P}\, NAME \\
\hline
\mathsf{disjoint}\, \langle \mathrm{dom}\, nameNOPE, \mathrm{dom}\, namePEN, \mathrm{dom}\, namePCN \rangle \\
topLevelEntityNaming = \mathrm{dom}\, nameNOPE \cup \\
\quad \mathrm{dom}\, namePEN \cup \mathrm{dom}\, namePCN
\end{array}
$$

The auxiliary function *mapNameToNode* maps the names of nodes to nodes based on the restriction that nodes are not the only entity thus named.

Definition 4.9

$$
\begin{array}{|l}
mapNameToNode : NAME \mathrel{+\!\!\!\rightarrow\!\!\!\!\!\rightarrow} NODE \\
\hline
\mathrm{dom}\, mapNameToNode = (\mathrm{ran}\, nameNOPE^{\sim}) \cup \\
\quad (\mathrm{ran}\, namePEN^{\sim}) \cup (\mathrm{ran}\, namePCN^{\sim}) \\
\forall\, name : NAME;\; node : NODE \bullet \\
\quad ((node, name) \in nameNOPE^{\sim} \Rightarrow \\
\quad\quad (name, node) \in mapNameToNode) \wedge \\
\quad ((node, name) \in namePEN^{\sim} \Rightarrow \\
\quad\quad (name, node) \in mapNameToNode) \wedge \\
\quad ((node, name) \in namePCN^{\sim} \Rightarrow \\
\quad\quad (name, node) \in mapNameToNode)
\end{array}
$$

Moreover, the auxiliary function *nodeName* determines whether a given name is that of a *NODE*.

Definition 4.10

$$
\begin{array}{|l}
nodeName : NODE \twoheadrightarrow\!\!\!\!\!\!+\; NAME \\
\hline
\forall\, name : NAME;\; node : NODE \bullet \\
\qquad (name \in \operatorname{dom} nameNOPE \Rightarrow (node, name) \in nodeName) \wedge \\
\qquad (name \in \operatorname{dom} namePEN \Rightarrow (node, name) \in nodeName) \wedge \\
\qquad (name \in \operatorname{dom} namePCN \Rightarrow (node, name) \in nodeName)
\end{array}
$$

The overall basic model assumes — for ease of notation — that the numbers of each node type are known.
For each node type, a partial injection defining the mapping between the NAME and specific node type is provided; the name spaces thus provided are disjoint.

Definition 4.11

$$
\begin{array}{|l}
\textit{GlobalModelState} \\
\hline
numNOPE : \mathbb{N} \\
numPEN : \mathbb{N} \\
numPCN : \mathbb{N} \\
\\
extantNOPE : \mathbb{F}\, NOPE \\
extantPEN : \mathbb{F}\, PEN \\
extantPCN : \mathbb{F}_1\, PCN \\
\hline
\end{array}
$$

Given the definition of *GlobalModelState*, a designated initial PCN is required. The following axiomatic definition 4.12 provides for such a named entity.

Definition 4.12

$$
\begin{array}{|l}
initialPCN : PCN
\end{array}
$$

Definition 4.13

GlobalModelInitialization

$\Delta GlobalModelState$

$extantNOPE' = \emptyset$
$extantPEN' = \emptyset$
$extantPCN' = \{initialPCN\}$

For changing the model configuration, a number of operation schemata are introduced. Conceptually, these are invoked implicitly as the system state changes.

The *IntroduceNodePEN* schema in definition 4.14 introduces a new node of type *PEN* into the global model and updates the global state accordingly along with the naming function *namePEN*.

Definition 4.14

IntroduceNodePEN

$\Delta GlobalModelState$
$injectedPEN? : PEN$

$injectedPEN? \notin extantPEN$

$extantPEN' = extantPEN \cup injectedPEN?$
$numPEN' = numPEN + 1$
$\exists\, name_{injectedPEN} : NAME \bullet$
$\quad namePEN' = namePEN \oplus$
$\qquad \{name_{injectedPEN} \mapsto injectedPEN?\}$

Correspondingly, the operation schema *RemoveNodePEN* in definition 4.15 models the removal of a node of type *PEN*.

Definition 4.15

$$
\begin{array}{l}
\textit{RemoveNodePEN} \\
\hline
\Delta GlobalModelState \\
removedPEN? : PEN \\
\hline
removedPEN? \in extantPEN \\
\\
extantPEN' = extantPEN \setminus removedPEN? \\
numPEN' = numPEN - 1 \\
\forall\, name_{removedPEN} : NAME \bullet \\
\quad namePEN' = namePEN \setminus \\
\qquad \{removedPEN? \mapsto name_{removedPEN}\} \\
\hline
\end{array}
$$

The *IntroduceNodePCN* schema in definition 4.16 provides the analogon to the *IntroduceNodePEN* schema given in definition 4.14 for nodes of type *PCN*.

Definition 4.16

$$
\begin{array}{l}
\textit{IntroduceNodePCN} \\
\hline
\Delta GlobalModelState \\
injectedPCN? : PCN \\
\hline
injectedPCN? \notin extantPCN \\
\\
extantPCN' = extantPCN \cup injectedPCN? \\
numPCN' = numPCN + 1 \\
\exists\, name_{injectedPCN} : NAME \bullet \\
\quad namePCN' = namePCN \oplus \\
\qquad \{name_{injectedPCN} \mapsto injectedPCN?\} \\
\hline
\end{array}
$$

Similarly, *RemoveNodePCN* given in definition 4.17 is the analogon to *RemoveNodePEN* of definition 4.15 ; however, an additional precondition is required to ensure that the final *PCN* is not removed from the model. It is not, however, required that this last *PCN* be the node designated *initialPCN*.

Definition 4.17

$$
\begin{array}{l}
\textit{RemoveNodePCN} \\
\hline
\Delta \textit{GlobalModelState} \\
\textit{removedPCN?} : \textit{PCN} \\
\hline
\textit{removedPCN?} \in \textit{extantPCN} \\
\textit{numPCN} \neq 1 \\
\\
\textit{extantPCN}' = \textit{extantPCN} \setminus \textit{removedPCN?} \\
\textit{numPCN}' = \textit{numPCN} - 1 \\
\forall\, \textit{name}_{\textit{removedPCN}} : \textit{NAME} \bullet \\
\quad \textit{namePCN}' = \textit{namePCN} \setminus \\
\qquad \{\textit{removedPCN?} \mapsto \textit{name}_{\textit{removedPCN}}\}
\end{array}
$$

Finally, the operation schemata *IntroduceNodeNOPE* given in definition 4.18 and *RemoveNodeNOPE* given in definition 4.19 provide the analoga to *IntroduceNodePEN* and *RemoveNodePEN* for nodes of type *NOPE*.

Definition 4.18

$$
\begin{array}{l}
\textit{IntroduceNodeNOPE} \\
\hline
\Delta GlobalModelState \\
injectedNOPE : NOPE \\
\hline
injectedNOPE? \notin extantNOPE \\
\\
extantNOPE' = extantNOPE \cup injectedNOPE \\
numNOPE' = numNOPE + 1 \\
\exists\, name_{injectedNOPE} : NAME \bullet \\
\quad nameNOPE' = nameNOPE \oplus \\
\qquad \{name_{injectedNOPE} \mapsto injectedPEN?\}
\end{array}
$$

Definition 4.19

$$
\begin{array}{l}
\textit{RemoveNodeNOPE} \\
\hline
\Delta GlobalModelState \\
removedNOPE? : NOPE \\
\hline
removedNOPE? \in extantNOPE \\
\\
extantNOPE' = extantNOPE \setminus removedNOPE? \\
numNOPE' = numNOPE - 1 \\
\forall\, name_{removedNOPE} : NAME \bullet \\
\quad nameNOPE' = nameNOPE \setminus \\
\qquad \{removedNOPE? \mapsto name_{removedNOPE}\}
\end{array}
$$

"Time" is a required concept within the model for two purposes. One is the establishment of a mechanism for specifying and identifying causality, the other is the specification of physical time and time intervals.

Physical time is commonly described as a relation $\mathbf{T}_{Physical}$ over a set and for which the following axioms hold:

T1 $\mathbf{T}_{Physical}$ is transitive ($\forall x, y, z : x <_{\mathbf{T}_{Physical}} y \wedge y <_{\mathbf{T}_{Physical}} z \Rightarrow x_{\mathbf{T}_{Physical}} z$)

T2 $\mathbf{T}_{Physical}$ is irreflexive ($\neg \exists x : x <_{\mathbf{T}_{Physical}} x$)

T3 $\mathbf{T}_{Physical}$ is linear ($\forall x, y : x \neq y \Rightarrow x <_{\mathbf{T}_{Physical}} y \Leftrightarrow \neg y <_{\mathbf{T}_{Physical}} x$)

T4 $\mathbf{T}_{Physical}$ is unbounded ($\forall x \exists y : y <_{\mathbf{T}_{Physical}} x \wedge \forall x \exists y : x <_{\mathbf{T}_{Physical}} y$)

T5 $\mathbf{T}_{Physical}$ is dense ($\forall x, y : x <_{\mathbf{T}_{Physical}} y \Rightarrow \exists z : x <_{\mathbf{T}_{Physical}} z <_{\mathbf{T}_{Physical}} y$)

However, axiom **T5** is not required for the purposes of the general model described here and would require a carrier set isomorphous to $\mathbb{R}$ [740].
As shown by Lamport [390], it is therefore sufficient to define time as an isomorphism to $\mathbb{N}$ with the $<$ relation used directly for notational convenience.

Definition 4.20

$$TIME == \mathbb{N}$$

As the problem domain is a distributed system, it can in some cases be necessary to explicitly reason about local clocks; the following definition therefore introduces the notion of local time. As before, the axioms **T1–T4** are fulfilled by providing an isomorphism onto $\mathbb{N}$ (via the abbreviation *TIME*).
Provided that the clock conditions C1 and C2 of [390] are maintained, a total order can be obtained from the local time by providing a homomorphous mapping between *sequenceCoordinate* and *LOCALTIME* for each *NODE* (cf. section 4.8.3). However, for the purposes of this model, the global time model can be used directly.

Definition 4.21

$$\begin{array}{|l}
LOCALTIME ::= localTime\langle\langle TIME \times NODE\rangle\rangle \\
_ <_{LocalTime} _ : LOCALTIME \leftrightarrow LOCALTIME \\
extractLocalTime : LOCALTIME \rightarrow TIME \\
\hline
\forall t : TIME;\ n : NODE \bullet extractLocalTime(localTime(t, n)) = t \\
\forall x, y : LOCALTIME \bullet x <_{LocalTime} y \Leftrightarrow \\
\quad extractTimeLocalTime\ x < extractTimeLocalTime\ y
\end{array}$$

Additionally, a time interval schema is given in definition 4.22.

Definition 4.22

$$
\begin{array}{|l}
\textit{TimeInterval} \\
\hline
timeStart : TIME \\
timeEnd : TIME \\
\hline
timeStart \leq timeEnd \\
\hline
\end{array}
$$

The following introduces basic types for the top-level subjects within the generic model of PEN operating systems as given sets; non-specific subjects are represented by a disjoint union free type.
The corresponding intended operational semantics are discussed in section 5.2.1; while the Z notation does not permit later redefinition, it should, however, be noted that the set of subjects is extensible.

Definition 4.23

$$[\,USER, PROCESS, APPLICATION, HOST,\\ NETWORK, LINK\_NODE, BUS\_NODE\,]$$

$$
\begin{aligned}
SUBJECT ::= {} & user\langle\langle USER\rangle\rangle \\
| {} & process\langle\langle PROCESS\rangle\rangle \\
| {} & application\langle\langle APPLICATION\rangle\rangle \\
| {} & host\langle\langle HOST\rangle\rangle \\
| {} & network\langle\langle NETWORK\rangle\rangle \\
| {} & linkNode\langle\langle LINK\_NODE\rangle\rangle \\
| {} & busNode\langle\langle BUS\_NODE\rangle\rangle
\end{aligned}
$$

Correspondingly, the sets of object entities within the PEN model are also defined in the form of given sets with non-specific objects again represented by a disjoint union free type. Files and directories represent a special case since these are considered sub-types.
As before, the corresponding intended operational semantics are discussed in section 5.2.2 and there is no a priori restriction on the extension of the set in definition 4.24.

Definition 4.24

$$[FILE, EXECUTABLE\_FILE, DATA\_FILE, DIRECTORY, CONNECTION, DATAGRAM, IPC\_MESSAGE, DEVICE, PHYSICAL\_LINK, MEMORY\_SEGMENT, VIRTUAL\_CIRCUIT]$$

$$\begin{aligned}
OBJECT ::= & \ file\langle\langle FILE\rangle\rangle \\
| & \ executableFile\langle\langle EXECUTABLE_FILE\rangle\rangle \\
| & \ dataFile\langle\langle DATA_FILE\rangle\rangle \\
| & \ directory\langle\langle DIRECTORY\rangle\rangle \\
| & \ connection\langle\langle CONNECTION\rangle\rangle \\
| & \ virtualCircuit\langle\langle VIRTUAL_CIRCUIT\rangle\rangle \\
| & \ datagram\langle\langle DATAGRAM\rangle\rangle \\
| & \ ipcMessage\langle\langle IPC_MESSAGE\rangle\rangle \\
| & \ device\langle\langle DEVICE\rangle\rangle \\
| & \ physicalLink\langle\langle PHYSICAL_LINK\rangle\rangle \\
| & \ memorySegment\langle\langle MEMORY_SEGMENT\rangle\rangle \\
FILE_TYPE ::= & \ executableSubtype\langle\langle EXECUTABLE_FILE\rangle\rangle \\
| & \ dataSubtype\langle\langle DATA_FILE\rangle\rangle \\
| & \ directorySubtype\langle\langle DIRECTORY\rangle\rangle
\end{aligned}$$

Definition 4.25 provides for the existence of a suitable naming functions for the various subject entities. As described before for the case of definition 4.23, the entity types are extensible although the Z notation itself does not provide a construct for this purpose.

Definition 4.25

$$
\begin{array}{|l}
nameUSER : NAME \rightarrowtail\!\!\!\!\!\!\!\!\;\;+\!\!\!\!\rightarrow USER \\
namePROCESS : NAME \rightarrowtail\!\!\!\!\!\!\!\!\;\;+\!\!\!\!\rightarrow PROCESS \\
nameAPPLICATION : NAME \rightarrowtail\!\!\!\!\!\!\!\!\;\;+\!\!\!\!\rightarrow APPLICATION \\
nameHOST : NAME \rightarrowtail\!\!\!\!\!\!\!\!\;\;+\!\!\!\!\rightarrow HOST \\
nameNETWORK : NAME \rightarrowtail\!\!\!\!\!\!\!\!\;\;+\!\!\!\!\rightarrow NETWORK \\
nameLINK\_NODE : NAME \rightarrowtail\!\!\!\!\!\!\!\!\;\;+\!\!\!\!\rightarrow LINK\_NODE \\
nameBUS\_NODE : NAME \rightarrowtail\!\!\!\!\!\!\!\!\;\;+\!\!\!\!\rightarrow BUS\_NODE \\
subjectNaming : \mathbb{P}\, NAME \\
\hline
\mathsf{disjoint}\ \langle\, \mathrm{dom}\, nameUSER, \mathrm{dom}\, namePROCESS, \\
\qquad\qquad \mathrm{dom}\, nameAPPLICATION, \mathrm{dom}\, nameHOST, \\
\qquad\qquad \mathrm{dom}\, nameNETWORK, \mathrm{dom}\, nameLINK\_NODE, \\
\qquad\qquad \mathrm{dom}\, nameBUS\_NODE \,\rangle \\
subjectNaming == \mathrm{dom}\, nameUSER \cup \mathrm{dom}\, namePROCESS \cup \\
\qquad\qquad \mathrm{dom}\, nameAPPLICATION \cup \mathrm{dom}\, nameHOST \cup \\
\qquad\qquad \mathrm{dom}\, nameNETWORK \cup \mathrm{dom}\, nameLINK\_NODE \cup \\
\qquad\qquad \mathrm{dom}\, nameBUS\_NODE \\
\mathrm{dom}\, nameSUBJECT = subjectNaming
\end{array}
$$

Correspondingly, definition 4.26 provides naming functions for object entities. To reflect the possibility that the various file types may in fact represent different interpretations of a common structure or type, the naming space for files is subdivided according to type information (cf. definition 5.13).

Definition 4.26

$$
\begin{array}{l}
nameFILE : NAME \rightarrowtail\!\!\!\!\!\!\!+\; FILE \\
nameEXECUTABLE\_FILE : NAME \rightarrowtail\!\!\!\!\!\!\!+\; EXECUTABLE\_FILE \\
nameDATA\_FILE : NAME \rightarrowtail\!\!\!\!\!\!\!+\; DATA\_FILE \\
nameDIRECTORY : NAME \rightarrowtail\!\!\!\!\!\!\!+\; DIRECTORY \\
nameVIRTUAL\_CIRCUIT : NAME \rightarrowtail\!\!\!\!\!\!\!+\; VIRTUAL\_CIRCUIT \\
nameDATAGRAM : NAME \rightarrowtail\!\!\!\!\!\!\!+\; DATAGRAM \\
nameIPC\_MESSAGE : NAME \rightarrowtail\!\!\!\!\!\!\!+\; IPC\_MESSAGE \\
nameDEVICE : NAME \rightarrowtail\!\!\!\!\!\!\!+\; DEVICE \\
namePHYSICAL\_LINK : NAME \rightarrowtail\!\!\!\!\!\!\!+\; PHYSICAL\_LINK \\
nameMEMORY\_SEGMENT : NAME \rightarrowtail\!\!\!\!\!\!\!+\; \\
\quad MEMORY\_SEGMENT \\
objectNaming : \mathbb{P}\, NAME \\
\hline
\mathsf{disjoint}\, \langle\, \mathrm{dom}\, nameEXECUTABLE\_FILE, \\
\qquad \mathrm{dom}\, nameDATA\_FILE, \\
\qquad \mathrm{dom}\, nameDIRECTORY \,\rangle \\
\mathrm{dom}\, nameFILE = \mathrm{dom}\, nameEXECUTABLE\_FILE \cup \\
\qquad \mathrm{dom}\, nameDATA\_FILE \cup \mathrm{dom}\, nameDIRECTORY \\
\mathsf{disjoint}\, \langle\, \mathrm{dom}\, nameFILE, \\
\qquad \mathrm{dom}\, nameVIRTUAL\_CIRCUIT, \\
\qquad \mathrm{dom}\, nameDATAGRAM, \\
\qquad \mathrm{dom}\, nameIPC\_MESSAGE, \\
\qquad \mathrm{dom}\, nameDEVICE, \\
\qquad \mathrm{dom}\, namePHYSICAL\_LINK, \\
\qquad \mathrm{dom}\, MEMORY\_SEGMENT \,\rangle \\
objectNaming == \mathrm{dom}\, nameFILE \cup \\
\qquad \mathrm{dom}\, nameVIRTUAL\_CIRCUIT \cup \\
\qquad \mathrm{dom}\, nameDATAGRAM \cup \\
\qquad \mathrm{dom}\, nameIPC\_MESSAGE \cup \\
\qquad \mathrm{dom}\, nameDEVICE \cup \\
\qquad \mathrm{dom}\, namePHYSICAL\_LINK \cup \\
\qquad \mathrm{dom}\, nameMEMORY\_SEGMENT
\end{array}
$$

Albeit part of the system model itself, the existence of meet and join relations is stipulated for notational convenience (cf. section 6.1).

Definition 4.27

$$
\begin{array}{l}
\_\nabla_{type}\_ : NAME \times NAME \rightarrow NAME \\
\_\Delta_{type}\_ : NAME \times NAME \rightarrow NAME \\
\_\nabla_{identity}\_ : NAME \times NAME \rightarrow NAME \\
\_\Delta_{identity}\_ : NAME \times NAME \rightarrow NAME
\end{array}
$$

Operation descriptors (see definition 5.9) are also modeled as given sets in definition 4.28.

Definition 4.28

$$[\,OPDESC\,]$$

Operations can be described as mappings between sets of subjects and objects, each of which are uniquely identified by an operation descriptor where for each operation descriptor, the number of arguments is well-defined.

Definition 4.29

$$
\begin{array}{l}
\underline{\quad OperationDescriptor \quad} \\
operationDescriptor : OPDESC \\
numSubjects : OPDESC \rightarrow \mathbb{N} \\
numObjects : OPDESC \rightarrow \mathbb{N}
\end{array}
$$

Operations therefore are modeled as operation descriptors conjoined with finite sequences of subjects and objects.

Definition 4.30

$$
\begin{array}{l}
\underline{\quad Operation \quad} \\
OperationDescriptor \\
subjects : \operatorname{seq} SUBJECT \\
objects : \operatorname{seq} OBJECT \\
\hline
numSubjects = \#subjects \\
numObjects = \#objects
\end{array}
$$

A hypothesis (referred to depending on context as a *PolicyRequest*, *PolicyReply*, *Fact*, or simply as *Hypothesis*, see definition 5.11) is then defined as the conjoining of an operation and a validity interval or, in the case of audit records, a time record of the event in question.

Definition 4.31

Hypothesis

Operation
TimeInterval

$$PolicyRequest == Hypothesis$$
$$PolicyReply == Hypothesis$$
$$Fact == Hypothesis$$

4.8.1 PEN Instances

The following model excerpt provides some of the essential properties of interest for the purposes of security policy enforcement in a policy-enforcing node.

Definition 4.32

PENInstanceState

$processes : \mathbb{F}\, PROCESS$
$users : \mathbb{F}\, USER$
$devices : \mathbb{F}\, DEVICE$
$memorySegments : \mathbb{F}\, MEMORY\_SEGMENT$
$ipcMessages : \mathbb{F}\, IPC\_MESSAGE$
$ecrm : ECRM$

$physicalLinks : \mathbb{F}\, PHYSICAL\_LINK$
$connections : \mathbb{P}\, CONNECTION$
$virtualCircuits : \mathbb{P}\, VIRTUAL\_CIRCUIT$
$datagrams : \mathbb{P}(VIRTUAL\_CIRCUIT \times \mathbb{P}\, DATAGRAM)$

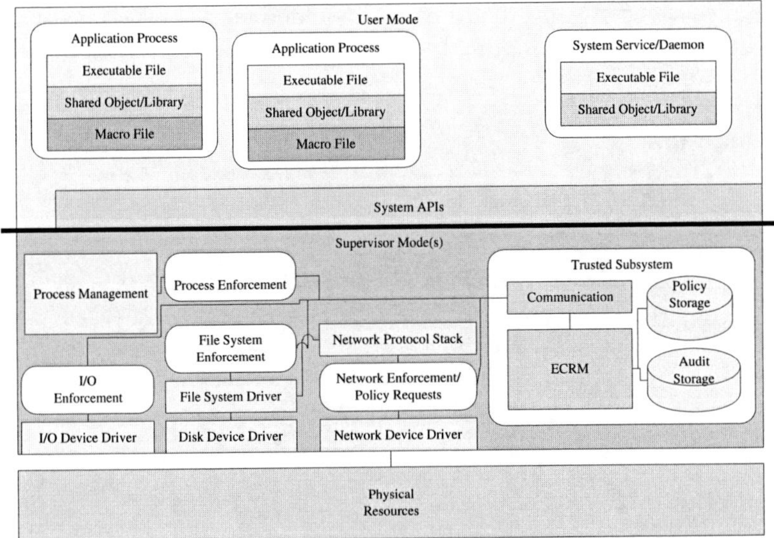

Figure 4.1: PEN Components

Definition 4.33 provides a model for an entity that is defined to permit the establishment of a communication channel to an ERM; since no further qualifications are necessary, this occurs in the form of a given set.

Definition 4.33

$[\, ERMLOCATOR \,]$

Similarly, definitions 4.34 and 4.35 stipulates the existence of security policy decision functions (SPDF, see definition 8.4 and section 3.2), which are identified by a mechanism.

Definition 4.34

$[\, SPDF\_MECHANISM \,]$

Definition 4.35 ensures that the result yielded by a SPDF contains, if positive, at least the original hypothesis, in accordance with definition 5.20.

Definition 4.35

$$
\begin{array}{l}
SPDF : (SPDF\_MECHANISM \times \mathbb{P}\, PolicyRequest \times \mathbb{P}\, Fact) \\
\quad \rightarrow \mathbb{P}\, PolicyReply \\
\hline
\forall\, policyReply : \mathrm{ran}\, SPDF;\ spdf : SPDF\_MECHANISM; \\
\quad policyRequests : \mathbb{P}\, PolicyRequest; \\
\qquad facts : \mathbb{P}\, Fact \mid \\
\qquad (spdf, policyRequests, facts) \mapsto policyReply \bullet \\
\qquad\quad policyReply \neq \{\} \Leftrightarrow \\
\qquad\quad policyRequest \subseteq policyReply
\end{array}
$$

Definition 4.36

$$
\begin{array}{l}
\underline{\ ECRMState\ } \\
nodeIdentity : NODE \\
knownERMInstances : \mathrm{iseq}\, ERMLOCATOR \\
localDecisionInstances : \mathrm{bag}\, SPDF\_MECHANISM \\
policyRepository : PolicyRepository \\
auditRepository : AuditRepository \\
localClock : LOCALTIME \\
timerRegulated : LOCALTIME \\
timerRegulatedInterval : TIME \\
\hline
knownERMInstances \neq \langle\rangle
\end{array}
$$

For the purposes of this discussion, the set of facts to be considered is a simple finite set where the validity interval has not yet expired, as given in definition 4.37.

Definition 4.37

$$\begin{array}{|l} \textit{PolicyRepository} \\ \hline \textit{localPolicyFacts} : \mathbb{F}\,\textit{Fact} \\ \hline \forall \textit{fact} : \textit{Fact} \mid \textit{fact} \in \textit{localPolicyFacts} \bullet \\ \quad \textit{fact.TimeInterval.timeEnd} \leq \\ \qquad \textit{extractLocalTime localClock} \\ \hline \end{array}$$

Similarly, definition 4.38 describes audit events as sets; however, in this case the validity interval determines the time during which an auditable event took place.

Definition 4.38

$$\begin{array}{|l} \textit{AuditRepository} \\ \hline \textit{localAuditRecord} : \mathbb{F}\,\textit{Fact} \\ \hline \end{array}$$

Definition 4.39 provides the initialization for the ECRM; definitions for the schema elements *nodeIdentity* and *knownERMInstances* are omitted and need to be suitably refined in a specific instantiation.

Definition 4.39

$$\begin{array}{|l} \textit{ECRMInit} \\ \hline \Delta \textit{ECRMState} \\ \hline \textit{localDecisionInstances}' = [\![\,]\!] \\ \textit{policyRepository}'.\textit{localPolicyFacts} = \emptyset \\ \textit{auditRepository}'.\textit{localAuditRecord} = \emptyset \\ \textit{localClock}' = \textit{localTime}(0, \textit{nodeIdentity}) \\ \textit{timerRegulatedInterval}' = 1 \textit{timerRegulated}' = \textit{localClock} \\ \hline \end{array}$$

Security policy decision function invocations based on information available in a local policy repository are encapsulated by the schema *ECRMLocalDecision* in definition 4.40. This description assumes the

existence of a set of security policy decision functions in accordance the signatures provided by definition 4.35; these need to be defined in a refinement step beyond the scope of this basic model.
For each decision taken, an audit record is created which records the timestamp of the decision, while the result obtained in the decision are conjoined with the set of facts held locally.

Definition 4.40

$$
\begin{array}{l}
\textit{ECRMLocalDecision} \\
\hline
\Delta ECRMState \\
decisionRequest? : PolicyRequest \\
decisionReply! : \mathbb{F}\ PolicyReply \\
requestRecord : Hypothesis \\
\hline
\forall\ decisionMechanism : SPDF\_MECHANISM \mid \\
\quad decisionMechanism \sqsubseteq localDecisionInstances \bullet \\
\quad SPDF(decisionMechanism, decisionRequest?, \\
\quad\quad policyRepository.localPolicyFacts) \\
\quad\quad \in decisionReply! \\
policyRepository'.localPolicyFacts = decisionReply! \cup \\
\quad policyRepository.localPolicyFacts \\
requestRecord' = decisionRequest \\
requestRecord'.TimeInterval.timeStart = \\
\quad extractLocalTimelocalClock \\
requestRecord'.TimeInterval.timeEnd = \\
\quad extractLocalTimelocalClock \\
auditRepository'.localAuditRecord = \\
\quad auditRepository.localAuditRecord \cup requestRecord' \\
\hline
\end{array}
$$

In analogy to definition 4.35, the *RemoteERMSPDF* functions according to definition 4.41 provide a security policy decision function signature for decisions taken at PCN nodes; for the purposes of this model, the mechanisms for locating a PCN based on an arbitrary *ERMLOCATOR*, communication with the PCN, and possibly transformation within the local model are not elaborated further.

Definition 4.41

$$
\begin{array}{|l}
RemoteERMSPDF : (ERMLOCATOR \times \\
\quad \mathbb{P}\, PolicyRequest \times \mathbb{P}\, Fact) \rightarrow \mathbb{P}\, PolicyReply \\
\hline
\forall\, policyReply : \operatorname{ran} RemoteERMSPDF; \\
\quad ermLocator : ERMLOCATOR;\ facts : \mathbb{P}\, Fact; \\
\quad policyRequests : \mathbb{P}\, PolicyRequest \mid \\
\quad (ermLocator, policyRequests, facts) \mapsto policyReply \bullet \\
\quad policyReply \neq \{\} \Leftrightarrow policyRequest \subseteq policyReply
\end{array}
$$

Furthermore, definition 4.42 provides the analogon for PCN-based decisions to the ECRM-based decisions given in definition 4.40.

Definition 4.42

$$
\begin{array}{|l}
\hline
\textit{ECRMRemoteDecision} \\
\Delta ECRMState \\
decisionRequest? : PolicyRequest \\
decisionReply! : \mathbb{F}\, PolicyReply \\
requestRecord : Hypothesis \\
\hline
\forall\, ermLocator : \operatorname{ran} knownERMInstances \bullet \\
\quad RemoteERMSPDF(ermLocator, decisionRequest?, \\
\qquad policyRepository.localPolicyFacts) \\
\qquad \in decisionReply! \\
policyRepository'.localPolicyFacts = \\
\quad decisionReply! \cup policyRepository.localPolicyFacts \\
requestRecord' = decisionRequest \\
requestRecord'.TimeInterval.timeStart = \\
\quad extractLocalTimelocalClock \\
requestRecord'.TimeInterval.timeEnd = \\
\quad extractLocalTimelocalClock \\
auditRepository'.localAuditRecord = \\
\quad auditRepository.localAuditRecord \cup requestRecord' \\
\hline
\end{array}
$$

Definition 4.43 conjoins the possible decision mechanisms by means of schema calculus. For reactive decisions, this is invoked by all operations of all enforcement modules (i.e. as a minimum all operations involved in the components shown in figure 4.1) as a precondition. For these, a priori unlimited in the number of instances, framing schemata for promotion are be required in addition to the external node modeling discussed in section 4.8.3.

Definition 4.43

$$ECRMDecision == ECRMLocalDecision \lor ECRMRemoteDecision$$

In order to support regulated operations, the ECRM needs to process non-reactive events.
This is modeled by the use of a real-time clock associated with the ECRM along with an arbitrary but fixed time interval for initiating processing as shown in definition 4.22 as well as by a signaling mechanism that regulates the inclusion of new policy requests.
The regulated operations of an ECRM are modeled as a queue of individual policy requests; this imposes an order on the requests that can be utilized for modeling certain policy requests with higher recurrence rates while retaining an overall deterministic evaluation order over policy requests.
For the purposes of this basic model, however, simple queuing is used.

Definition 4.44

ECRMRegulatedRequestQueue

$$regulatedRequestQueue : \mathrm{seq}\ PolicyRequest$$

Definition 4.45

$$
\begin{array}{l}
\mathit{adjustRegulatedQueue} : \\
\quad \mathrm{seq}\, \mathit{PolicyRequest} \rightarrow \mathrm{seq}\, \mathit{PolicyRequest} \\
\mathit{adjustRegulatedQueueElement} : \\
\quad \mathit{PolicyRequest} \rightarrow \mathrm{seq}\, \mathit{PolicyRequest} \\
\hline
\forall \mathit{queueElement} : \mathit{PolicyRequest} \bullet \\
\quad \mathit{queueElement.TimeInterval.timeEnd} \leq \\
\quad \mathit{extractTimeLocalTime}\ \mathit{localClock} \Rightarrow \\
\qquad \mathit{adjustRegulatedQueueElement}\ \mathit{queueElement} = \\
\qquad\quad \langle \mathit{queueElement} \rangle \wedge \\
\quad \neg(\mathit{queueElement.TimeInterval.timeEnd} \leq \\
\quad \mathit{extractTimeLocalTime}\ \mathit{localClock}) \Rightarrow \\
\qquad \mathit{adjustRegulatedQueueElement}\ \mathit{queueElement} = \langle\ \rangle \\
\\
\mathit{adjustRegulatedQueue} \langle\ \rangle = \langle\ \rangle \\
\forall \mathit{queue} : \mathrm{seq}\, \mathit{PolicyRequest} \bullet \mathit{adjustRegulatedQueue}\ \mathit{queue} = \\
\quad \mathit{adjustRegulatedQueueElement}\ \mathrm{head}\ \mathit{queue} \frown \mathrm{tail}\ \mathit{queue}
\end{array}
$$

Definition 4.46

$$
\begin{array}{l}
\underline{\mathit{ECRMRegulatedRequest}} \\
\Delta \mathit{ECRMState} \\
\hline
\mathit{localClock} \geq \mathit{timerRegulated} \\
\mathit{regulatedRequestQueue} \neq \langle\ \rangle \\
\\
\mathit{ECRMDecision}[\mathrm{head}\ \mathit{adjustRegulatedQueue} \\
\quad \mathit{regulatedRequestQueue} / \mathit{decisionRequest?}] \\
\mathit{regulatedRequestQueue}' = \mathrm{tail}\ \mathit{adjustRegulatedQueue} \\
\quad \mathit{regulatedRequestQueue} \frown \\
\quad \mathit{adjustRegulatedQueueElement} \\
\qquad \mathrm{head}\ \mathit{regulatedRequestQueue} \\
\mathit{timerRegulated}' = (\mathrm{first}\ \mathit{localClock} + \\
\quad \mathit{timerRegulatedInterval} \mapsto \mathrm{second}\ \mathit{localClock})
\end{array}
$$

Finally, definition 4.47 conjoins the sub-schemata for the ECRM model.

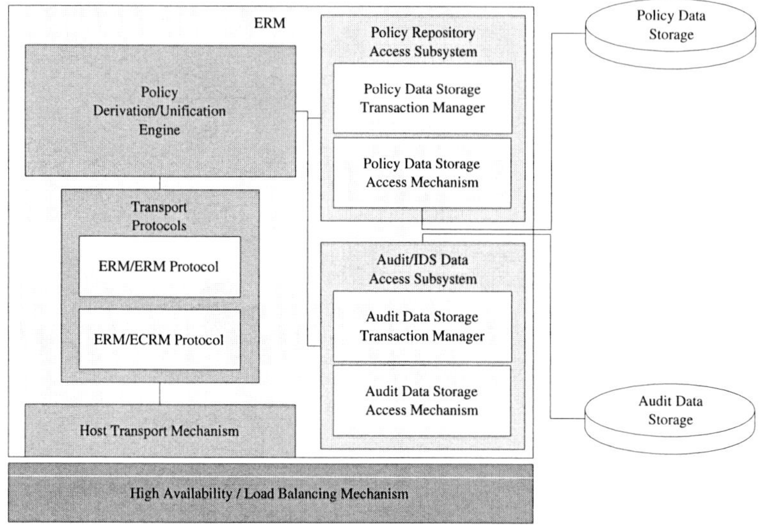

Figure 4.2: ERM Components

Definition 4.47

$$ECRM == ECRMDecision \land ECRMRegulatedRequest \land ECRMRegulatedRequest$$

4.8.2 PCN Instances

Policy controller nodes can be considered specialized instances of policy-enforcing nodes in that there may no remote decision instance, although individual PCN may refer to and cache policy decisions of other PCN.

Moreover, PCN also differ from PEN in that policy elements can be both injected and removed and policy elements can be both queried from PEN and other PCN and propagated to PEN and other PCN instances; for details see [118].

Figure 4.2 on page 104 provides a schematic overview of the components contained in a PCN.

4.8.3 Modeling Concurrent Node Behavior

The following discussion does not by itself represent valid Z notation; instead, in the interest of brevity schema names have been replaced by syntactical variables throughout the discussion.
Under this assumption, regardless of the type of node, it can be derived easily that the specification of the state of a node can be structured without loss of generality in such a way that internal state, input variables, and output variables are separated:

$$State == InternalState \wedge InputVariables \wedge OutputVariables$$

Moreover, all operations of a node can be represented in the following form, for arbitrary values of i and assuming a mapping between the $Operation_i$ and the operation schemata of the node:

$$\begin{array}{|l} \hline Operation_i \\ \Delta State \\ \hline \ldots \\ \hline \end{array}$$

The environment of a node — specifically its input variables — may change at any time; if a node performs finite state transition repetitions, this is referred to as *stuttering*; while a specification can be derived that is stutter-free [2, 3, 4], it is typically desirable to permit such behavior.
Hence, a generic node operation *Stutter* can be defined, which is again mapped to the specific operation schema of a node:

$$\begin{array}{|l} \hline Stutter \\ \Xi InternalState \\ \Xi OutputVariables \\ \Delta InputVariables \\ \hline \end{array}$$

Furthermore, a node specification can without loss of generality be structured in such a way that an initialization operation *Initialization*, which is again mapped to the specific initialization schema of a node, provides the initial state of the node. Hence, the set of operations for a specification of a node with n operations can be defined as

$$Operations == \{ Initialization, Operation_1, \ldots, Operation_n, Stutter \}$$

For the specification of a node, a sequence of states can be used to model behavior, where

$$\begin{array}{l} StateSequence : \mathbb{N} \rightarrow State \\ projectionIn : State \rightarrow InputVariables \\ projectionOut : State \rightarrow OutputVariables \\ \hline projectionIn = \{State \bullet (\theta State, \theta InputVariables)\} \\ projectionOut = \{State \bullet (\theta State, \theta OutputVariables)\} \end{array}$$

which yields a sequence $StateSequence^{in}$ containing the input values during the sequence of states in *StateSequence* and, correspondingly, the output values in the sequence $StateSequence^{out}$.

$$\begin{array}{l} \_^{in} : (\mathbb{N} \rightarrow State) \rightarrow (\mathbb{N} \rightarrow InputVariables) \\ \_^{out} : (\mathbb{N} \rightarrow State) \rightarrow (\mathbb{N} \rightarrow OutputVariables) \\ \hline \forall StateSequence : (\mathbb{N} \rightarrow State) \bullet \\ \quad (StateSequence^{in} = projectionIn \circ StateSequence \land \\ \quad StateSequence^{outn} = projectionOut \circ StateSequence) \end{array}$$

The concatenation of state bindings, *ConcatenatedStateBindings*, permits reasoning over states occurring in sequence.

$$\begin{array}{l} ConcatenatedStateBindings : State \times State \rightarrow \Delta State \\ \hline ConcatenatedStateBindings = \\ \quad \{\Delta State \bullet ((\theta State, \theta State'), \theta(\Delta State))\} \end{array}$$

The global model of a reactive (sequential) system node requires the specification of all possible states after a finite sequence of compositions of operations. Analogous to a construct from finite automata theory and assuming a sequence of operations *operationSequence*, an arbitrary but fixed coordinate within the *sequenceCoordinate* sequence, and an input flow *inputFlow*, a set of all states which can be reached by performing the first *sequenceCoordinate* + 1 operations of *operationSequence* with input values of

$$inputFlow(0), \ldots, inputFlow(sequenceCoordinate)$$

can be constructed [773, 56] as

$$ReachableStates(operationSequence, inputFlow)(sequenceCoordinate + 1).$$

$$
\begin{array}{|l}
ReachableStates : (\mathbb{N} \rightarrow Operations) \times \\
\quad (\mathbb{N} \rightarrow InputVariables) \rightarrow (\mathbb{N}\, \mathbb{P}\, State) \\
\hline
\forall\, operationSequence : (\mathbb{N} \rightarrow Operations);\ inputFlow : \\
\quad (\mathbb{N} \rightarrow InputVariables) \bullet \\
\qquad (ReachableStates(operationSequence, inputFlow)0 = \\
\qquad State \wedge (\forall\, sequenceCoordinate : \mathbb{N} \bullet \\
\qquad\quad (ReachableStates(operationSequence, \\
\qquad\qquad inputFlow))(sequenceCoordinate + 1) = \\
\qquad\quad \{state_{succ} : State \mid \\
\qquad\quad (\exists\, state_{pred} : (ReachableStates(operationSequence, \\
\qquad\qquad inputFlow)sequenceCoordinate \bullet \\
\qquad\qquad ConcatenatedStateBindings(state_{pred}, \\
\qquad\qquad state_{succ}) \in operationSequence \\
\qquad\qquad sequenceCoordinate \wedge \\
\qquad\qquad projectionIn(state_{succ} = \\
\qquad\qquad\quad inputFlow(sequenceCoordinate + 1))\}))))
\end{array}
$$

Again, using the concatenation of states, a dependency on input flows can be established as shown by the *PrecedingStates* function [55].

$$
\begin{array}{l}
PrecedingStates : \mathbb{P}\,\Delta State \rightarrow \mathbb{P}\,State \\
\hline
\forall\, preState : \mathbb{P}\,\Delta State \bullet PrecedingStates\ preState = \\
\quad \{state_{pred} : State \mid \\
\qquad \exists\, state_{succ} : State \bullet ConcatenatedStateBindings \\
\qquad\quad (state_{pred}, state_{succ}) \in preState\}
\end{array}
$$

Given the preceding definitions, the sequences of operations which can be performed in a reactive system node irrespective of individual, nondeterministically chosen states using the function *StateCones* [57].

$$
\begin{array}{l}
StateCones : (\mathbb{N} \rightarrow InputVariables) \rightarrow \mathbb{P}(\mathbb{N} \rightarrow Operations) \\
\hline
\forall\, inputFlow : \mathbb{N} \rightarrow InputVariables \bullet StateCones(inputFlow) = \\
\quad \{operationSequence : \mathbb{N} \rightarrow Operations \mid \\
\qquad operationSequence\ 0 = Initialization\ \wedge \\
\quad (\forall\, sequenceCoordinate : \mathbb{N}_1 \bullet \\
\qquad (ReachableStates(operationSequence, inputFlow)) \\
\qquad\quad sequenceCoordinate \subseteq \\
\qquad\qquad PrecedingStates(operationSequence, \\
\qquad\qquad sequenceCoordinate))\}
\end{array}
$$

Finally — again in analogy to finite automata — a Z machine as defined by *ZM* defines all permitted operations, states, input and output sequences [55, 57].

$$
\begin{array}{|l}
\hline
\textit{ZM} \\
\textit{operationSequence} : \mathbb{N} \rightarrow \textit{Operations} \\
\textit{StateSequence} : \mathbb{N} \rightarrow \textit{State} \\
\textit{inputFlow} : \mathbb{N} \rightarrow \textit{InputVariables} \\
\textit{outputFlow} : \mathbb{N} \rightarrow \textit{OutputVariables} \\
\hline
\textit{operationSequence} \in \textit{StateCones}(\textit{inputFlow}) \wedge \\
\quad (\forall\, \textit{sequenceCoordinate} : \mathbb{N} \bullet \\
\qquad \textit{ConcatenatedStateBindings}(\\
\qquad\quad \textit{StateSequence}(\textit{sequenceCoordinate}), \\
\qquad\quad \textit{StateSequence}(\textit{sequenceCoordinate} + 1)) \in \\
\qquad \textit{operationSequencesequenceCoordinate}) \\
\textit{StateSequence}^{in} = \textit{inputFlow} \\
\textit{StateSequence}^{out} = \textit{outputFlow} \\
\hline
\end{array}
$$

The formulation of the Z machine schema *ZM* thus can be considered a safety condition; while it is also possible to impose similar liveness conditions with the help of weak and strong fairness operators on operations in *Operation*, this is not considered in this discussion [15, 16, 642, 2, 55, 56].

Moreover, the overall system thus can be modeled by N Z machines ZM_i, each specifying a node $node_i$ where $i \in 1 \ldots N$. The following discussion assumes as before that a syntactical mapping between the indexed operations and sets used below and the actual sets, types, and schemata exist.

The environment of the nodes $node_i$ derives from a subset of other nodes in the model and can, assuming a fixed but arbitrary node $node_i$ which is related by input and output flows to k_i other nodes, be described by a set of inputs $node_{i_{jm}}$:

$$\{node_{i_{1_1}}, \ldots, node_{i_{k_i}}\} \subset \{node_1, \ldots, node_N\} \setminus \{node_i\}$$

Given these node identities, the Z machines must be conjoined and the relations over the respective input and output flows established. The input flow $inputFlow_i$ for a given $node_i$ are computed from the output flows of the nodes constituting the environment of $node_i$; for this purpose the family of abstract auxiliary functions $FlowMapping_i$ is introduced which determines the $inputFlow_i$ based on the respective $outputFlow_{i_{1_1}}, \ldots, outputFlow_{i_{k_i}}$ of the nodes $node_{i_{1_1}}, \ldots, node_{i_{k_i}}$. The $FlowMapping_i$ yield, for each element of the sequence, an input derived from the output variables of the state sequences of other nodes.

$$FlowMapping_i : OutputVariables_{i_{1_1}} \times \cdots \times OutputVariables_{i_{k_i}} \times InputVariables \rightarrow InputVariables_i$$

The output flow *outputFlow* of the entire model can thus be computed by another abstract auxiliary function *OutputMapping*.

$$OutputMapping : OutputVariables_1 \times \cdots \times OutputVariables_N \rightarrow OutputVariables$$

Moreover, conjoining the Z machines of the nodes $\{node_1, \ldots, node_N\}$ and providing relations between the input and output flows using the abstract auxiliary functions $FlowMapping_i$ and the abstract auxiliary function *OutputMapping* results in a closed system that can, again, be considered as a Z machine for further composition. In the following schema, signature definitions for indexed *inputFlow* and *outputFlow* are omitted for brevity and are assumed to be subsumed by the non-indexed signature definition:

ConcurrentZMachines

$$\bigwedge_{p=1}^{N} ZM_p$$
$$inputFlow : \mathbb{N} \rightarrow InputVariables$$
$$outputFlow : \mathbb{N} \rightarrow OutputVariables$$

$$\begin{aligned}
&\forall i : 1 \ldots N;\ sequenceCoordinate : \mathbb{N} \bullet \\
&\quad (inputFlow_i sequenceCoordinate = \\
&\qquad FlowMapping_i(outputFlow_{i_{1_1}} \\
&\qquad\quad sequenceCoordinate, \ldots \\
&\qquad\quad outputFlow_{i_{k_i}} \\
&\qquad\quad sequenceCoordinate, \\
&\qquad\quad inputFlow\ sequenceCoordinate) \wedge \\
&\quad outputFlow\ sequenceCoordinate = \\
&\qquad OutputMapping(\\
&\qquad\quad outputFlow_1 sequenceCoordinate, \ldots \\
&\qquad\quad outputFlow_N sequenceCoordinate))
\end{aligned}$$

As before, fairness conditions could be defined analogously.

CHAPTER

FIVE

Security Policies

In this chapter the mechanisms for specifying and deriving statements of the security policy as well as the underlying model are described.
Section 5.2 contains definitions for later use in section 5.3 which introduces the definition of the security policy.
It is in the selection of the primitives of section 5.2 that the conscious restriction to the description of operating system security manifests itself.
A result on the expressiveness of the policy mechanism is obtained in section 5.5.
Section 5.4 contains the definition of the security model.
In section 5.6 the application of a security policy to a hypothesis along with the requisite algorithm for the application of multiple policies is defined.

5.1 Elements of Security Policy Statements

It should be noted that the following definitions are merely elements of a formal theory, namely $\mathcal{PC}$; while names are suggestive, the following does not constitute an interpretation in and of itself (see chapter 6).

Definition 5.1 *A **subject** is a constant denoted as* σ.

For each subject σ^i*, there exist suitable constants* $C^T_{\sigma^i}$, $C^I_{\sigma^i}$ *and two predicates* subject_type($\sigma^i, C^T_{\sigma^i}$) *and* subject_identity($\sigma^i, C^I_{\sigma^i}$) *such that an individual subject is uniquely identified by*

$$\text{subject\_type}(\sigma^i, C^T_{\sigma^i}) \wedge \text{subject\_identity}(\sigma^i, C^I_{\sigma^i}).$$

The set of all subjects is denoted as $\mathbf{\Sigma}$.

Definition 5.2 *Each subject* σ *is embedded in a lattice* $\mathcal{L}^T_\sigma$ *imposed by* subject_subset_type *on* $\mathbf{\Sigma}$. *For all* σ^i *there is a constant* C *and a predicate* subject_subset_type(σ^i, C) *such that* C *is an element of* $\mathcal{L}^T_\sigma$ *which contains* σ^i.
The function symbol Subject_subset_type(A, B) *yields the lub of the elements of* $\mathcal{L}^T_\sigma$.

Definition 5.3 *Each subject* σ *is embedded in a lattice* $\mathcal{L}^I_\sigma$ *imposed by* subject_subset_identity *on* $\mathbf{\Sigma}$. *For all* σ^i *there is a constant* C *and a predicate* subject_subset_identity(σ^i, C) *such that* C *is an element of* $\mathcal{L}^I_\sigma$ *which contains* σ^i.
The function symbol Subject_subset_identity(A, B) *yields the lub of the elements of* $\mathcal{L}^I_\sigma$.

Intuitively, a set of subject types within the lattice $\mathcal{L}^T_\sigma$ can therefore be represented by choosing suitable elements σ^i and C for the parameters of subject_subset_type. Correspondingly, a set of subject instances in $\mathcal{L}^I_\sigma$ can be identified by choosing suitable elements σ^i and C for the parameters of subject_subset_identity. The simplest lattice algebra satisfying the requirements is obtained by defining subject_subset_identity and subject_subset_type as the set inclusion relation on $\mathcal{P}(\mathbf{\Sigma})$ according to observation 2.17. However, despite the suggestive name, only the general lattice algebraic properties are required for definitions 5.2 and 5.3.

Definition 5.4 Σ *is a vector of subjects* σ_i.

Definition 5.5 *An **object** is a constant denoted as ω.*
For each object ω^i, there exist suitable constants $C^T_{\omega^i}$, $C^I_{\omega^i}$ and two predicates $\text{object\_type}(\omega^i, C^T_{\omega^i})$ *and* $\text{object\_identity}(\omega^i, C^I_{\omega^i})$ *such that an individual object is uniquely identified by*

$$\text{object\_type}(\omega^i, C^T_{\omega^i}) \wedge \text{object\_identity}(\omega^i, C^I_{\omega^i}).$$

The set of all objects is denoted as $\mathbf{\Omega}$.

Definition 5.6 *Each object ω is embedded in a lattice $\mathcal{L}^T_\omega$ imposed by* object_subset_type *on* $\mathbf{\Omega}$.
For all ω^i there is a constant C and a predicate $\text{object\_subset\_type}(\omega^i, C)$ *such that C is an element of $\mathcal{L}^T_\omega$ which contains ω^i.*
The function symbol $\text{Object\_subset\_type}(A, B)$ *yields the lub of the elements of $\mathcal{L}^T_\omega$.*

Definition 5.7 *Each object ω is embedded in a lattice $\mathcal{L}^I_\omega$ imposed by* object_subset_identity *on* $\mathbf{\Omega}$. *For all ω^i there is a constant C and a predicate* $\text{object\_subset\_identity}(\omega^i, C)$ *such that C is an element of $\mathcal{L}^I_\omega$ which contains ω^i.*
The function symbol $\text{Object\_subset\_identity}(A, B)$ *yields the lub of the elements of $\mathcal{L}^I_\omega$.*

Intuitively, a set of object types within the lattice $\mathcal{L}^T_\omega$ can therefore be represented by choosing suitable elements ω^i and C for the parameters of object_subset_type. Correspondingly, a set of subject instances in $\mathcal{L}^I_\omega$ can be identified by choosing suitable elements ω^i and C for the parameters of object_subset_identity. The simplest lattice algebra satisfying the requirements is obtained by defining object_subset_identity and object_subset_type as the set inclusion relation on $\mathcal{P}(\mathbf{\Omega})$ according to observation 2.17. However, despite the suggestive name, only the general lattice algebraic properties are required for definitions 5.6 and 5.7.

Definition 5.8 *Ω is a vector of subjects ω_i.*

Definition 5.9 *ξ is an* ***operation descriptor*** *$\xi \in \Xi$ where Ξ is the set of well formed operation descriptors.*
For all ξ there exists a predicate

$$\text{operation}_{\xi}(\sigma_1, \ldots, \sigma_i, \omega_1, \ldots, \omega_j)$$

where i and j depend on ξ.

Definition 5.10 *λ is an* ***operation lifetime*** *$\lambda \in \mathbf{\Lambda}$. The set of valid operation lifetimes $\mathbf{\Lambda}$ is defined as ordered pairs (λ^s, λ^e) where λ^s and λ^e are elements of a set on which an order is defined for which $\lambda^s \leq \lambda^e$ holds.*

Definition 5.11 *Requests for policy decisions (called a* ***policy request*** *or simply a* ***hypothesis****) can be formulated as a 4-tuple $(\Sigma, \Omega, \xi, \lambda)$.*

5.2 Modeling Primitives

As before, some caution must be paid to the fact that the following definitions are elements of a formal theory $\mathcal{PC}$ and do not in and of themselves carry any semantic meaning despite suggestive naming which is provided only in the definition of an interpretation. For each element, a description is given which provides guidance for mapping the primitives to the elements of an interpretation.
The choice of predicates — given below without loss of generality for the case of operating systems — for individual operations and the corresponding embedding in a lattice emanates from an analysis of underlying concepts in the domain of general purpose operating systems and, orthogonally, the algebraic embedding of these concepts [78, 548, 434, 273, 236].
While omitted in this discussion, the embedding of operations within lattice algebraic structures for obtaining a semantic hierarchy can proceed analogously.

5.2.1 Subjects

Definition 5.12 *The following predicates are defined for subjects; S is a metavariable denoting an element of* **Σ**.

subject(S) *This predicate holds for each constant for which one of the predicates* user(S), process(S), application(S), host(S), network(S), link_node(S), *or* bus_node(S) *holds.*

user(S) *An interpretation should map this primitive to an entity which corresponds to an independent agent not subject to algorithmic control.*

process(S) *An interpretation should map this primitive to an entity which consists of a collection of resources and the ability to execute operations.*

application(S) *An interpretation should map this primitive to an entity which consists of a collection of resources and algorithmic patterns that may be different for each specific instance of the primitive.*

host(S) *An interpretation should map this primitive to an individual node of a distributed system which is operating under local control and which processes security policy data locally.*

network(S) *An interpretation should map this primitive to a set of individual nodes of a distributed system in which membership in the set can be determined by a predicate over each element of any such set.*

link_node(S) *An interpretation should map this primitive to an endpoint of a network link where the set of endpoints is a priori indeterminate and may change without an individual link node obtaining notification of changes of the existence of other link node.*

bus_node(S) *An interpretation should map this primitive to an endpoint of a link connecting a node to other endpoints where the endpoint has the means to enumerate other endpoints and obtain knowledge of endpoints being attached or detached to the common link.*

5.2.2 Objects

Definition 5.13 *The following predicates are defined for objects. O is a metavariable denoting an element of* $\mathbf{\Omega}$.

object(S) *This predicate holds for each constant for which one of the predicates* file(O), data_file(O), executable_file(O), connection(O), virtual_circuit(O), datagram(O), physical_link(O), *or the predicate* memory_segment(O) *holds.*

file(O) *An interpretation should map this primitive to a resource consisting of nonvolatile storage.*

executable_file(O) *An interpretation should map this primitive to a resource consisting of nonvolatile storage which can be interpreted in such a way that a primitive for which the* process(S) *predicate holds can execute the information contained therein.*

data_file(O) *An interpretation should map this primitive to a resource consisting of nonvolatile storage for which the* executable_file(O) *predicate does not hold.*

directory(O) *An interpretation should map this primitive to a resource consisting of nonvolatile storage and identifying a possibly empty set of entities* O_i *for whose elements the* file(O_i) *predicate holds.*

connection(O) *An interpretation should map this primitive to an exclusively used direct connection between a subject and another subject node without transiting intermediate nodes.*

virtual_circuit(O) *An interpretation should map this primitive to a connection between a subject and another subject which may transition across one or more intermediate nodes and which may be multiplexed with other objects for which the* virtual_circuit(O) *or* connection(O) *predicates hold.*

datagram(O) *An interpretation should map this primitive to a resource which may transmit data from a subject to another subject.*

ipc_message(O) *An interpretation should map this primitive to a resource which may transmit data from a subject to another subject.*

device(O) *An interpretation should map this primitive to a resource which is associated with an actual physical or abstract device, access to which is mediated.*

physical_link(O) *An interpretation should map this primitive to a physical connection between the subject and another subject.*

memory_segment(O) *An interpretation should map this primitive to a resource consisting of volatile storage which can be accessed by multiple subjects.*

5.2.3 Operations

Definition 5.14 *The following operation predicates identifying the right to perform an operation are defined:*

create_file(S, O_1, O_2) *An interpretation should map this primitive to an operation which creates a resource uniquely identified by the metavariable O_2 for which the predicate* file(O_2) *holds. The creator of the resource is identified by the metavariable S for which the predicate* subject(S) *holds. The entity identified by O_2 is subordinate to another entity identified by the metavariable O_1 for which the predicate* directory(O_1) *holds.*

read_file(S, O) *An interpretation should map this primitive to an operation which makes the information contained in the resource identified by the metavariable O for which the predicate* file(O) *holds available to the entity identified by the metavariable S for which the predicate* subject(S) *holds.*

write_file(S, O) *An interpretation should map this primitive to an operation which transfers information under the control of the entity identified by the metavariable S for which the predicate* subject(S) *holds to the resource identified by the metavariable O for which the predicate* file(O) *holds.*

overwrite_file(S, O) *An interpretation should map this primitive to an operation which transfers information under the control of the entity identified by the metavariable S for which the predicate* subject(S) *holds to the resource identified by the metavariable O for which the predicate* file(O) *holds while possibly removing information present in O prior to the operation.*

append_file(S, O) *An interpretation should map this primitive to an operation which transfers information under the control of the entity identified by the metavariable S for which the predicate* subject(S) *holds to the resource identified by the metavariable O for which the predicate* file(O) *holds without removing information present in O prior to the operation.*

delete_file(S, O) *An interpretation should map this primitive to an operation in which the the resource identified by the metavariable O for which the predicate* file(O) *holds is destroyed by the entity identified by the metavariable S for which the predicate* subject(S) *holds.*

read_file_information(S, O) *An interpretation should map this primitive to an operation which makes information on the resource identified by the metavariable O for which the predicate* file(O) *holds and which is mutually disjoint with information for which the operation identified by the predicate* read_file(S, O) *holds available to the entity identified by the metavariable S for which the predicate* subject(S) *holds.*

write_file_information(S, O) *An interpretation should map this primitive to an operation which transfers information under the control of the entity identified by the metavariable S for which the predicate* subject(S) *holds to the resource identified by the metavariable O for which the predicate* file(O) *holds and which is mutually disjoint with information for which the operation identified by the predicate* write_file(S, O) *holds.*

read_directory(S, O) *An interpretation should map this primitive to an operation which transfers information on a set of objects O_i for which the predicates* file(O_i) *hold and where the predicate* directory(O)

holds; the information transferred for the set of objects O_i is mutually disjoint with information for which the operation identified by the predicates read_file(S, O_i) *hold and is made available to the entity identified by the metavariable S for which the predicate* subject(S) *holds.*

create_directory(S, O_1, O_2) *An interpretation should map this primitive to an operation which creates an entity which is uniquely identified by the metavariable O_2 for which the predicate* directory(O_2) *holds and which is subordinate to an entity identified by the metavariable O_1 for which in turn* directory(O_1) *holds. The creator of the resource O_2 is identified by the metavariable S for which the predicate* subject(S) *holds.*

search_directory(S, O_1, O_2) *An interpretation should map this primitive to an operation which transfers information on an object identified by the metavariable O_2 for which the predicate* file(O_2) *or the predicate* directory(O_2) *holds and where the predicate* directory(O_1) *holds; the information transferred for the entity O_2 is mutually disjoint with information for which the operation identified by the predicate* read_file(S, O_2) *holds and is made available to the entity identified by the metavariable S for which the predicate* subject(S) *holds.*

delete_directory(S, O) *An interpretation should map this primitive to an operation in which the the resource identified by the metavariable O for which the predicate* directory(O) *holds is destroyed by the entity identified by the metavariable S for which the predicate* subject(S) *holds.*

create_memory_segment(S, O) *An interpretation should map this primitive to an operation which creates a resource which is uniquely identified by the metavariable O and for which the* memory_segment(O) *predicate holds. The creator of the resource is identified by the metavariable S for which the predicate* subject(S) *holds.*

read_memory_segment(S, O) *An interpretation should map this primitive to an operation which makes the information contained in the resource identified by the metavariable O for which the predicate* memory_segment(*O*) *holds available to the entity identified by the metavariable S for which the predicate* subject(*S*) *holds.*

write_memory_segment(*S*, *O*) *An interpretation should map this primitive to an operation which transfers information under the control of the entity identified by the metavariable S for which the predicate* subject(*S*) *holds to the resource identified by the metavariable O for which the predicate* file(*O*) *holds.*

delete_memory_segment(*S*, *O*) *An interpretation should map this primitive to an operation in which the the resource identified by the metavariable O for which the predicate* memory_segment(*O*) *holds is destroyed by the entity identified by the metavariable S for which the predicate* subject(*S*) *holds.*

open_device(*S*, *O*) *An interpretation should map this primitive to an operation which assigns a resource uniquely identified by the metavariable O for which the predicate* device(*O*) *holds to an entity identified by the metavariable S for which the predicate* subject(*S*) *holds.*

read_device_data(*S*, *O*) *An interpretation should map this primitive to an operation which makes the information associated with the resource identified by the metavariable O for which the predicate* device(*O*) *holds available to the entity identified by the metavariable S for which the predicate* subject(*S*) *holds.*

write_device_data(*S*, *O*) *An interpretation should map this primitive to an operation which transfers information under the control of the entity identified by the metavariable S for which the predicate* subject(*S*) *holds to the resource identified by the metavariable O for which the predicate* device(*O*) *holds.*

read_device_control(*S*, *O*) *An interpretation should map this primitive to an operation which makes information on the resource identified by the metavariable O for which the predicate* device(*O*) *holds and which is mutually disjoint with information for which the operation identified by the predicate* read_device_data(S, O) *holds available to the entity identified by the metavariable S for which the predicate* subject(S) *holds.*

write_device_control(S, O) *An interpretation should map this primitive to an operation which transfers information under the control of the entity identified by the metavariable S for which the predicate* subject(S) *holds to the resource identified by the metavariable O for which the predicate* device(O) *holds and which is mutually disjoint with information for which the operation identified by the predicate* write_device_data(S, O) *holds.*

close_device(S, O) *An interpretation should map this primitive to an operation in which the the resource identified by the metavariable O for which the predicate* device(O) *holds is disassociated with the entity identified by the metavariable S for which the predicate* subject(S) *holds.*

send_ipc(S_1, S_2, O) *An interpretation should map this primitive to an operation in which the the resource identified by the metavariable O for which the predicate* ipc_message(O) *holds can be used to transmit information under the control of the entity identified by the metavariable S_1 for which the predicate* subject(S_1) *holds to the entity identified by the metavariable S_2 for which the predicate* subject(S_2) *holds.*

receive_ipc(S_1, S_2, O) *An interpretation should map this primitive to an operation in which the the resource identified by the metavariable O for which the predicate* ipc_message(O) *holds may be used to transmit information under the control of the entity identified by the metavariable S_2 for which the predicate* subject(S_2) *holds to the entity identified by the metavariable S_1 for which the predicate* subject(S_1) *holds.*

send_datagram(S_1, S_2, O) *An interpretation should map this primitive to an operation in which the the resource identified by the metavariable O for which the predicate* datagram(O) *holds can be used to* *transmit information under the control of the entity identified by the metavariable* S_1 *for which the predicate* subject(S_1) *holds to the entity identified by the metavariable* S_2 *for which the predicate* subject(S_2) *holds.*

receive_datagram(S_1, S_2, O) *An interpretation should map this primitive to an operation in which the the resource identified by the metavariable* O *for which the predicate* datagram(O) *holds may be used to transmit information under the control of the entity identified by the metavariable* S_2 *for which the predicate* subject(S_2) *holds to the entity identified by the metavariable* S_1 *for which the predicate* subject(S_1) *holds.*

create_virtual_circuit(S_1, S_2, O) *An interpretation should map this primitive to an operation which creates a resource uniquely identified by the metavariable* O *for which the predicate* virtual_circuit(O) *holds. The creator of the resource is identified by the metavariable* S_1 *for which the predicate* subject(S_1) *holds; the primitive also defines a relation on the entities* S_1 *and* S_2 *where the predicate* subject(S_2) *also holds.*

send_virtual_circuit(S, O) *An interpretation should map this primitive to an operation in which the the resource identified by the metavariable* O *for which the predicate* virtual_circuit(O) *holds can be used to transmit information under the control of the entity identified by the metavariable* S *for which the predicate* subject(S) *holds and which must be equal to the parameter* S_1 *of the* create_virtual_circuit *predicate which the parameter* O *identifies to the entity identified by the parameter* S_2 *of the predicate* create_virtual_circuit *which the parameter* O *identifies.*

receive_virtual_circuit(S, O_1, O_2) *An interpretation should map this primitive to an operation in which the the resource identified by the metavariable* O_1 *for which the predicate* virtual_circuit(O) *holds makes information available to the entity* S *for which the predicate* subject(S) *holds and which must be equal to the parameter* S_2 *of the predicate* create_virtual_circuit *which the parameter* O_2 *identifies;*

the information source is equal to the entity identified by parameter S_1 *of the predicate* create_virtual_circuit *which the parameter* O_2 *identifies.*

delete_virtual_circuit(S, O) *An interpretation should map this primitive to an operation in which the the resource identified by the metavariable* O *for which the predicate* virtual_circuit(O) *holds is destroyed by the entity identified by the metavariable* S *for which the predicate* subject(S) *holds.*

5.2.4 Derivative Instances

Definition 5.15 *The following predicate is defined entirely in terms of interpreted entities and therefore does not need to be subject to interpretation itself assuming standard logic connectives are used by the interpretation:*

derived_or_self(X, Y) *The interpretation must ensure that this predicate is true for any two constants* X *and* Y *if and only if*

- *X and Y are identical ground expressions*
- *The predicates* subject(X), subject(Y), *and* subject_subset_type(X, Y) *hold.*
- *The predicates* object(X), object(Y), *and* object_subset_type(X, Y) *hold.*
- *The predicates* subject(X), subject(Y), *and* subject_subset_identity(X, Y) *hold.*
- *The predicates* object(X), object(Y), *and* object_subset_identity(X, Y) *hold.*

5.2.5 Sequences

The elements of the formal theory do not impose temporal constraints. However, a it is desirable to introduce a mechanism for expressing the temporal composition of operations [552]. For this purpose the concept of time must be introduced.

Definition 5.16 *A* ***timeline*** *is a first-order linear time structure* $\mathcal{M} = (\Phi, \psi, \mathrm{l})$ *where* Φ *is a set of states* ϕ *,* ψ *is an infinite sequence of states* $\mathbb{N} \rightarrow \Phi$*, and* l *is a labeling of each state with the interpretation of all symbols* $S \rightarrow \Phi$*.*
The notation $\mathcal{M}, \psi \models A$ *is used to signify semantically that in structure* $\mathcal{M}$*, the well-formed formula* A *is true of timeline* ψ*.* $\models$ *is defined inductively on the structure of the well-formed formulae of the theory:* $\psi \models A$ *if and only if* $A \in \mathrm{l}(\phi_0)$*. Logical connectives are defined as usual. The construct* ψ^i *represents the suffix path* $\phi_i, \phi_{i+1}, \ldots$*.*

Definition 5.17 *The following primitive operations are used to describe temporal relations where* A, B *are well-formed formulae which must not contain quantifiers.*

precedes(A, B) *This primitive represents the notion of "A precedes B". The semantics to be assigned by an interpretation must be that* $\psi \models$ precedes(A, B) *if and only if* $(\forall j) x^j \models B$ *implies* $(\exists k < j) x^k \models A$

until(A, B) *This primitive represents the notion of "A until B". The semantics to be assigned by an interpretation must be that* $\psi \models$ until(A, B) *if and only if* $(\exists j) x^j \models B$ *and* $(\forall k < j) x^j \models A$*.*

It should be noted that the mechanism introduced above does not include axioms for reasoning on the temporal structure. Hence, it does not represent a full temporal logic. The rationale for this restriction stems from the problems of efficient deduction and decidability inherent in the inclusion of deductive mechanisms and quantifiers. As a result, the expressiveness compared to a full temporal logic is severely curtailed.
However, it should be noted that tense logic (or temporal, i.e. most modal logic with some exceptions such as propositional dynamic logic that are presumably of limited interest) can be efficiently translated into first order logic [17, 639, 321]. While this would in some sense permit both a more intuitive formulation of policy and a cleaner definition of semantics, it arguably also increases overall complexity without adding to the expressiveness (see section 5.5).

5.3 The Security Policy

The term security policy is, unfortunately, in use in a variety of interpretations and some care must be taken not to confuse the formal definition with existing preconceptions.

Definition 5.18 *A **security policy** is a consistent set of constants, function symbols and predicate symbols in $\mathcal{PC}$.*

Intuitively, a security policy is a set of rules which determine operations performed by subjects on objects; the rules of a security policy will eventually apply to the elements of definition 5.11.
However, it is not required that each rule directly refers to the elements described in section 5.1.
It should be noted that the policy must be formulated as a set of valid predicates; the consistency requirement in conjunction with the use of $\mathcal{PC}$ rules out the specification of a policy as positive rules and exceptions as would be possible using non-monotonic logic [462].

5.4 Security Models

The following definition of a security model can be considered incomplete in that an most security models must rely on a set of modeling primitives such as the ones defined in definitions 5.12, 5.13, and 5.14. However, it is conceivable that some security models do not require those primitives or are defined in terms of different primitives; definition 5.19 takes this possibility into account.

Definition 5.19 *A **security model** is a set of constants, function symbols and predicate symbols in $\mathcal{PC}$.*
*A security policy is said to be in **compliance** with a security model if the conjunction of each predicate in the security policy with all predicates in the security model is valid.*
*The predicates of the security model are also referred to as **constraint predicates**.*

The constraint predicates are merely an artifact of the formulation of the security model in $\mathcal{PC}$; an equivalent but more efficient construct can be found by realizing the security model as a set of meta--operations on security policy predicates. Such meta-operations can be used to eliminate or restrict policy predicates which are non--compliant syntactically.

5.5 Expressiveness of the Security Policy

The computability of each existing security model by a security policy according to definition 5.18 is obvious.

Theorem 5.1 *Any security model that can be calculated by a machine working on finite data in accordance with a finite program of instructions can be implemented by a security policy according to definition 5.18.*

Proof The proof of this theorem is a direct consequence of the Church-Turing thesis. Since it can be shown that $\mathcal{PC}$ is equivalent to a Turing machine with an initial tape, the theorem follows. □

5.6 Security Policy Application

Algorithm 5.1 ***Policy element derivation algorithm***
Let the predicates of the security policy P according to definition 5.18 be enumerated and represented by the metavariables P_i. Let the parameters of each predicate be enumerated and represented by the metavariables P_i^j and P_i^k, respectively. Let the number of parameters for each predicate be denoted by P_i^p.
For each predicate P_i perform the following steps:

1. *Set $j = 1$*
2. *If the predicate* subject(P_i^j) *holds, add a predicate to P for all constants S_m for which the predicates* subject_subset_type(S_m, P_i^j) *and*

subject_subset_identity(S_m, P_i^j) *hold where parameter* P_i^j *is substituted with* S_m.

3. *Set* $k = 1$

4. *If the predicate* object(P_i^k) *holds, add a predicate to P for all constants* O_n *for which the predicates* object_subset_type(O_n, P_i^j) *and* object_subset_identity(S_n, P_i^j) *hold where parameter* P_i^j *is substituted with* O_n.

5. *If* $k = p$, *go to step 6; otherwise set* $k = k + 1$ *and go to step 4.*

6. *If* $j = p$, *stop; otherwise set* $j = j + 1$ *and go to step 2.*

It should be noted that algorithm 5.1 is not intended for direct execution; depending on the lattices $\mathcal{L}_\sigma^T$, $\mathcal{L}_\sigma^I$, $\mathcal{L}_\omega^T$, and $\mathcal{L}_\omega^I$ chosen, significantly more efficient idempotent realizations can be obtained.

Algorithm 5.2 ***Hypothesis parameter projection algorithm***
Let the subjects of a hypothesis according to definition 5.11 be enumerated and represented by the metavariables S_m.
Let the objects of a hypothesis according to definition 5.11 be enumerated and represented by the metavariables O_n.
Let the parameters of the operation descriptor ξ *of a hypothesis be enumerated and represented by the metavariables* P_p.

1. *Set* $i = 1$.

2. *Set* $j = 1$.

3. *Set* $k = 1$.

4. *If* $j \geq m$, *go to step 6; otherwise substitute the parameter* P_i *with* S_j.

5. *Set* $i = i + 1$, *set* $j = j + 1$, *and go to step 4.*

6. *If* $k \geq n$, *stop; otherwise substitute the parameter* P_i *with* O_k.

7. *Set* $i = i + 1$, *set* $k = k + 1$, *and go to step 6.*

Definition 5.20 *The **application** of a security policy to a hypothesis according to definition 5.11 is the unification of the hypothesis with the elements of the security policy and the elements derived from these according to algorithm 5.1 after application of algorithm 5.2 and is denoted with the symbol ∇, i.e.*

$$\{(\Sigma^\nabla, \Omega^\nabla, \xi^\nabla, \lambda^\nabla)_k\} = \nabla(\Sigma, \Omega, \xi, \lambda).$$

or, for short

$$\{(\Sigma, \Omega, \xi, \lambda)_k^\nabla\} = \nabla(\Sigma, \Omega, \xi, \lambda).$$

If $|\{(\Sigma, \Omega, \xi, \lambda)_k^\nabla\}| = 1$, then this is denoted as $(\Sigma, \Omega, \xi, \lambda)^\nabla$ or $(\Sigma^\nabla, \Omega^\nabla, \xi^\nabla, \lambda^\nabla)$ as a shorthand.
*Elements of $\{(\Sigma, \Omega, \xi, \lambda)_k^\nabla\}$ for which $\lambda_k^s < \lambda_p^s$ or $\lambda_k^e > \lambda_p^e$ for the policy element with which it was unified are removed from $\{(\Sigma, \Omega, \xi, \lambda)_k^\nabla\}$. The elements are verbally described as **reply tuples**.*

It should be noted that the ∇ operator is ab initio independent of the formal theory established in previous sections.

Definition 5.21 *A **negative reply tuple** is denoted as $(\Sigma^\nabla, \Omega^\nabla, \xi^\perp, \lambda^\nabla)$. The application of any policy decision on a negative reply tuple is again the negative reply tuple:*

$$\forall \nabla : \nabla(\Sigma^\nabla, \Omega^\nabla, \xi^\perp, \lambda^\nabla) = (\Sigma^\nabla, \Omega^\nabla, \xi^\perp, \lambda^\nabla).$$

*If the application of a policy decision yields a result different from $(\Sigma^\nabla, \Omega^\nabla, \xi^\perp, \lambda^\nabla)$ the hypothesis is said to be **valid** under the policy.*

Definition 5.22 *If, after application of a security policy to a hypothesis according to definition 5.20, $\{(\Sigma, \Omega, \xi, \lambda)_k^\nabla\}$ is not equal to $\{(\Sigma^\nabla, \Omega^\nabla, \xi^\perp, \lambda^\nabla)\}$, $\{(\Sigma, \Omega, \xi, \lambda)_k^\nabla\}$ contains elements whose unification with the hypothesis* **H** *yields* **H**.

Definition 5.23 *A policy request is* ***granted*** *(or alternatively, the hypothesis is valid) if and only if* $\exists\,(\Sigma,\Omega,\xi,\lambda)_i^{\nabla} \in \{(\Sigma,\Omega,\xi,\lambda)^{\nabla}\}$ *such that* $(\Sigma,\Omega,\xi,\lambda)_i^{\nabla} \neq (\Sigma^{\nabla},\Omega^{\nabla},\xi^{\perp},\lambda^{\nabla})$.

Multiple policies can be applied to a hypothesis.

Definition 5.24 *The application of multiple policies is denoted as*

$$\{(\Sigma,\Omega,\xi,\lambda)_k'\} = \nabla_1 \circ \nabla_2 \circ \cdots \circ \nabla_n(\Sigma,\Omega,\xi,\lambda)$$

or, in cases where the set of policies and their sequence applied is obvious, as

$$\{(\Sigma,\Omega,\xi,\lambda)_k'\} = \nabla^*(\Sigma,\Omega,\xi,\lambda).$$

The sequence of applying policies is left-associative.

Algorithm 5.3 ***Multiple policy application algorithm***

1. *Set* $i = 1$, *set* $\{\mathbf{H}\}$ *(the set of hypotheses) to the original hypothesis* $\{H\}$.
2. *Apply* ∇_i *to each element of* $\{\mathbf{H}\}$ *and let the set union of the results be* $\{\mathbf{H}\}_i'$.
3. *Remove all elements of* $\{\mathbf{H}\}_i'$ *for which* $\xi = \xi^{\perp}$.
4. *If* $i = n$ *and* $\bigcup_{k=1}^{n}\{\mathbf{H}\}_k' = \emptyset$, *set the reply to* $\{(\Sigma^{\nabla},\Omega^{\nabla},\xi^{\perp},\lambda^{\nabla})\}$ *and stop.*
5. *If* $i = n$ *set the reply to* $\bigcup_{k=1}^{n}\{\mathbf{H}\}_k' \cup H$ *and stop.*
6. *Set* $\{\mathbf{H}\} = \bigcup_{k=1}^{i}\{\mathbf{H}\}_k' \cup H$, *set* $i = i + 1$, *and go to step 2.*

Theorem 5.2 *Algorithm 5.3 yields either a set of reply tuples* $\{(\Sigma,\Omega,\xi,\lambda)_k^{\nabla}\}$ *that either valid under at least one policy applied and can be unified with the hypothesis or the set* $\{(\Sigma^{\nabla},\Omega^{\nabla},\xi^{\perp},\lambda^{\nabla})\}$.

Proof Proof follows by induction.

Basis Case $i = 1$. If, after application of step 2 $\{(\Sigma^\nabla, \Omega^\nabla, \xi^\perp, \lambda^\nabla)\}$ is obtained, steps 3 and 4 will ensure that the result is

$$\{(\Sigma^\nabla, \Omega^\nabla, \xi^\perp, \lambda^\nabla)\}$$

as well, and the theorem holds.

Otherwise after application of step 2 $\{\mathbf{H}\}'_1$ is valid under ∇_1 according to definitions 5.20 and 5.21, hence $\bigcup_{k=1}^{1}\{\mathbf{H}\}'_k \neq \emptyset$ after step 3 and step 5 is executed.

By definition 5.20 H must be unifiable with $\bigcup_{k=1}^{1}\{\mathbf{H}\}'_k$, therefore the theorem holds after step 5.

Induction Step The elements of $\bigcup_{k=1}^{i-1}\{\mathbf{H}\}'_k$ are consistent with at least one policy according to the induction hypothesis.

If, after application of step 2 $\{(\Sigma^\nabla, \Omega^\nabla, \xi^\perp, \lambda^\nabla)\}$ is obtained for $\{\mathbf{H}\}'_i$, step 3 will ensure that no element that is not valid under the policy ∇_i is in $\{\mathbf{H}\}'_i$ and hence in $\bigcup_{k=1}^{i}\{\mathbf{H}\}'_k$.

If in step 4 none of the policy applications has yielded a valid reply, then $\bigcup_{k=1}^{n}\{\mathbf{H}\}'_k = \emptyset$, and thus $\{(\Sigma^\nabla, \Omega^\nabla, \xi^\perp, \lambda^\nabla)\}$ results and step 5 is never reached since the algorithm terminates.

Otherwise in step 5 $\bigcup_{k=1}^{n}\{\mathbf{H}\}'_k$ contains only valid tuples. The set union of this set with H is also valid because of definition 5.22 since any element whose unification with H yields H is consistent with H.

Any hypothesis added to $\{\mathbf{H}\}$ is consistent with those of previous steps and with H, hence valid hypotheses generated from these will also be valid.

□

Observation 5.1 *The set of derived tuples under a set of security policies* $\{\nabla_i\}$ *for a given hypothesis depends on the sequence of application of the members of* $\{\nabla_i\}$.

5.7 Metapolicies

Multiple security policies according to definition 5.18 can be applicable at any point in time. These must meet the requirements set forth in definition 5.25.

Definition 5.25 *Given a set of security policies* $\mathbf{\Psi} = \{\Psi_1, \ldots, \Psi_n\}$ *where all elements of* $\mathbf{\Psi}$ *are to be applied to a given state,* $\bigcup_{k=1}^{n} \mathbf{\Psi}$ *must be consistent.*

Proving the consistency property cannot be automated in the general case as a direct consequence of Rice's theorem. Except in trivial instances or given considerable resources that are applicable only in case of largely static policies, this is also presumably not feasible using manual proof techniques.
A pragmatic approach is to ensure consistency by the imposition of a separation of authority for the definition of policy elements, e.g. in parallel to a policy domain in accordance with definition 6.9.
Given a set of security policies that satisfy the requirement of definition 5.25, the individual policies can be considered atomic elements over which an algebra can be defined as shown in definition 5.26 that can yield new consistent policies.

Definition 5.26 *Given a set of atomic security policies* $\mathbf{\Psi} = \{\Psi_1, \ldots, \Psi_n\}$ *and a well-formed formula in* $\mathcal{PC}\ \varpi$*, the following operators over policies exist:*

$\Psi_i \cup \Psi_j$ *This is the union of all elements of the policies.*

This results in any operation permitted under either policy being admissible.

$\Psi_i \cap \Psi_j$ *This is the set intersection of all elements of the policies where for all well-formed formulae* $C_{\Psi_i} \in \Psi_i$ *and* $C_{\Psi_j} \in \Psi_j$ *the element is included in the result if and only if*

$$\text{derived\_or\_self}(C_{\Psi_i}, C_{\Psi_j}) \wedge \text{derived\_or\_self}(C_{\Psi_j}, C_{\Psi_i})$$

holds.

$\Psi_i - \Psi_j$ *This is the set difference of all elements of the policies where for all well-formed formulae $C_{\Psi_i} \in \Psi_i$ and $C_{\Psi_j} \in \Psi_j$ the element is included in the result if and only if*

$$\neg\ \text{derived\_or\_self}(C_{\Psi_j}, C_{\Psi_i})$$

holds.

$\Psi_i \mid \varpi$ *This is the scoping operator where for each of the function symbols and predicate symbols in Ψ_i, the symbol is transformed by a conjunction with ϖ.*

$\neg\ \Psi_i$ *This is the policy negation operator where for each of the function symbols and predicate symbols in Ψ_i, the symbol is transformed by negation.*

The following identities can be derived trivially:

Observation 5.2 1. *Union inclusion:*

$$\Psi_i \subseteq (\Psi_j \cap \Psi_k)$$

2. *Associativity of unions and intersections:*

   (a)
$$(\Psi_i \cup \Psi_j) \cup \Psi_k = \Psi_i \cup (\Psi_j \cup \Psi_k)$$

   (b)
$$(\Psi_i \cap \Psi_j) \cap \Psi_k = \Psi_i \cap (\Psi_j \cap \Psi_k)$$

3. *Commutativity of unions and intersections:*

   (a)
$$\Psi_i \cup \Psi_j = \Psi_j \cup \Psi_i$$

   (b)
$$\Psi_i \cap \Psi_j = \Psi_j \cap \Psi_i$$

4. *Distributivity of unions and intersections:*

(a)

$$\Psi_i \cup (\Psi_j \cap \Psi_k) = (\Psi_i \cup \Psi_j) \cap (\Psi_i \cup \Psi_k)$$

(b)

$$\Psi_i \cap (\Psi_j \cup \Psi_k) = (\Psi_i \cap \Psi_j) \cup (\Psi_i \cap \Psi_k)$$

5. *Difference operator properties:*

(a)

$$\Psi_i - \Psi_i = \emptyset$$

(b)

$$\Psi_i - (\Psi_j \cup \Psi_k) = \Psi_i - \Psi_j - \Psi_k$$

(c)

$$\neg\, (\Psi_i - \Psi_j) = \neg\, \Psi_i - \neg\, \Psi_j$$

6. *Negation operator properties:*

(a)

$$\neg\, \neg\, \Psi_i = \Psi_i$$

(b)

$$\neg\, (\Psi_i \cup \Psi_j) = \neg\, \Psi_i \cup \neg\, \Psi_j$$

(c)

$$\neg\, (\Psi_i \cap \Psi_j) = \neg\, \Psi_i \cap \neg\, \Psi_j$$

CHAPTER

SIX

Interpretation

Section 5.2 provides the elements of a formal theory which can be used in the definition of a security policy; as indicated by the choice of primitives defined there, the emphasis in the discussion is on operating system security although the mechanisms are sufficiently general for other areas.
Therefore an interpretation according to definition 2.22 of the primitives for use in well-formed formulae is required. This interpretation must then be augmented by interpretation elements, if any, specific to a security model according to definition 5.19.

6.1 Subject Type and Identity

One of the goals of this discussion is to ensure the consistent application of security policies regardless of which environment, i.e. which interpretation, it occurs under.
A prerequisite for this is the identification of subject type and identity such that algorithm 5.1 can be applied efficiently and that hierarchies of such subjects can be established such that derivative predicates can be applied to subordinate subjects.

Definition 6.1 *The lattice* $\mathcal{L}_\sigma^T$ *is defined by the* subject_subset_type *relation on the elements of* $\mathcal{P}(\Sigma)$. $\forall X, Y \in \mathcal{P}(\Sigma)$, subject_subset_type *is defined as* $X \subseteq Y$.

Similarly, hierarchies for subject identities can be used by security models for efficient representation and in the creation of hierarchical security models.

Definition 6.2 *The lattice* $\mathcal{L}_\sigma^I$ *is defined by the* subject_subset_identity *relation on the elements of* $\mathcal{P}(\Sigma)$. $\forall X, Y \in \mathcal{P}(\Sigma)$, subject_subset_identity *is defined as* $X \subseteq Y$.

Each element of these lattices can be represented as a path in a Hasse diagram. Definitions 6.3 and 6.4 provide an implementable representation for both subject types and identity lattice relations.

Definition 6.3 *The* ***subject type vector*** *is a vector of numerical elements for which a bijective mapping exists to the elements of* $\mathcal{L}_\sigma^T$.
Let the elements of the vector be numbered $X_1, \ldots, X_n$. *For each* X_i *and all elements of the vector* X_j *where* $i < j$, X_i *is the lub of* X_i, X_j.

Definition 6.4 *The* ***subject identity vector*** *is a vector of numerical elements for which a bijective mapping exists to the elements of* $\mathcal{L}_\sigma^I$.
Let the elements of the vector be numbered $X_1, \ldots, X_n$. *For each* X_i *and all elements of the vector* X_j *where* $i < j$, X_i *is the lub of* X_i, X_j.

Observation 6.1 *The numerical representation of each vector element in definitions 6.3 and 6.4 is unique.*

Proof Any object can be enumerated as a Gödel number of a Turing machine with finite data and a finite, possibly empty program of instructions. □
For both the type and identity of a subject, an interpretation may provide an interpretation-specific naming schemata if and only if bijective function is defined by the interpretation which maps the interpretation-specific naming schemata to the schemata of definitions 6.3 and 6.4, respectively.

6.2 Object Type and Identity

The requirements for the identification of object type and identity are analogous to those delineated for subjects in section 6.1, hence the following definitions.

Definition 6.5 *The lattice $\mathcal{L}^T_\omega$ is defined by the* object_subset_type *relation on the elements of $\mathcal{P}(\boldsymbol{\Omega})$. $\forall X, Y \in \mathcal{P}(\boldsymbol{\Omega})$,* object_subset_type *is defined as $X \subseteq Y$.*

Definition 6.6 *The lattice $\mathcal{L}^I_\omega$ is defined by the* object_subset_identity *relation on the elements of $\mathcal{P}(\boldsymbol{\Omega})$. $\forall X, Y \in \mathcal{P}(\boldsymbol{\Omega})$,* object_subset_identity *is defined as $X \subseteq Y$.*

The lattice relation representations for object types and identities are again defined analogous to the definitions for subjects.

Definition 6.7 *The* ***object type vector*** *is a vector of numerical elements for which a bijective mapping exists to the elements of $\mathcal{L}^T_\omega$.*
Let the elements of the vector be numbered $X_1, \ldots, X_n$. For each X_i and all elements of the vector X_j where $i < j$, X_i is the lub of X_i, X_j.

Definition 6.8 *The* ***object identity vector*** *is a vector of numerical elements for which a bijective mapping exists to the elements of $\mathcal{L}^I_\omega$.*
Let the elements of the vector be numbered $X_1, \ldots, X_n$. For each X_i and all elements of the vector X_j where $i < j$, X_i is the lub of X_i, X_j.

Observation 6.1 holds accordingly for the vectors in definition 6.7 and 6.8; for both the type and identity of an object, an interpretation may provide an interpretation-specific naming schemata if and only if bijective function is defined by the interpretation which maps the interpretation-specific naming schemata to the schemata of definitions 6.7 and 6.8, respectively.

6.3 Imposition of a Policy Hierarchy

As noted in observation 5.1, for the set of derived tuples under a set of security policies for a given hypothesis to be deterministic, the sequence of policy application must — while in itself completely arbitrary — be well-defined.
This is somewhat problematic in that any such ordering imposes a boundary condition on the expressiveness of the security models and policies and the reason why it is defined here rather than in chapter 5.
The mechanism for imposing a sequence described here is a hierarchy.

Definition 6.9 *A* ***policy domain*** *is a set of subjects* $\{\sigma_i\}$ *and objects* $\{\omega_j\}$ *for which an instance, the* ***domain authority****, may issue reply tuples.*

Definition 6.10 *A policy domain may contain zero or more* ***sub-domains*** *and is called a* ***parent domain****. A sub-domain consists of a subset of the sets of subjects of the parent domain* $\{\sigma_i^S\} \subseteq \{\sigma_i^P\}$ *and a subset of objects of the parent domain* $\{\omega_i^S\} \subseteq \{\omega_i^P\}$*. The policy domains are the vertices of a tree, i.e. a directed, connected, and acyclical graph with precisely one vertex of in-degree zero, the root [244]. The resulting graph is called the* ***policy domain tree****.*

To ensure efficient identification of policy domains, the following definitions impose additional restrictions on the set inclusion relations of definition 6.10.

Definition 6.11 *For each element* X_i *of a subject identity vector, there exists a corresponding unique vertex in the policy domain tree.*

Definition 6.12 *For each element* X_i *of an object identity vector, there exists a corresponding unique vertex in the policy domain tree.*

6.4 Interpretation Outline: Microsoft Windows 2000

Realizing any security policy requires the existence of an interpretation according to definition 2.22. For this, a domain $\mathcal{D}$ must be given along with a mapping of each predicate symbol, function symbol, and constant.
To ensure that a given interpretation is a model of a theory requires rigorous proof, yet the scope of the effort for specifying an operating system with the desired functionality as well as proving the correspondence between the formal specification and the implementation including all underlying components down to the gate level appears to be truly daunting.
However, as outlined in chapter 1, this is — while certainly a theoretically worthwhile goal — not the objective pursued in this discussion. Rather, it is to provide the mechanisms described here on actively deployed systems. While it may be argued that the potential assurance provided by an interpretation from a system with source code availability is greater, few systems provide satisfactory design evidence to provide a significant increase in assurance ([500, 647, 499] notwithstanding).
Therefore, the following is to be considered a sample outline of an interpretation based on published material, behavioral analysis, and to a certain extent reverse engineering of a system for which neither detailed specification, nor source code, nor evidence of correspondence between specification and source code was available, namely the Microsoft Windows 2000 family of operating systems.
It should be noted that the system-specific interpretation must be augmented with interpretation definitions for network elements and definitions covering physical connectivity.

6.4.1 Architectural Considerations

The Microsoft Windows 2000 operating system (the following applies with some minor modifications to all versions of Windows NT

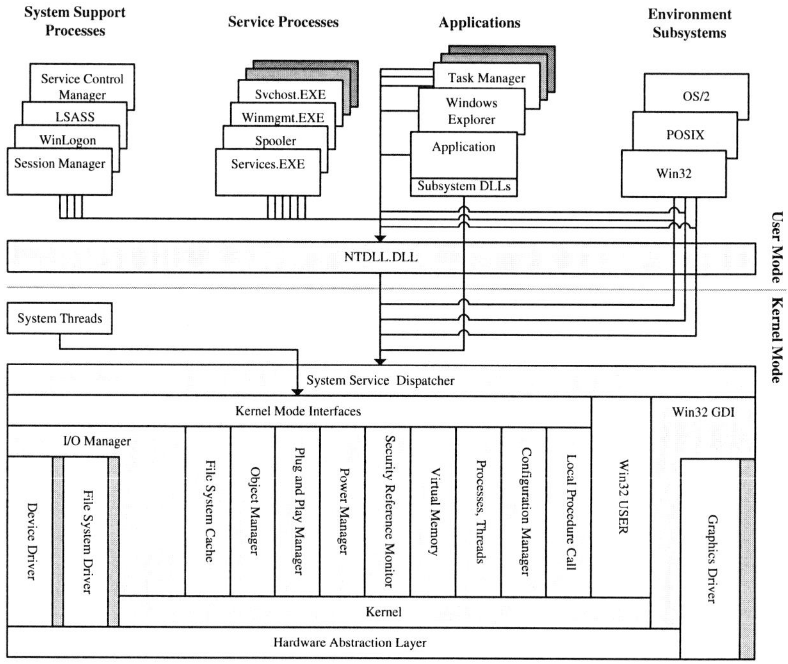

Figure 6.1: Windows 2000 Architectural Overview

ranging from 3.1 to XP; version from the initial 3.1 release through 3.51 employed a different architecture for graphics device drivers, some components in the kernel related to plug-and-play and power management were added after version 4.0) is a traditional monolithic operating system employing two levels of hardware protection: a user mode for regular applications, system services, system support functions, and environmental subsystems and a supervisor or kernel mode. A simplified schema of the system components is shown in figure 6.1 on page 142 [157, 675, 676].

The system provides an environmental subsystem mechanism which exposes a certain interface to processes in user mode; all of these subsystems — the dominant of which is the Win32 API — do, however, at some point map to a native interface.

The native API is the equivalent of Unix system calls or OpenVMS system services. A number of Win32 calls are merely forwarding their arguments to corresponding native API functions while others map more complex operations to primitives provided by the native API [503, 645].

While there exist two interfaces with different prefixes (`Zw` and `Nt`), the entry points these interfaces map to are in fact the same code. The difference between these entries is that user mode programs calls (with the `Nt` prefix) are accessed via `NTDLL.DLL`, where a change to supervisor mode occurs.

As a result of this call, the system service dispatcher records the previous state of the processor in the `ETHREAD` structure before calling the actual executive service residing in `NTOSKRNL.EXE`. Kernel mode code links directly against `NTOSKRNL.EXE`, the `Zw` prefix entry point re-enters supervisor mode and proceeds through the system service dispatcher bypassing object access control mechanisms. The kernel depicted as such in figure 6.1 is not a separate entity; it is also contained in the `NTOSKRNL.EXE` along with the executive. It contains scheduling and synchronization primitives (called kernel objects) as well as mechanisms for trap handling.

Some of this depends on specific hardware; most of the hardware dependencies are, however, encapsulated in a separate layer, the HAL. The exception to this by design are graphics device drivers; however, any code in kernel mode technically has the ability to bypass the HAL.

The executive itself contains several components in addition to the services mentioned above, including native API functions not exposed to environmental subsystems, functions and symbols exposed for use by device drivers, file system drivers, and file system filter drivers (see chapters 9 and 10) as well as kernel support functions for use by the I/O and memory managers.

The executive contains the following major subsystems:

File System Cache The Cache Manager implements the read and write caching strategy for file system access; it is closely interrelated to the virtual memory manager since Windows 2000 implements a unified file system cache and virtual memory architecture.

Object Manager This subsystem creates, manages, and destroys executive objects used for representing all operating system resources in a centralized manner.

Plug and Play Manager With this subsystem newly introduced in Windows 2000 a mechanism is implemented which identifies devices and their resource requirements and in response loads the required drivers and allocates the resources requested by such a device.

Power Manager This subsystem generates power management I/O notifications for device drivers and coordinates events related to power consumption.

Security Reference Monitor The reference monitor verifies ACLs for system objects on access and can create audit events.

Virtual Memory This subsystem realizes separate address spaces for processes each backed by a demand paging mechanism (anticipatory paging in conjunction with the Cache Manager).

Process and Thread Manager While the actual objects for managing and scheduling processes and threads are implemented as kernel objects, this subsystem adds additional semantics and functionality by encapsulating the pertinent kernel objects.

Configuration Manager This subsystem (also referred to as the Registry) handles both transient and persistent configuration information for both the operating system itself and also for user mode processes.

Local Procedure Call Facility This subsystem realizes an IPC mechanism which is used extensively throughout the Windows 2000 operating system.

Of particular interest for the following discussion is the Object Manager and the resources implemented by it. It provides a centralized naming schema for all resources and a centralized access mechanism which is also used by the native Windows 2000 security mechanisms. The Object Manager deals only with executive objects; this, however, is not a limitation as kernel objects are never used outside the `NTOSKRNL.EXE` module and are always encapsulated in executive objects; access from the outside is mediated through the use of handles. Objects are structured into a header containing generic information for use by the Object Manager itself and a type-specific body. The object header contains the following information:

Object name An unique name for the object within its object directory.

Object directory A structured name creating a hierarchical name space; adding the object name to the object directory of a given object results in a node-wide unique name for the object.

Security descriptor This entry contains the ACL entries for security enforcement by the Windows 2000 Security Reference Monitor.

Quota charges Resource charges levied against the process the object is assigned to on opening a handle to the object.

Open handle count This entry contains the number of open handles to the object.

Open handles list This entry contains a pointer to a list of processes which retain open handles to the object.

Object type This entry contains a pointer to a type object containing attributes specific to a given type of object.

Reference count The reference count is the number of kernel-mode references to the address of the object itself.

The type object attributes are given below:

Type name This entry contains the name of the type of object. Windows 2000 defines the following object types and names (not all of which are exposed beyond the executive):

1. `Adapter`
2. `Callback`
3. `Controller`
4. `Desktop`
5. `Device`
6. `Directory`
7. `Driver`
8. `Event`
9. `EventPair`
10. `File`
11. `IoCompletion`
12. `Job`
13. `Key`
14. `Mutant`
15. `Port`
16. `Process`
17. `Profile`
18. `Section`
19. `Semaphore`
20. `SymbolicLink`
21. `Thread`
22. `Timer`
23. `Token`
24. `Type`
25. `WaitablePort`
26. `WindowStation`
27. `WmiGuid`

Pool type This field indicates whether objects of this type should be allocated from non-paged pool or if they may be allocated from paged pool.

Default quota charges This field contains the values to be charged to the process quotas for paged and non-paged objects, respectively.

Access types This field indicates the access types a caller may request when opening a handle to an object of this type.

Generic access rights mapping This field contains a mapping between generic access rights defined by Windows 2000 (i.e. "read", "write", "terminate", and "all") and type-specific access rights, if any.

Synchronization Thie field indicates whether a thread object may wait on objects of this type.

Methods This entry contains a list of function pointers to functions to be called each time when performing certain generic operations on an object of this type.

Given the brief exposition above, it is both necessary and sufficient for an interpretation for the Windows 2000 operating system to operate on subjects and objects as defined by the executive and to identify sets of operations for mappings of the operation predicate symbols.

6.4.2 Subjects

6.4.2.1 user(S)

These entities represents a human user. It also encompasses stimuli to the system which are not attributable to other subject entities but which can be identified and presumably authenticated. Within a Windows 2000 system, this concept is partially represented as unique user account SIDs. The restriction to identifiable human users implies that the following SIDs must not be mappable to a constant for which the predicate user(S) holds: `SeNullSid`, `SeWorldSid`, `SeLocalSid`, `SeCreatorOwnerSid`, `SeCreatorGroupSid`, `SeNtAuthoritySid`, `SeDialupSid`, `SeNetworkSid`, `SeBatchSid`,

SeInteractiveSid, SeLocalSystemSid, SeAliasAdminsSid, SeAliasUsersSid, SeAliasGuestsSid, SeAliasPowerUsersSid, SeAliasAccountOpsSid, SeAliasSystemOpsSid, SeAliasPrintOpsSid, and SeAliasBackupOpsSid.

6.4.2.2 process(S)

These entities are mapped to the EPROCESS executive object. Windows 2000 provides each process with an address space backed by demand paging whose separation from other processes in user mode is provided by the underlying hardware protection mechanisms. As the name implies, the UniqueProcessId field provides an identification mechanism within the name space of a given nodes processes. Threads (represented by ETHREAD executive object) are not mapped to the process(S) predicate since they are subsumed by the EPROCESS object which contains a list of kernel thread objects (KTHREAD) in the KPROCESS kernel object contained in the EPROCESS object; reverse mapping is possible. If a security model implements information flow controls including covert timing channels, a separate additional thread predicate becomes necessary.

6.4.2.3 application(S)

Applications are not represented by a native primitive within Windows 2000. Rather, these entities are composite constructs. Since the system supports shared executable objects, the pertinent components must be identified. The relevant list structure pointed to by ModuleLoaderInfoHead is part of the PEB which is in turn associated with the EPROCESS executive object.
The native function NtCreateProcess, which is involved in both initial creation of processess and subsequent dynamic loading of shared executable objects eventually calls on the services of the Loader for initialization of the executable object images.
The final step to be completed before reaching entry points is a routine called LdrpRunInitializeRoutines which is called within the context of the process to be initialized and iterates over the elements

of `ModuleLoaderInfo` to initialize executable objects that have not been initialized before.
At this stage the newly initialized elements must be copied from the `PEB`.
The reason for this is that the `PEB` is writable in the process context itself and thus can be manipulated by code running in user mode.
By updating the module list (which is in the form of a path in the Windows 2000 file system and must be mapped to unique constants to which executable_file(O) applies) immediately prior to the transfer of control, the integrity of the new entries (and only the new entries) in `ModuleLoaderInfo` is assured.
Moreover, the load order is not equivalent to the initialization order, raising the possibility of symbol name and initialization conflicts; by using the well-defined initialization order, this issue is avoided.
Identification of the individual elements can proceed by applying a cryptographic hash function to each element; the interpretation--specific identification is then the vector of hashed elements in initialization order including information on the size of the vector.

6.4.2.4 host(S)

These entities are mapped to individual nodes running a single instance of Windows 2000.
An unique identifier for the system must either be provided externally or the system SID can be used for this purpose.

6.4.2.5 network(S)

These entities are not covered by the operating system interpretation and must be augmented.

6.4.2.6 link_node(S)

Link nodes are not represented by a native primitive within Windows 2000. They are, however, implicitly represented by a specific instance

of the `Device` executive object. The relevant API for this occurrence is the NDIS, version 5 of which was introduced by Windows 2000. Such device objects are created by the system NDIS library on behalf of NDIS NICs or directly by a NDIS Miniport driver by calling the `NdisMRegisterDevice`.
Due to the addition of upper-level protocol task offloading and fast packet forwarding mechanisms provided, it is the lowest level at which the properties required for the link_node(S) predicate can be provided by an interpretation.

6.4.2.7 bus_node(S)

Bus nodes are directly represented by a primitive in the form of bus drivers.
This is a class of driver newly introduced with Windows 2000; since older revisions do not support this mechanism and the possibility of using monolithic legacy drivers exist, these are also subsumed under this primitive.
In the case of legacy drivers individual drivers to be used must have the capabiltiy to enumerate endpoints, obtain and provide information on changes in the nodes.
Drivers supporting only a single node can also be subsumed under this category. The bus driver mechanism itself is also divided into two categories.
At the lowest abstraction layer are hardware bus drivers; this category is augmented by PnP hardware bus drivers controlling dynamically configurable I/O buses.
The driver collaborates with the Plug and Play Manager to configure and reconfigure system hardware resources (such as mappings for device memory and IRQs) for all child devices connected to the I/O bus or buses it controls.
Common to all these instances is that they are represented by a specific instance of the `Device` executive object and can thus be identified uniquely by the interpretation-specific identification mechanism.

6.4.3 Objects

6.4.3.1 file(*O*)

Files are distinct from other objects in that they have representations in memory and persistent external representations. However, the relevant instance for consideration in the interpretation is the in-- memory representation.
The entities are mapped to the `FILE_OBJECT` objects as represented by the `File` executive object[158, 493]. There exists another object within the file system implementation that can be considered representative of the file primitive, namely the file control block (FCB) .
There can exist several `FILE_OBJECT` instances in memory for a given physical file representation; however, there is only one FCB for such a representation.
Since FCB objects do not carry all necessary information for the relevant operations and are file-system dependent, they do not qualify for mapping to the file primitive.

6.4.3.2 data_file(*O*)

These entities are mapped to objects to which the file primitive is mapped and to which the primitive executable_file is not mapped.

6.4.3.3 executable_file(*O*)

These entities are mapped to objects for which, given a section object containing the data of the object accessed with `SECTION_MAP_EXECUTE` privileges, the executive function `ZwCreateProcess` can either identify the program as a PE [476] executable file or find a supporting image (`POSIX.EXE` for POSIX, `OS2.EXE` for OS/2 executable images, `NTVDM.EXE` for MS-DOS and 16-Bit Windows applications, and `CMD.EXE` for scripts for which the file ending is `.BAT` or `.CMD`) by matching the object content with the expected format.

6.4.3.4 directory(*O*)

These entities are mapped to two separate entities. First, they are mapped to `Directory` executive objects. The abstraction provided by the system extend the directory notion beyond nonvolatile storage objects; this is a semantically consistent extension of the requirements for the interpretation. Nonvolatile storage objects are the second type of entity, they are represented by `File` executive objects for which the corresponding `FILE_BASIC_INFORMATION` structure contains the attribute `FILE_ATTRIBUTE_DIRECTORY`.

6.4.3.5 connection(*O*)

There is no direct construct to which this primitive can be mapped. However, the required mapping can be provided by `Device` executive objects referring to display drivers, port drivers, and miniport drivers provided that the `Device` object refers to a connected physical device.

6.4.3.6 virtual_circuit(*O*)

There exist several types of connection oriented primitives for communication between nodes to which this primitive maps. First, named pipes which are represented internally by the operating system as file objects. The identification of a given object to which the primitive file maps as a named pipe and hence as a virtual_circuit occurs if the `ZwCreateNamedFile` function is called to yield the handle to the file object whose namespace is managed by the `NPFS.SYS` driver. Second, mailslots which are also represented internally by the operating system as file objects.
The identification of a given object to which the primitive file maps as a and hence as a virtual_circuit occurs if the `ZwCreateMailslotFile` function is called to yield the handle to the file object whose namespace is managed by the `MSFS.SYS` driver. Third, virtual circuits to which virtual_circuit is mapped are established using the `MSAFD.SYS` file system driver [346] also represented internally by the operating

system as file objects and including WinSock, SPX over IPX, NetBIOS, AppleTalk, ATM, and IrDA infrared sockets.

6.4.3.7 datagram(O)

There exists no native to which this primitive can be mapped. Instead, it must be derived from interpreted data as provided by one of several driver objects, namely `MSFS.SYS`, `NPFS.SYS`, `MSAFD.SYS`, and particularly `NDIS.SYS`. The latter is required as all of the remaining drivers transmit data via this object and components may issue direct requests for data transmission to this driver.

6.4.3.8 physical_link(O)

These entities are not covered by the operating system interpretation and must be augmented.

6.4.3.9 memory_segment(O)

These entities are mapped to `Section` executive objects representing volatile storage which may be accessed by several processes; these are the only objects for which user processes may share access.

6.4.4 Operations

It should be noted that not all of the functions discussed here are documented as interfaces of the operating system or even exported to any of the subsystem personalities.

6.4.4.1 create_file(S, O_1, O_2)

This primitive is mapped to the native function `ZwCreateFile` provided that access is performed with the `CreateDisposition` set to `FILE_CREATE` and the object does not exist, set to `FILE_OPEN_IF` if the object does not exist, or set to `FILE_OVERWRITE_IF` if the object

does not exist, or FILE_SUPERSEDE if the object exists. For this mapping to apply the CreateOptions parameter must not contain the flag FILE_DIRECTORY_FILE.
The identity of O_1 is provided either by the handle to the File executive object RootDirectory in OBJECT_ATTRIBUTES if that handle is not NULL or from the prefix of the fully resolved name without any symbolic links remaining of the system-specific namespace up to the last separator in the field ObjectName of the OBJECT_ATTRIBUTES structure pointed to by the parameter ObjectAttributes.
The identity of O_2 is derived from the fully resolved name without any symbolic links remaining as the element after the last separator (if any) in the field ObjectName of the structure OBJECT_ATTRIBUTES pointed to by the parameter ObjectAttributes.
The names obtained for O_1, O_2 must be mapped accordingly.
The abstraction provided by the system extend the directory notion beyond nonvolatile storage objects; this is a semantically consistent extension of the requirements for the interpretation.

6.4.4.2 read_file(S, O)

This primitive is mapped to the ZwReadFile and ZwReadFileScatter native functions as well as to the ZwMapViewOfSection native function. In all cases, the identity of O is provided by the handle to the File executive object in the parameter FileHandle. The data read from the identified object falls under the area of control of the process subject; this applies to other subject types by extension.
Prior establishment of the File executive object and its mapping in the interpretation by means of ZwCreateFile or ZwOpenFile is implicitly assumed; the access characteristics from the parameter DesiredAccess to these calls are not relevant for consideration in the evaluation of this primitive.
In the case of ZwMapViewOfSection the Section executive object identified by the SectionHandle handle must have been obtained by either a call to the native function ZwCreateSection with the same parameter and the parameter DesiredAccess having the flag SECTION_MAP_READ or SECTION_MAP_EXECUTE set or by a call to the

native function ZwOpenSection where the parameter DesiredAccess contained the set flags SECTION_MAP_READ or SECTION_MAP_EXECUTE and the parameter FileHandle identical to that used in the subsequent function call.

6.4.4.3 write_file(S, O)

This primitive is mapped to the ZwWriteFile, ZwWriteFileGather native functions, and to the function ZwMapViewOfSection. In both cases, the identity of O is provided by the handle to the File executive object in the parameter FileHandle. The data written to the identified object falls under the area of control of the process subject; this applies to other subject types by extension.
Prior establishment of the File executive object and its mapping in the interpretation by means of ZwCreateFile or ZwOpenFile is implicitly assumed; the access characteristics from the parameter DesiredAccess to these calls are not relevant for consideration in the evaluation of this primitive.
In the case of ZwMapViewOfSection the Section executive object identified by the SectionHandle handle must have been obtained by either a call to the native function ZwCreateSection with the same parameter and the parameter DesiredAccess having the flag SECTION_MAP_WRITE set or by a call to the ZwOpenSection native function where the parameter DesiredAccess contained the set flag SECTION_MAP_WRITE and the parameter FileHandle identical to that used in the subsequent function call.

6.4.4.4 overwrite_file(S, O)

There is no directly mappable function for this primitive; it must be modeled by a sequence of calls beginning with a call to ZwCreateFile, ZwOpenFile, ZwCreateSection, or ZwOpenSection for the object identified by O as represented by the File executive object handle passed in to the FileHandle parameters of these functions provided that the CreateDisposition set to FILE_OVERWRITE, FILE_OVERWRITE_IF, or

FILE_SUPERSEDE. Subsequent calls are those identified by the predicate write_file(S, O) with identical S and O. In the cases of the operations on section objects associated with the file handles, the preceding calls must have been made with the flags DesiredAccess having the flag SECTION_MAP_WRITE set.

6.4.4.5 append_file(S, O)

There is no directly mappable function for this primitive; it must be modeled by a sequence of calls beginning with a call to ZwCreateFile, ZwOpenFile, ZwCreateSection, or ZwOpenSection for the object identified by O as represented by the File executive object handle passed in to the FileHandle parameters of these functions with the flags for DesiredAccess set to the logical disjunction of SYNCHRONIZE and FILE_APPEND_DATA and no other flags set in that parameter. In the cases of the operations on section objects associated with the file handles, the preceding calls must have been made with the flags DesiredAccess having the flag SECTION_MAP_WRITE set.
The second element of the sequence consists of calls to the native functions ZwWriteFile and ZwWriteFileGather with O set to the same File executive object handle and the parameter ByteOffset set to NULL.

6.4.4.6 delete_file(S, O)

This primitive is mapped to two native functions. The first function is ZwDeleteFile where the identity of O is derived from the fully qualified native name space name ObjectName contained in the OBJECT_ATTRIBUTES structure referred to by the ObjectAttributes parameter. The second function is ZwSetInformationFile where the object O is identified as represented by the File executive object handle passed in to the FileHandle parameter if the parameter FileInformationClass is set to FileDispositionInformation.

6.4.4.7 read_file_information(S, O)

This primitive is mapped to four native functions. The first function is `ZwQueryEaFile` where the object O is identified as represented by the `File` executive object handle passed in to the `FileHandle` parameter. The second function is `ZwQueryAttributesFile` where the identity of O is derived from the fully qualified native name space name `ObjectName` contained in the `OBJECT_ATTRIBUTES` structure referred to by the parameter `ObjectAttributes`. The third function is `ZwQueryFullAttributesFile` where the identity of O is derived from the fully qualified native name space name `ObjectName` contained in the `OBJECT_ATTRIBUTES` structure referred to by the parameter `ObjectAttributes`.
The fourth function is `ZwQueryInformationFile` where the object O is identified as represented by the `File` executive object handle passed in to the `FileHandle` parameter.

6.4.4.8 write_file_information(S, O)

This primitive is mapped to two native functions. The first function is `ZwSetEaFile` where the object O is identified as represented by the `File` executive object handle passed in to the `FileHandle` parameter. The second function is `ZwSetInformationFile` where the object O is identified as represented by the `File` executive object handle passed in to the `FileHandle` parameter.

6.4.4.9 read_directory(S, O)

This primitive is mapped to two native functions. The first function is `ZwQueryDirectoryFile` where the object O is identified as represented by the `File` executive object handle passed in to the `FileHandle` parameter.
The second function is `ZwQueryOleDirectoryFile` where the object O is identified as represented by the `File` executive object handle passed in to the `FileHandle` parameter. The second function is not

supported by any of the native file systems. The abstraction provided by the system extend the directory notion beyond nonvolatile storage objects; this is a semantically consistent extension of the requirements for the interpretation; this is provided by the native function `ZwQueryDirectoryObject` where the object O is identified as represented by the `Directory` executive object handle passed in to the `DirectoryHandle` parameter.

6.4.4.10 create_directory(S, O_1, O_2)

This primitive is mapped to two native functions. The first native function is applicable to nonvolatile storage and is `ZwCreateFile` provided that access is performed with the `CreateDisposition` parameter set to `FILE_CREATE` and the object does not exist, set to `FILE_OPEN_IF` if the object does not exist, in case the object does not exist if it is set to `FILE_OVERWRITE_IF`, or `FILE_SUPERSEDE` if the object exists. For this mapping to apply the `CreateOptions` parameter must contain the flag `FILE_DIRECTORY_FILE`
The identity of O_1 is provided either by the handle to the `File` executive object `RootDirectory` in `OBJECT_ATTRIBUTES` if that handle is not `NULL` or from the prefix of the fully resolved name without any symbolic links remaining of the system-specific namespace up to the last separator in the field `ObjectName` of the structure `OBJECT_ATTRIBUTES` pointed to by the parameter `ObjectAttributes`.
The identity of O_2 is derived as the element of the fully resolved name without any symbolic links remaining after the last separator (if any) in the field `ObjectName` of the `OBJECT_ATTRIBUTES` structure pointed to by the parameter `ObjectAttributes`.
The abstraction provided by the system extend the directory notion beyond nonvolatile storage objects; this is a semantically consistent extension of the requirements for the interpretation; this is provided by the native function `ZwCreateDirectoryObject` whose parameter `ObjectAttributes` must be filled with the identities for O_1 in the handle to the executive object `RootDirectory` in `OBJECT_ATTRIBUTES` if that handle is not `NULL` or the prefix of the system-specific namespace up to but not including the last separator in the field `ObjectName` of

the OBJECT_ATTRIBUTES structure; the identity of O_2 is derived as the element after the last separator (if any) in the field ObjectName of the OBJECT_ATTRIBUTES structure.
The names obtained for O_1, O_2 must be mapped accordingly in each case.

6.4.4.11 search_directory(S, O_1, O_2)

This primitive does not have a direct correspondence. Instead it is mapped to two qualified functions. The first function is the ZwQueryDirectoryFile native function where O_1 is identified as represented by the File executive object handle passed in to the FileHandle parameter; the parameter FileInformation referring to the FILE_BASIC_INFORMATION data structure either directly or as part of another FILE_INFORMATION_CLASS must have the attribute FILE_ATTRIBUTE_DIRECTORY set.
The object O_2 is mapped to the mapping of the FileName parameter which must be set to a value not containing the characters * and ?. The second function is the ZwQueryOleDirectoryFile native function where O_1 is also identified as represented by the File executive object handle passed in to the FileHandle parameter and which must have the attribute FILE_ATTRIBUTE_DIRECTORY set in the FILE_BASIC_INFORMATION structure referred to directly or indirectly by the parameter FileInformation. The object O_2 is mapped to the mapping of the FileName parameter which must be set to a value not containing the characters * and ?. The second function is not supported by any of the native file systems. The abstraction provided by the system extend the directory notion beyond non-volatile storage objects; this is a semantically consistent extension of the requirements for the interpretation; this is provided by the native function ZwQueryDirectoryObject where the object O is identified as represented by the Directory executive object handle passed in to the DirectoryHandle parameter. However, this function is semantically identical to the primitive as applied to the mapped Directory executive object using the read_directory(S, O) predicate.

6.4.4.12 delete_directory(S, O)

This primitive is mapped to two native functions. The first function is `ZwDeleteFile` where the identity of O is derived from the fully qualified native name space name `ObjectName` contained in the `OBJECT_ATTRIBUTES` structure referred to by the `ObjectAttributes` parameter which must refer to an object which must have the attribute `FILE_ATTRIBUTE_DIRECTORY` set in the corresponding structure `FILE_BASIC_INFORMATION`.
The second function is `ZwSetInformationFile` where the object O is identified as represented by the `File` executive object handle passed in to the `FileHandle` parameter if the `FileInformationClass` parameter is set to `FileDispositionInformation` and which must have the attribute `FILE_ATTRIBUTE_DIRECTORY` set in the structure `FILE_BASIC_INFORMATION` referred to directly or indirectly by the `FileInformation` parameter.

6.4.4.13 create_memory_segment(S, O)

This primitive is mapped to the native function `ZwCreateSection` where the identity of O is mapped to the `SectionHandle` return parameter.

6.4.4.14 read_memory_segment(S, O)

This primitive is mapped to the native function `ZwMapViewOfSection` with the identity of O derived from the parameter `SectionHandle` and the identity of S explicitly set by the parameter `ProcessHandle`. As a precondition, the `Section` executive object identified by the `SectionHandle` handle must have been obtained by either a call to the native function `ZwCreateSection` with the same parameter and the parameter `DesiredAccess` having the flag `SECTION_MAP_READ` or `SECTION_MAP_EXECUTE` set or by a call to the native function `SectionHandle` where the parameter `DesiredAccess` contained the set flags `SECTION_MAP_READ` or `SECTION_MAP_EXECUTE`.

6.4.4.15 write_memory_segment(S, O)

This primitive is mapped to the native function ZwMapViewOfSection with the identity of O derived from the parameter SectionHandle and the identity of S explicitly set by the parameter ProcessHandle. As a precondition, the Section executive object identified by the SectionHandle handle must have been obtained by either a call to the native function ZwCreateSection with the same parameter and the parameter DesiredAccess having the flag SECTION_MAP_WRITE set or by a call to the native function SectionHandle where the parameter DesiredAccess contained the set flags SECTION_MAP_WRITE.

6.4.4.16 delete_memory_segment(S, O)

This primitive is mapped to the ZwUnmapViewOfSection native function with the identity of O derived from the parameter BaseAddress and the identity of S explicitly set by the parameter ProcessHandle. This will not directly result in the destruction of the object, but removes the object from the area of control of the identified subject.

6.4.4.17 open_device(S, O)

This primitive is mapped to two native functions ZwCreateFile and ZwOpenFile, provided that the object referenced is a Device executive object. The identity of O is derived from the fully resolved name without any symbolic links remaining in the field ObjectName of the OBJECT_ATTRIBUTES structure pointed to by the parameter ObjectAttributes.
The CreateDisposition must not be set to FILE_CREATE and if the object referenced by the ObjectName field of the OBJECT_ATTRIBUTES structure does not exist, it must not be set to one of the values FILE_OPEN_IF, FILE_SUPERSEDE, or FILE_OVERWRITE_IF.

6.4.4.18 read_device_data(S, O)

This primitive is mapped to the ZwReadFile, ZwReadFileScatter native functions as well as to the native function ZwMapViewOfSection. In all cases, the identity of O is provided by the handle to the File executive object in the parameter FileHandle and must represent a Device executive object. The data read from the identified object falls under the area of control of the process subject; this applies to other subject types by extension.
Prior establishment of the File executive object and its mapping in the interpretation by means of ZwCreateFile or ZwOpenFile as mapped to the open_device(S, O) primitive is implicitly assumed; the access characteristics from the DesiredAccess parameter to these calls are not relevant for consideration in the evaluation of this primitive.
In the case of ZwMapViewOfSection the Section executive object identified by the SectionHandle handle must have been obtained by either a call to the native function ZwCreateSection with the same parameter and the parameter DesiredAccess having the flag SECTION_MAP_READ or SECTION_MAP_EXECUTE set or by a call to the native function ZwOpenSection where the parameter DesiredAccess contained the set flags SECTION_MAP_READ or SECTION_MAP_EXECUTE and the parameter FileHandle identical to that used in the subsequent function call.

6.4.4.19 write_device_data(S, O)

This primitive is mapped to the ZwWriteFile, ZwWriteFileGather native functions, and the function ZwMapViewOfSection. In both cases, the identity of O is provided by the handle to the File executive object in the parameter FileHandle and must represent a Device executive object. The data written to the identified object falls under the area of control of the process subject; this applies to other subject types by extension.
Prior establishment of the File executive object, its reference to a Device object, and its mapping in the interpretation by means of

ZwCreateFile or ZwOpenFile is implicitly assumed; the access characteristics from the DesiredAccess parameter to these calls are not relevant for consideration in the evaluation of this primitive.
In the case of ZwMapViewOfSection the Section executive object identified by the SectionHandle handle must have been obtained by either a call to the native function ZwCreateSection with the same parameter and the parameter DesiredAccess having the flag SECTION_MAP_WRITE set or by a call to the ZwOpenSection native function where the parameter DesiredAccess contained the set flag SECTION_MAP_WRITE and the parameter FileHandle identical to that used in the subsequent function call.

6.4.4.20 read_device_control(S, O)

This primitive is mapped to six native functions. The first function is ZwQueryEaFile where the object O is identified as represented by the File executive object handle passed in to the FileHandle parameter which must refer to a Device executive object. The second function is ZwQueryAttributesFile where the identity of O is derived from the fully qualified native name space name ObjectName not including unresolved symbolic links contained in the OBJECT_ATTRIBUTES structure referred to by the parameter ObjectAttributes.
The third function is ZwQueryFullAttributesFile where the identity of O is derived from the fully qualified native name space name ObjectName not including unresolved symbolic links contained in the OBJECT_ATTRIBUTES structure referred to by the parameter ObjectAttributes.
The fourth function is ZwQueryInformationFile where the object O is identified as represented by the File executive object handle passed in to the FileHandle parameter which must refer to a Device executive object.
The fifth function is ZwDeviceIoControlFile where the identity of O is provided by the handle to the File executive object in the parameter FileHandle and must represent a Device executive object. The parameter IoControlCode must not contain the flags FILE_WRITE_ACCESS or FILE_ANY_ACCESS.

The sixth function is `ZwFsControlFile` where the identity of O is provided by the handle to the `File` executive object in the parameter `FileHandle` and must represent a `Device` executive object.
The parameter `IoControlCode` must not have either of the flags `FILE_WRITE_ACCESS` or `FILE_ANY_ACCESS` set.

6.4.4.21 write_device_control(S, O)

This primitive is mapped to four native functions. The first function is `ZwSetEaFile` where the object O is identified as represented by the `File` executive object handle passed in to the `FileHandle` parameter which must refer to a `Device` executive object. The second function is `ZwSetInformationFile` where the object O is identified as represented by the `File` executive object handle passed in to the `FileHandle` parameter which must refer to a `Device` executive object. The third function is `ZwDeviceIoControlFile` where the identity of O is provided by the handle to the `File` executive object in the parameter `FileHandle` and must represent a `Device` executive object. The parameter `IoControlCode` must contain the flags `FILE_WRITE_ACCESS` or `FILE_ANY_ACCESS`. The fourth function is `ZwFsControlFile` where the identity of O is provided by the handle to the `File` executive object in the parameter `FileHandle` and must represent a `Device` executive object. The parameter `IoControlCode` must contain the flags `FILE_WRITE_ACCESS` or `FILE_ANY_ACCESS`.

6.4.4.22 close_device(S, O)

This primitive is mapped to the native function `ZwClose` where the object O is identified as represented by the `File` executive object handle passed in to the `Handle` parameter which must refer to a `Device` executive object.

6.4.4.23 send_ipc(S_1, S_2, O)

This primitive is mapped to three native functions. It is assumed that prior to these, the IPC mechanism (the LPC facility) is established

between the subjects implicitly identified by S_1 and S_2. As the LPC mechanism is bidirectional, only the general association is possible at the time the identities are implicitly established using `ZwCreatePort` or `ZwCreateWaitablePort` for one subject and `ZwConnectPort` or `ZwSecureConnectPort` for the other identity. In both cases the result of the respective calls is a handle to a `Port` or `WaitablePort` executive object. The first function is `ZwRequestPort` where the identity of the subject S_1 is mapped to the identity of the caller and the identity of the subject S_2 is mapped to the other subject holding a handle to the `Port` or `WaitablePort` executive object identified by the parameter `PortHandle` and the identity of the object O is mapped to the `PORT_MESSAGE` entity pointed to by the parameter `RequestMessage`. The second function is `ZwRequestWaitReplyPort` where again the of the subject S_1 is mapped to the identity of the caller and the identity of the subject S_2 is mapped to the other subject holding a handle to the `Port` or `WaitablePort` executive object identified by the parameter `PortHandle` and the identity of the object O is mapped to the `PORT_MESSAGE` entity pointed to by the parameter `ReplyMessage`. The third function is `ZwWriteRequestData` where again the of the subject S_1 is mapped to the identity of the caller and the identity of the subject S_2 is mapped to the other subject holding a handle to the `Port` or `WaitablePort` executive object identified by the parameter `PortHandle` and the identity of the object O is mapped to the `PORT_MESSAGE` entity pointed to by the parameter `Message`.
Due to an intermixing of semantically disjoint primitives, it is necessary to obtain the primitive receive_ipc(S_1, S_2, O) in conjunction with this primitive.

6.4.4.24 receive_ipc(S_1, S_2, O)

This primitive is mapped to seven native functions. It is assumed that prior to these, the IPC mechanism (the LPC facility) is established between the subjects implicitly identified by S_1 and S_2. As the LPC mechanism is bidirectional, only the general association is possible at the time the identities are implicitly established using `ZwCreatePort` or `ZwCreateWaitablePort` for one subject and `ZwConnectPort` or

ZwSecureConnectPort for the other identity. In both cases the result of the respective calls is a handle to a Port or WaitablePort executive object.

The first function is ZwReplyPort where the identity of the subject S_1 is mapped to the identity of the caller and the identity of the subject S_2 is mapped to the other subject holding a handle to the Port or WaitablePort executive object identified by the parameter PortHandle and the identity of the object O is mapped to the PORT_MESSAGE entity pointed to by the parameter ReplyMessage.

The second function is ZwReplyWaitReplyPort where the identity of the subject S_1 is mapped to the identity of the caller and the identity of the subject S_2 is mapped to the other subject holding a handle to the WaitablePort executive object identified by the parameter PortHandle and the identity of the object O is mapped to the PORT_MESSAGE entity pointed to by the parameter ReplyMessage.

The third function is ZwReplyWaitReceivePort where the identity of the subject S_1 is mapped to the identity of the caller and the identity of the subject S_2 is mapped to the other subject holding a handle to the Port or WaitablePort executive object identified by the parameter PortHandle and the identity of the object O is mapped to the PORT_MESSAGE entity pointed to by the parameter Message; the fourth function is ZwReplyWaitReceivePortEx with mappings analogous to ZwReplyWaitReceivePort.

The fifth function is ZwReadRequestData where the identity of the subject S_1 is mapped to the identity of the caller and the identity of the subject S_2 is mapped to the other subject holding a handle to the Port or WaitablePort executive object identified by the parameter PortHandle and the identity of the object O is mapped to the PORT_MESSAGE entity pointed to by the parameter Message.

The sixth function is ZwRequestWaitReplyPort where the identity of the subject S_1 is mapped to the identity of the caller and the identity of the subject S_2 is mapped to the other subject holding a handle to the Port or WaitablePort executive object identified by the parameter PortHandle and the identity of the object O is mapped to the PORT_MESSAGE entity pointed to by the parameter

ReplyMessage. The seventh function is ZwQueryInformationPort where the identity of the subject S_1 is mapped to the identity of the caller and the identity of the subject S_2 is mapped to the other subject holding a handle to the Port or WaitablePort executive object identified by the parameter PortHandle and the identity of the object O is mapped to the PORT_INFORMATION_CLASS entity pointed to by the parameter PortInformation with specific instantiation for the parameter PortInformationClass.
Due to an intermixing of semantically disjoint primitives, it is necessary to obtain the primitive send_ipc(S_1, S_2, O) in conjunction with this primitive.

6.4.4.25 send_datagram(S_1, S_2, O)

This primitive is mapped to two native functions. The first function is NdisSend where the identity of the subject S_1 is implicitly mapped to the identity of the caller and the identity of the subject S_2 is mapped to the protocol-specific address contained in the payload contained in the data structure pointed to by the parameter PNDIS_PACKET. There must exist a mapping of the protocol-specific sender address contained in the payload to the entity S_1 or a derivative entity of S_1. The identity of the object O is mapped to the entity referred to by the parameter PNDIS_PACKET. The second function is NdisSendPackets where the identity of the subject S_1 is implicitly mapped to the identity of the caller. The primitive is mapped recursively as described above to each of the packets pointed in the sequence pointed to by the parameter PacketArray and enumerated by the parameter NumberOfPackets

6.4.4.26 receive_datagram(S_1, S_2, O)

This primitive is mapped to the NdisGetReceivedPacket native function where the identity of the subject S_2 is implicitly mapped to the identity of the caller and the identity of the subject S_1 is mapped to the protocol-specific address contained in the payload contained in the data structure pointed to by the return value of type

PNDIS_PACKET. There must exist a mapping of the protocol-specific sender address contained in the payload to the entity S_2 or a derivative entity of S_2. The identity of the object O is mapped to the entity referred to by the return value PNDIS_PACKET.

6.4.4.27 create_virtual_circuit(S_1, S_2, O)

This primitive is mapped to two native functions and a sequence of predicates. The first function is NdisClMakeCall where the identity of the subject S_1 is implicitly mapped to the identity of the caller and the identity of the subject S_2 is derived from the interpretation-specific identity, which is contained in the CO_MEDIA_PARAMETERS structure, itself part of the data structure CO_CALL_PARAMETERS pointed to by the parameter CallParameters.
The second function is NdisCmDispatchIncomingCall, in which the identity of the subject S_1 is implicitly mapped to the identity of the caller and the identity of the subject S_2 is derived from the interpretation-specific identity contained in the CO_MEDIA_PARAMETERS structure, itself contained in the CO_CALL_PARAMETERS structure pointed to by the parameter CallParameters. In both cases the calls to the functions must have been preceded by a call to NdisCoCreateVc by the same entity S_1 is mapped to which is the source of the mapping of the object O, which is mapped to the entity of type NDIS_HANDLEreferred to by the parameter NdisVcHandle.
The primitive is also mapped to a sequence of interpretation-specific and connectionless datagram exchange mechanisms for which the send_datagram(S_1, S_2, O) and receive_datagram(S_1, S_2, O) predicates apply and for which the interpretation must provide an entity to which the object O can be mapped; the entity to which S_2 is mapped is also interpretation-specific. The sequence and additional property predicates for the sequence must be augmented.
In each case virtual_circuit(O) is valid.

6.4.4.28 send_virtual_circuit(S, O)

This primitive is mapped to three native functions. The first is `NdisCoSendPackets` where the identity of S is implicitly mapped to the caller and the identity of O is one for which the predicate virtual_circuit(O) applies and was established in the interpretation by a call to `NdisCoCreateVc`.
The second and third function are required for interpretation-specific connectionless datagram based virtual circuits whose semantics must be augmented. The second function is `NdisSend` where the identity of the subject S is implicitly mapped to the identity of the caller. For the identity of O there must exist a mapping of the interpretation--specific entity to which O is mapped and to which virtual_circuit(O) applies. The third function is `NdisSendPackets` where the identity of the subject S_1 is implicitly mapped to the identity of the caller. The primitive is mapped recursively as described above to each of the packets pointed in the sequence pointed to by the parameter `PacketArray` and enumerated by the parameter `NumberOfPackets`.

6.4.4.29 receive_virtual_circuit(S, O_1, O_2)

This primitive is mapped to the `NdisGetReceivedPacket` native function where the identity of the subject S is implicitly mapped to the caller, which must be equal to the parameter S_2 of the predicate create_virtual_circuit which the parameter O_2 identifies; the information source is equal to the entity identified by parameter S_1 of the predicate create_virtual_circuit which the parameter O_2 identifies. For the identity of O_1 there must exist a mapping of the interpretation-specific entity to which O_1 is mapped and to which virtual_circuit(O_1) applies.

6.4.4.30 delete_virtual_circuit(S, O)

The primitive is also mapped to a sequence of interpretation-specific and connectionless datagram exchanges for which the primitives send_datagram(S_1, S_2, O) and receive_datagram(S_1, S_2, O) apply and for which the interpretation must provide an entity to which the object O can be mapped; the entity to which S_2 is mapped is also interpretation-specific. The sequence and additional property predicates for the sequence must be augmented.

CHAPTER

SEVEN

Security Models

The term **security model**[1] is generally used to describe a set of requirements a system must fulfill to meet the IA criteria for a given information system. The justification for abstraction from general security policies inherent in the term model can be derived directly from the results of Denning, Denning, Garland, Harrison, and Ruzzo discussed in section 3.1.

Of the IA criteria given in definition 1.1, the confidentiality aspect is primarily considered here, although integrity and availability considerations also must be part of a comprehensive security model.

Confidentiality has been the main focus in the research on security models, motivated primarily by governmental requirements [515]. In this commonly used more restrictive usage of the term, security models specify restrictions on object access or information flow which enforce confidentiality requirements.

Notes on Transformation into the Security Policy Calculus

In the interest of brevity, a number of simplifications and syntactical shorthands compared to a first order theory — which could be trans-

[1] Irrespective of definition 5.19

formed directly into the syntax used for the mechanical derivation of security policy decisions [282] — are used.

The security models discussed in this chapter include dynamic elements (e.g. the enter operation of the HRU model). As a result, such operations could lead to contradictions within a naïve global system model, which cannot be dealt with adequately in a monotonic theory such as $\mathcal{PC}$; such constraints are not given explicitly in the following discussion. Instead, whenever transitions between the sets of formulae valid for (open) intervals t_1 and t_2 (where $t_1 < t_2$) have to be denoted, a decoration similar to that used by the Z notation is used [679] in that a well-formed formula A valid at t_1 and replaced in t_2 is denoted as A'.

However, as can be seen easily, the conjunction of time intervals containment as defined in chapter 4 with any formula representing such a dynamic operation permits a monotonic formulation.

In addition, references to subjects and objects generally need to be replaced with $\forall$-quantized formulae over the predicate derived_or_self according to definition 5.15 (if applicable, similar predicates can be substituted for operations). This has also been omitted in the following descriptions in the interest of legibility. Similarly, set notation has been used where appropriate, to obtain a valid first order theory formula such expressions must be replaced by adding the requisite number of facts and verifying membership predicates for a given constant. To illustrate this (particularly in section 7.4, as the discussion of section 7.2.1 is rather trivial), the notational convention for using meta-expressions from section 2.2 has again been used.

7.1 Access Control Models

Access control based on a system modeled in terms of subjects and objects was part of the ADEPT-50 system formulated by Weissman and Schaefer [761, 628]; this model was refined several times, first by Lampson [396] in the context of the Cosine Task Force on Operating Systems [530] as well as by Graham and Denning and developed independently for database management systems by Conway *et al.*

[145] whose decision rules (which may be either data dependent or independent) can be viewed as a generalization of the model proposed by Lampson.

The commonality of these models lies in the representation by means of a state machine consisting of a triple $V = \langle S, O, M \rangle$ where S is a set of subjects s_i, O a set of objects o_j, and M a two-dimensional matrix M specifying the access rights for each subject to an object in the form of a vector $a_{i,j} \in A$ where A is a finite set of access rights. A security model partitions the set of all possible states of $\langle S, O, M \rangle$ into secure and non-secure states.

$$M = \begin{bmatrix} a_{s_1,o_1} & \cdots & a_{s_1,o_n} \\ \vdots & \ddots & \vdots \\ a_{s_m,o_1} & \cdots & a_{s_m,o_n} \end{bmatrix}$$

Objects or object types are associated with monitors which prevent subjects s_i from accessing an object o_j if the requisite right is not contained in $a_{i,j}$.

It is further assumed in the models that the state machine starts from a state that is considered secure in terms of the model.

This model was further refined by Harrison, Ruzzo, and Ullman [293] in the HRU model as described in section 3.1. In this model, the primitive operations give rise to explicit transitions within the state model and are described explicitly in terms of operations.

Harrison *et al.* also showed that the safety problem is decidable and $\mathcal{PSPACE}$-complete for systems without create requests and is decidable for monotonic (here: not containing destroy or delete operations) and monoconditional (i.e. containing only operations with at most one conditional clause[2]).

Additionally. Lipton and Snyder showed that while the safety problem is decidable for finite sets S, it is intractable [419].

The transformation into first order theory of HRU model derivates as well as simple access control lists is trivial.

[2]Permitting more than one conditional clause even while retaining the monotonicity constraint results in the system becoming undecidable [292].

An alternative approach to access control was proposed by Jones, Lipton, and Snyder with the Take-Grant graph model.
This model is also based on a state machine; however, secure states are described by a directed graph G where vertices of the graph represent subjects $s_i \in S$ and objects $o_j \in O$ (where $S \cap O = \emptyset$) and edges represent access rights with two special rights describing the transfer of authority, take(o) denoting the right to take any of the rights associated with o and grant(o) the right to share rights with o. State transitions are modeled as graph rewriting rules; the safety of a configuration can be computed from an initial graph by showing that all graphs that can be derived from the initial graph fulfill the safety condition; Jones *et al.* showed that even with the number of subjects and objects unbounded, safety is decidable, and that the computational complexity is linear in the size of the initial configuration [349, 674].
Using the state machine model, one can see that there exists a relation over the state space where the set of all states S can be subdivided into unauthorized states under a given security model P, $P \subseteq S$. However, for the set of states that can be reached under a given security mechanism $R = P$ need not hold; if that is the case a system is called precise [171].
General access control models are additionally subdivided into two types:

Definition 7.1 ***Discretionary Access Control*** *or DAC mechanisms are procedures and mechanisms which enforce the specified mediation on objects associated with a user at the discretion of an individual user by permitting user-definable parameterization of mediation.*
Mandatory Access Control *or MAC mechanisms are procedures and mechanisms which enforce the specified mediation on subjects, objects, and operations without the possibility of parameterization through individual users or other subjects.*

7.2 Information Flow Control Models

Access control models represent an intuitive approach and are compact in description to permit high assurance in their specification. However, while access control models can be used to describe information flow, no explicit formulation of the flow or dissemination of information is made in these models resulting in undesirable imprecision. The underlying intuition for information flows is that a flow from A to B occurs if the behavior of an entity B can be influenced by events occurring in entity A.

Early models were influenced by the requirements for defense applications, particularly the multilevel security (MLS) policy described in definition 7.2 [515, 737, 735, 736].

Definition 7.2 *In the multilevel security policy, each object o is assigned a 2-tuple* $\langle A, C \rangle$ *called the security class where A denotes an authority level and C a non-hierarchical category.*
There are eight authority levels

1. *Unclassified*
2. *Sensitive but Unclassified*
3. *Confidential*
4. *Secret*
5. *Secret (Code Word)*
6. *Top Secret*
7. *Top Secret SIOP-ESI*
8. *Top Secret (Code Word)*

on which a partial order ($\leq_A$) *is defined in the numerical order given here with (1) denoting the lower bound*[3]. *"Code Word" classifications impose an additional requirement in that only subjects for which the general authority level is equal or higher and all code words for a given object match the authority level have access to the object. For the purposes of this definition, code words are equivalent to non-hierarchical categories.*

[3]For practical purposes, this is typically further qualified by the type of background investigation, access to sensitive compartmented information (SCI), and whether the individual has passed a Counter-Intelligence or a Full Scope polygraph test [520, 521, 522].

The partial order on the finite set of non-hierarchical categories ($\leq_C$) is defined analogous to that of code word authority levels; given classes $\langle A, C\rangle$ and $\langle A', C'\rangle$, the partial order on classes ($\leq$) is defined as

$$\langle A, C\rangle \leq \langle A', C'\rangle \Leftrightarrow A \leq_A A' \wedge C \leq_C C' \tag{7.1}$$

7.2.1 The Bell-LaPadula Model

The Bell-LaPadula model (BLP) is presumably the most widely used model for mandatory access control (The TCSEC addressed only policies for which a consensus regarding implementation existed at the time, other policies were not considered even though requirements existed; e.g. although Director of Central Intelligence Directive (DCID) 1/7 [182] predates the TCSEC, the policies it specifies are not adequately covered by the TCSEC); it implements the MLS policy according to definition 7.2[4]. The model can be viewed as an intermediate model that is based on concepts in access control yet realizes information flow controls; although this is not made explicit in the original publications of the model [62, 63, 64] it has been shown later to be equivalent [283, 698] to noninterference.

Compared to the HRU model, the BLP contains several restrictions. The sets S and O are not permitted to change between states (i.e. the operations *delete* r from $A[s, o]$, *create subject* s, *create object* o, *destroy subject* s, and *destroy object* o of HRU are disallowed), and A is restricted to the rights *read*, *write*, *append*, and *execute* (only the former two are considered).

The model assumes a fixed Lattice $\mathcal{L}$ of security levels with the partial order defined by $\geq$ and defines a total function $f : S \cup O \rightarrow L$ yielding the security level of its argument (i.e. a subject or object) in a given state.

An additional restriction compared to the HRU model is that the state machine is represented as a tuple $V = (F, M)$ (where F is the range of

[4]MAC and MLS are frequently used interchangeably due to the dominance of the Bell-LaPadula model; this is incorrect.

S and O under f). A system under BLP consists of an initial state v_0, a set of operations R, and a state transition function $t : \langle F, M \rangle \times R \rightarrow V$. The security model of BLP introduces three definitions (7.3 through 7.5) providing necessary and sufficient conditions for the MLS policy. For ease of notation, set-theoretic and relational expressions are used in the formalization instead of a pure first-order notation.

Definition 7.3 *A state (F, M) is **read secure** if and only if*

$$\forall s \in S. \forall o \in O : read \in M[s, o] \Rightarrow f(s) \geq f(o)$$

*This property is commonly called **simple security**.*

Definition 7.4 *A state (F, M) is **write secure** if and only if*

$$\forall s \in S. \forall o \in O : write \in M[s, o] \Rightarrow f(o) \geq f(s)$$

*This property is commonly called the ***-property**.*

Definition 7.5 *A state (F, M) is **state secure** if and only if it is read secure and write secure.*

This permitted the formulation of the basic security theorem (BST):

Theorem 7.1 ***Basic Security Theorem***
A system (v_0, R, t) is secure if and only if v_0 is state secure and for every state v reachable from v_0 by executing a finite but arbitrary sequence of requests $r \in R$, if $t(v, c) = v^*$ (where $v = (F, M)$ and $v^* = (F^*, M^*)$), for each $s \in S$ and $o \in O$:

- $(read \in M^*[s, o]) \wedge (read \notin M[s, o]) \Rightarrow f^*(s) \geq f^*(o)$
- $(read \in M[s, o]) \wedge (f^*(s) \not\geq f^*(o)) \Rightarrow read \notin M^*[s, o]$
- $(write \in M^*[s, o]) \wedge (write \notin M[s, o]) \Rightarrow f^*(o) \geq f^*(s)$
- $(write \in M[s, o]) \wedge (f^*(o) \not\geq f^*(s)) \Rightarrow write \notin M^*[s, o]$

For a proof of theorem 7.1 refer to [64].
However, as shown by McLean [465], the BLP model contains tacit assumptions about the operations performed by subjects. By adding

explicit constraints on permissible operations, this can be remedied. For this purpose, a new function $c : S \cup O \rightarrow P(S)$ is introduced which specifies the set of subjects P permitted to change the security level of its argument [467]. The function t is replaced with a function $t' : (S \times V \times R) \rightarrow V$, which includes information on the subject inducing the state change. Definitions 7.6 and 7.7 introduce such constraints and create a lattice of models including a model where no security level may change, a constraint referred to as **tranquility**.

Definition 7.6 *A transition function $t^\star$ is **transition secure** if and only if, for $v = (f, m)$ and $v^* = (f^*, m^*)$,*

$$t^\star(s, v, r) = v^* \Rightarrow \forall x \in S \cup O : f(x) \neq f^*(x) \Rightarrow s \in c(x).$$

Definition 7.7 *A system $(v_0, R, t^\star)$ is secure if and only if*

1. *v_0 and all states reachable from v_0 by a finite sequence of one or more requests from R are secure according to definition 7.1, and*
2. *$t^\star$ is transition secure according to definition 7.6.*

7.2.2 Noninterference, Nondeducibility, and Beyond

While, as noted in section 7.2.1, BLP can be considered an information flow control model in terms of its results, it predates a rigid formulation of information-theoretic properties. Early formulations are due to Fenton [205] and Walter [754, 755], whereas Denning formulated the information flow within the context of a lattice structure providing a more elegant algebraic model than the earlier research [174, 169, 170] based on augmenting the run-time information flow control proposed e.g. by Fenton in his Data Mark Machine with compile-time controls. Information flow was primarily considered (and in some cases used eponymously with) covert channel analysis [397, 662, 661, 478, 718, 719, 775, 275, 745, 426], a step that had to be performed manually in the BLP model.
Cohen derived information flow from functional semantics based on the work of Denning and introduced inductive proofs of information

flow security [139], a result that was later refined by Feiertag [201] within the context of the PSOS work at SRI (cf. section 3.5.1) into a mechanism for creating theorems on information flows that could be proven automatically. Similar work was done by Popek and Farber [557].

Goguen and Meseguer subsequently formulated **noninterference** for deterministic systems again based on inductive information flows in the form of execution traces (histories) [255, 458, 345] within the framework of the PSOS project (see section 3.5.1). The trace model used here is only one of several possible formalisms for analysis, others include the direct use of temporal logic [211] or process algebrae [45, 592, 591]. While simpler and more compact in its presentation, the trace model does not readily yield a formulation in $\mathcal{PC}$; however, while a formulation in temporal logic could be transformed into $\mathcal{PC}$ [321], the efficiency and efficacy of this approach is unclear.

In the noninterference model the information flow directed from a subject s is given by a function $out(s, h.read(s))$, where $h.read$ is a trace of the system whose last input is $read(s)$, i.e. an information flow towards the subject s. Confidentiality can thus be defined as **purges** of traces, where a purge removes operations performed by a subject whose security level in a lattice model is not dominated by that of s.

Definition 7.8 *Let clearance be a function from the set of subjects to security levels* $S \rightarrow \langle A, C\rangle$ *and let* purge *be a function subjects* $\times$ *traces* $\rightarrow$ *traces such that*

1. *For all subjects* s, $\text{purge}(s, \langle\rangle) = \langle\rangle$ *where* $\langle\rangle$ *is the empty trace.*
2. *For all subjects* s_1, s_2 *and operations* op_i, $\text{purge}(s_1, h.op_i(s_2)) = \text{purge}(s_1, h).op_i(s_2)$ *if* $op_i(s_2)$ *is an input flow operation performed by subject* s_2 *and* $clearance(s_2) \leq_C clearance(s_1)$.
3. *For all subjects* s_1, s_2 *and operations* op_i, $\text{purge}(s_1, h.op_i(s_2)) = \text{purge}(s_1, h)$ *if* $op_i(s_2)$ *is an input flow operation performed by* s_2 *and* $\neg clearance(s_2) \leq_C clearance(s_1)$.

A system satisfies noninterference if and only if for all subjects s_i, *for all traces* $t \in T$ *and all output flow operations* op_j, $out(s_i, t, op_j(s_i)) = out(s_i, \text{purge}(s_i, t).op_j(s_i))$.

Definition 7.8 is non-constructive; Goguen and Meseguer subsequently developed a series of **unwinding conditions** [256] sufficient for determining noninterference in state machines, a result that was later refined by McLean [468]. Millen showed that noninterference for deterministic systems provides perfect protection against covert channels [478], although for this to occur, time dependencies must also be considered part of the alphabet over which noninterference is determined so as to eliminate covert timing channels as well. Several alternative formalisms for noninterference analysis have been put forth [214], e.g. the use of CSP as proposed by Ryan [589, 426], mostly due to different models for computation.

Noninterference can, however, as shown by Sutherland [689], be too stringent; instead the more general concept of **nondeducibility** is introduced, i.e. given a set of possible state spaces W and two functions f_1 and f_2 $\text{dom} f_1 = W$ and $\text{dom} f_2 = W$, information flows (is deducible) from f_1 to f_2 if and only if there exists some possible state space w and some element $z \in \text{ran} f_2$ such that z is achieved by f_2 in some possible state space and in every possible state space w' such that $f_1^{w'} = f_2^{w}, f_2^{w'} \neq z$[5] [770, 14, 331]. Nondeducibility is, however, somewhat problematic in that several undesirable trace configurations fulfill the nondeducibility requirement as first shown by McCullough [458].

Similarly, McCullough also showed that nondeducibility did not hold under composition [460, 461] and proposed a restricted model, generalized noninterference (i.e. given any acceptable trace t and another trace t' formed by insertion or removal of a non-empty set of elements from t, there exists an acceptable trace t'' such that a a non-empty set of elements is inserted into t'' and t'' meets the generalized noninterference requirement), to remedy the problem of trace configurations.

However, this model does not compose either although – by further restricting generalized noninterference – McCullough was able to create a composable variant of noninterference [461].

[5]Intuitively, information flows from f_1 to f_2 if knowing the value of f_1 eliminates even a single possible value of f_2 from consideration.

Since the above is applicable for deterministic models, some interest has focused on non-deterministic (probabilistic) models. Examples of such models include the Flow Model defining **separability** by McLean [466], which requires for each pair of traces t_1, t_2 that interleaving a restricted subsequence of t_1 with an unrestricted subsequence of t_2 must itself result in a valid trace, and P-Restrictiveness by Gray [274, 275, 276], see [770, 334, 469, 689, 592, 591].
Similarly for possibilistic non-deterministic models, there also exist a number of frameworks that permit description and reasoning over several of the types of information flow control models discussed below, notably [441, 792, 471].
For distributed systems that must be considered independent nodes, the problem of transitive information flows (or, respectively, intransitive information flow policies) must also be considered; a number of approaches have been proposed, e.g. by Rushby [584] who introduced an intransitive ipurge function analogous to the purge function of definition 7.8; a similar approach was proposed by Pinsky [549].

However, these approaches are largely limited to deterministic systems and hence limited in their applicability. Roscoe and Goldsmith [582] introduced a formalism based on communicating sequential processes (CSP) that can be used to reason over intransitive flow control policies, while Mantel proposed a similar mechanism and framework based on trace models [442, 443].
As noted before, these approaches are, however, as yet suitable mainly for static verification of security policies (cf. observation 3.1) while their efficacy in dynamic environments is as yet unclear.

7.3 Integrity and Availability Models

Whereas most formal security model work concentrated on the confidentiality aspect of information assurance, integrity and availability are also necessary for information operations. Denial of service problems were first considered by Gligor [250] in terms of maximum waiting times. Yu and Gligor further refined this in terms of safety and liveness conditions by means of temporal logic [785]. Millen provided an alternative model based on finite state machines [479] for this approach.

Availability, particularly in the context of distributed systems, poses similar requirements as fault tolerance in general (and in particular tolerance to Byzantine [395] faults); while this is an active research area [125], it is not considered further in this dissertation.

Integrity protection has been the subject of research particularly in the application domain of database security [175, 337, 599].

For general purpose computer systems — apart from seminal work by Biba [75] and early work by Lipner [418] modeling integrity in analogy to BLP — the canonical example of an integrity policy is the model proposed by Clark and Wilson [135], which is closely patterned after procedures used in the physical world for ensuring integrity of data and transactions in the presence of adversaries, e.g. restricting the set of possible operations on data items or requiring cooperation between multiple designated subjects for performing certain operations.

While McLean provides a partial formal interpretation of this model [467], the model itself is not adequately formalized for automated verification [6].

Amoroso [21] introduced a model formalizing the notions of subject priority and object criticality; similarly, Fantechi *et al.* have proposed an integrity model permitting multiple levels of criticality [199] based on earlier work by Totel *et al.* [713]; the model permits (non-decidable) reasoning by means of temporal logic over a process algebra.

As with availability policies, while apparently a pragmatically satisfactory formal model for integrity has not been found and requires additional research, such models are not considered further in this dissertation.

Another approach to modeling integrity was proposed by Glasgow and MacEwen [247, 246] based on obligations as formulated in modal logic and applying the possible-worlds semantics as proposed by

Kripke [379]; this permits the formulation of models and policies such as the one proposed by Clark and MacEwen even for distributed systems.

7.4 Role-Based Access Control Models

Role-based access control (RBAC) can be considered to be derived from privilege grouping (e.g. group permissions) [50, 703, 380] and separation of duty mechanisms [135, 597, 108] and were formalized first by Ferraiolo and Kuhn and Ting *et al.* [207, 704].
The concepts were further formalized by a number of authors [512, 206, 598, 600]; this has led to the emergence of a consensus on elementary attributes of RBAC mechanisms. The core concepts are sets of operations, representing permissible actions associated with roles and subjects which are grouped into roles, thereby introducing an indirection step from direct permission assignments.
RBAC has become an attractive research subject and yielded several implementations; a conference dedicated to the subject exists since 1995 in SACMAT (formerly RBAC) [783, 780, 781, 782, 570, 596, 595].
As shown by Osborn *et al.*, RBAC can also be used itself to model discretionary and mandatory (particularly MLS) security models [526] and also lends itself well to automated reasoning approaches [446].
The proposed NIST standard [209, 208] identifies four model components, one (Core RBAC) of which is required in any RBAC model.
For this model, five sets are defined: subjects (*SUBJECTS*), roles (*ROLES*), objects (*OBJECTS*), sessions (*SESSIONS*), and operations (*OPERATIONS*). Over these, relations are defined as illustrated in figure 7.1. As in section 7.2.1, set-theoretic and relational expressions and abbreviations are used in the formalization to permit a compact notation.

Definition 7.9 *A role-based access control model can be modeled by the following sets and relations:*

1. *A set of subjects SUBJECTS,*

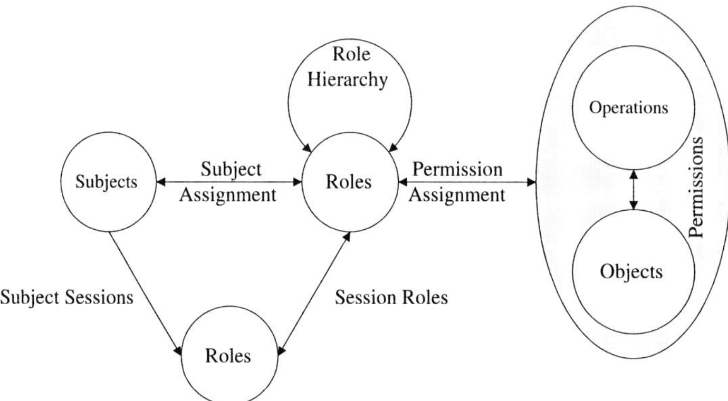

Figure 7.1: Hierarchical Role-Based Access Model sets, relations, and functions

2. *a set of roles ROLES,*
3. *a set of objects OBJECTS,*
4. *a set of sessions SESSIONS, and*
5. *a set of operations OPERATIONS.*
6. *The set of permissions PERMISSIONS is defined as a subset of the power set of the 2-tuple* $\mathbb{P}(OPERATIONS \times OBJECTS)$.
7. *The set of subject assignments SA is defined as a subset of the power set of the 2-tuple* $\mathbb{P}(SUBJECTS \times ROLES)$, *for which a set comprehension assigned_subjects is defined as*

$$assigned\_subjects(r) = \{ s \in SUBJECTS \mid (s, r) \in SA \}.$$

8. *The set of permission assignments PA is defined as a subset of the power set of the 2-tuple* $\mathbb{P}(PERMISSIONS \times ROLES)$, *for which a set comprehension assigned_permissions is defined as*

$$assigned\_permissions(r) = \{ p \in PERMISSIONS \mid (p, r) \in PA \}.$$

9. *A predicate subject_sessions(subject) mapping a subject onto a set of sessions from* $\mathbb{P}\, SESSIONS$ *and, correspondingly, an additional predicate session_subjects(session) as the inverse thereof.*

10. *A set comprehension predicate session_roles(s) mapping a session onto a set of roles from* $\mathbb{P}$ *ROLES such that*

$$session\_roles(s) = \{\, r \in ROLES \mid (session\_subjects(s), r) \in SA \,\}.$$

11. *The set of permissions for a given session as a predicate*

$$session\_permissions(s) = \bigcup_{r \in session\_roles(s)} assigned\_permissions(r)$$

Role hierarchies can be modeled by three additional elements:

1. *A set of role hierarchies RH obtained by defining a partial order on the set ROLES × ROLES by means of a relation predicate* $role\_containment(r_1, r_2)$*:*

$$role\_containment(r_1, r_2) \Rightarrow assigned\_subjects(r_2) \subseteq assigned\_subjects(r_1).$$

2. *A set comprehension predicate authorized_permissions mapping roles onto a set of permissions under the constraint of the role hierarchy:*

$$authorized\_permissions(r) = \{\, p \in PERMISSIONS, r' \in ROLES \mid role\_containment(r, r') \land (p, r') \in PA \,\}$$

Dynamic rules for maintaining the state of the RBAC facts can now be introduced; only the rules required for Core RBAC are considered here. Moreover, also following [209], only operations involving single subjects and objects are considered; operations involving multiple subjects and objects can be modeled analogously. The naming schema of [209] for the operations has been retained wherever possible.

The *RBAC_add_subject* operation in definition 7.10 updates the model by adding a subject to the set of subjects (which would occur in $\mathcal{PC}$ by adding a fact with a membership predicate for the given constant) as well as updating the function *subject_sessions* (again, since functions may be considered as relations over a denumerable domain and range, the same model as for sets applies).

Definition 7.10

$$\begin{aligned}
&RBAC_add_subject(subject) \Leftrightarrow \\
&\quad (subject \notin SUBJECTS \land \\
&\quad SUBJECTS' = SUBJECTS \cup \{ subject \} \\
&\quad subject_sessions' = subject_sessions \cup \{ subject \mapsto \emptyset \})
\end{aligned}$$

Analogously, definition 7.11 provides for the removal of a subject from the set of subjects; consistency constraints require that the *SA* relation and *assigned_subjects* function are updated. Sessions associated with a subject are also removed if the subject is no longer part of the model.

Definition 7.11

$$\begin{aligned}
&RBAC_delete_subject(subject) \Leftrightarrow \\
&\quad (subject \in SUBJECTS \land \\
&\quad \forall\, session \in SESSIONS : \\
&\qquad session \in subject_sessions(subject) \Rightarrow \\
&\qquad RBAC_delete_session(session) \land \\
&\quad SA' = SA \setminus \{ r \in ROLES \mid subject \mapsto r \} \land \\
&\quad assigned_subjects' = \{r \in ROLES \mid \\
&\qquad r \mapsto (assigned_subjects(r) \setminus \{ subject \})\} \land \\
&\quad SUBJECTS' = SUBJECTS \setminus \{ subject \})
\end{aligned}$$

Definition 7.12 adds a new role to the model and updates the functions *assigned_subjects* and *assigned_permissions* accordingly.

Definition 7.12

$$\begin{aligned}
&RBAC_add_role(role) \Leftrightarrow \\
&\quad (role \notin ROLES \land \\
&\quad ROLES' = ROLES \cup \{ role \} \land \\
&\quad assigned_subjects' = assigned_subjects \cup \{ role \mapsto \emptyset \} \land \\
&\quad assigned_permissions' = assigned_permissions \cup \{ role \mapsto \emptyset \})
\end{aligned}$$

The operation *RBAC_delete_role* in definition 7.13 permits the removal of a role from the RBAC model. This also requires that any functions and sets referring to this role, namely any sessions, the *PA* set, and the *assigned_subjects* and *assigned_permissions* functions must be updated accordingly by their removal.

Definition 7.13

$$
\begin{aligned}
&RBAC\_delete\_role(role) \Leftrightarrow \\
&\quad (role \in ROLES \wedge \\
&\quad \forall session \in SESSIONS : role \in session\_roles(session) \Rightarrow \\
&\qquad RBAC\_delete\_session(session) \wedge \\
&\quad SA' = SA \setminus \{ subject \in SUBJECTS \mid subject \mapsto role \} \wedge \\
&\quad assigned\_subjects' = assigned\_subjects \setminus \\
&\qquad \{ role \mapsto assigned\_subjects(role) \} \wedge \\
&\quad PA' = PA \setminus \{ operation \in OPERATIONS, object \in OBJECTS \mid \\
&\qquad (operation, object) \mapsto role \} \wedge \\
&\quad assigned\_permissions' = assigned\_permissions \setminus \\
&\qquad \{ role \mapsto assigned\_permissions(role) \} \wedge \\
&\quad ROLES' = ROLES \setminus \{ role \})
\end{aligned}
$$

Definition 7.14 describes an operation permitting the assignment of a subject to a role.

Definition 7.14

$$
\begin{aligned}
&RBAC\_assign\_subject(subject, role) \Leftrightarrow \\
&\quad (subject \in SUBJECTS \wedge \\
&\quad role \in ROLES \wedge \\
&\quad (subject \mapsto role) \notin SA \\
&\quad SA' = SA \cup \{ subject \mapsto role \} \wedge \\
&\quad assigned\_subjects' = assigned\_subjects \setminus \\
&\qquad \{ role \mapsto assigned\_subjects(role) \} \cup \\
&\qquad \{ role \mapsto (assigned\_subjects(role) \cup \{ role \}) \})
\end{aligned}
$$

Conversely, definition 7.15 describes the operation to dissassociate a subject from a role and also enforces the removal of sessions in which the subject is involved from the model.

Definition 7.15

$$
\begin{aligned}
& RBAC\_deassign\_subject(subject, role) \Leftrightarrow \\
& \quad (subject \in SUBJECTS \wedge \\
& \quad role \in ROLES \wedge \\
& \quad (subject \mapsto role) \in SA \wedge \\
& \quad \forall\, session \in SESSIONS : session \in subject\_sessions(subject) \wedge \\
& \qquad role \in session\_roles(session) \Rightarrow \\
& \qquad\quad RBAC\_delete\_session(session) \wedge \\
& \quad SA' = SA \setminus \{ subject \mapsto role \} \wedge \\
& \quad assigned\_subjects' = assigned\_subjects \setminus \\
& \qquad \{ role \mapsto assigned\_subjects(role) \} \cup \\
& \qquad \{ role \mapsto (assigned\_subjects(role) \setminus \{ subject \})\})
\end{aligned}
$$

In a pure RBAC model, permissions are assigned solely to roles and not to individual subjects; the only mechanism to grant such permissions is provided by the operation *RBAC_grant_permission* as given in definition 7.16. For this purpose, the function *assigned_permissions* and the set *PA* are modified.

Definition 7.16

$$
\begin{aligned}
& RBAC\_grant\_permission(object, operation, role) \Leftrightarrow \\
& \quad ((operation, object) \in PERMISSIONS \wedge \\
& \quad role \in ROLES \wedge \\
& \quad PA' = PA \cup \{ (operation, object) \mapsto role \} \wedge \\
& \quad assigned\_permissions' = assigned\_permissions \setminus \\
& \qquad \{ role \mapsto assigned\_permissions(role) \} \cup \\
& \qquad \{ role \mapsto (assigned\_permissions(role) \cup \\
& \qquad \{ (operation, object) \}) \})
\end{aligned}
$$

In symmetry to definition 7.16, the revocation of permissions from those assigned to a role is given by the *RBAC_revoke_permission* operation given in definition 7.17.

Definition 7.17

$$
\begin{array}{l}
RBAC\_revoke\_permission(object, operation, role) \Leftrightarrow \\
\quad ((operation, object) \in PERMISSIONS \wedge \\
\quad role \in ROLES \wedge \\
\quad ((operation, object) \mapsto role) \in PA \wedge \\
\quad PA' = PA \setminus \{ (operation, object) \mapsto role \} \wedge \\
\quad assigned\_permissions' = assigned\_permissions \setminus \\
\qquad \{ role \mapsto assigned\_permissions(role) \} \cup \\
\qquad \{ role \mapsto (assigned\_permissions(role) \setminus \\
\qquad \{ (operation, object) \}) \})
\end{array}
$$

The concept of a *session* is introduced to provide the possibility of one subject being simultaneously assigned to multiple roles, the activated role set defined by the function *active_role_set*. Definition 7.18 therefore provides a mapping between a subject and the activated role set and updates the set *SESSIONS* and the function *subject_sessions* accordingly.

Definition 7.18

$$
\begin{array}{l}
RBAC\_create\_session(subject, active\_role\_set, session) \Leftrightarrow \\
\quad (subject \in SUBJECTS \wedge \\
\quad active\_role\_set \subseteq \{ r \in ROLES \mid (subject \mapsto r) \in SA \} \wedge \\
\quad session \notin SESSIONS \wedge \\
\quad SESSIONS' = SESSIONS \cup \{ session \} \wedge \\
\quad subject\_sessions' = subject\_sessions \setminus \\
\qquad \{ subject \mapsto subject\_sessions(subject) \} \cup \\
\qquad \{ subject \mapsto (subject\_sessions(subject) \cup \{ session \}) \} \wedge \\
\quad session\_roles' = session\_roles \cup \{ session \mapsto active\_role\_set \})
\end{array}
$$

The disassociation of a subject with an active role set and subsequent removal of the session from the model is provided by the operation *RBAC_delete_session* as given in definition 7.19.

Definition 7.19

$$
\begin{array}{l}
RBAC\_delete\_session(subject, session) \Leftrightarrow \\
\quad (subject \in SUBJECTS \land \\
\quad session \in SESSIONS \land \\
\quad session \in subject\_sessions(subject) \land \\
\quad subject\_sessions' = subject\_sessions \setminus \\
\qquad \{ subject \mapsto subject\_sessions(subject) \} \cup \\
\qquad \{ subject \mapsto (subject\_sessions(subject) \setminus \{ session \}) \} \land \\
\quad session\_roles' = session\_roles \setminus \\
\qquad \{ session \mapsto session\_roles(session) \} \land \\
\quad SESSIONS' = SESSIONS \setminus \{ session \})
\end{array}
$$

Complementing definition 7.18, the addition of a role to an existing session is provided by the *RBAC_add_active_role* operation in definition 7.20.

Definition 7.20

$$
\begin{array}{l}
RBAC\_add\_active\_role(subject, session, role) \Leftrightarrow \\
\quad (subject \in SUBJECTS \land \\
\quad session \in SESSIONS \land \\
\quad role \in ROLES \land \\
\quad session \in subject\_sessions(subject) \land \\
\quad (subject \mapsto role) \in SA \land \\
\quad role \notin session\_roles(session) \land \\
\quad session\_roles' = session\_roles \setminus \\
\qquad \{ session \mapsto session\_roles(session) \} \cup \\
\qquad \{ session \mapsto (session\_roles(session) \cup \{ role \}) \})
\end{array}
$$

Complementary to definition 7.20, the *RBAC_drop_active_role* operation provides for the removal of a role from the set of roles contained by a session as given in definition 7.21.

Definition 7.21

$$
\begin{aligned}
&RBAC\_drop\_active\_role(subject, session, role) \Leftrightarrow \\
&\quad (subject \in SUBJECTS \land \\
&\quad session \in SESSIONS \land \\
&\quad role \in ROLES \land \\
&\quad session \in subject\_sessions(subject) \land \\
&\quad role \in session\_roles(session) \land \\
&\quad session\_roles' = session\_roles \setminus \\
&\qquad \{ session \mapsto session\_roles(session) \} \cup \\
&\qquad \{ session \mapsto (session\_roles(session) \setminus \{ role \}) \})
\end{aligned}
$$

The decision of a hypothesis for a given policy can therefore be given simply as shown in definition 7.22 by the operation *RBAC_decision*. This resolves the indirection by the concomitant association between the subject and the role and the association between the role and the permissions assigned to the role.

Definition 7.22

$$
\begin{aligned}
&RBAC\_decision(subject, session, operation, object) \Leftrightarrow \\
&\quad (session \in SESSIONS \land \\
&\quad operation \in OPERATIONS \land \\
&\quad object \in OBJECTS \land \\
&\quad \exists r \in ROLES : r \in session\_roles(session) \land \\
&\qquad ((operation, object) \mapsto r) \in PA
\end{aligned}
$$

CHAPTER

EIGHT

The Externally Controlled Reference Monitor Mechanism

The reference monitor as a structuring mechanism for operating system design was proposed by the Anderson panel [26] based on earlier work by Schell [743, 630] ; similar concepts (albeit at a somewhat lower abstraction level) were simultaneously and independently pursued by Graham and Denning [271].

As defined in a note in [26], the reference monitor concept was *"the notion that all references by any program to any program, data or device are validated against a list of authorized types of reference based on user and/or program function"*

It has since been used as a guiding principle for the design of secure operating systems or in adding security facilities to existing systems, arguably due to it being a simple yet powerful abstraction, but undoubtedly also with the aid of the TCSEC [733] which mandated the use of the reference monitor concept for systems conforming to the classes B2 and higher.

A standard reference monitor as shown in figure 8.1 implements a local security policy by verifying that the authentication and authorization policy (typically implemented as a database or as individual access control lists) permits access (this represents a reference in the terminology of [26]) of a subject (also referred to as principal) to an object. In addition, the reference monitor must audit each opera-

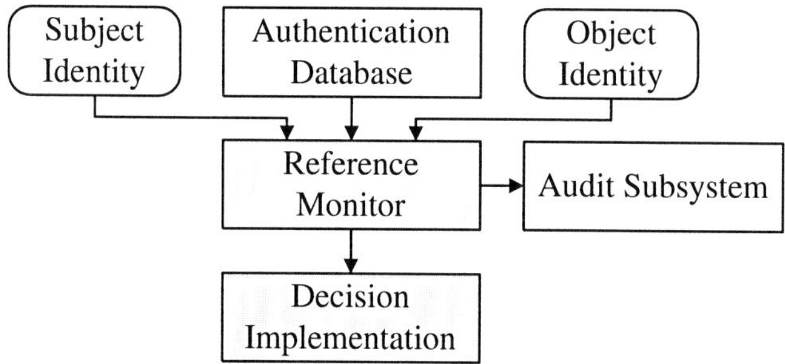

Figure 8.1: Reference monitor model proposed by the Anderson panel

tion and its outcome. The core requirements for a reference monitor were outlined in [26] as follows (notes set in italics are not part of the quotation):

1. The reference validation mechanism must be tamper proof. *This is also referred to as the* ***Isolation Property***
2. The reference validation mechanism must always be invoked. *This is also referred to as the* ***Completeness Property***
3. The reference validation mechanism must be small enough to be subject to analysis and tests to assure that it is correct. *This is also referred to as the* ***Verifiability Property***

The verifiability property reflects the etymology of the word "monitor" in that this term was initially used in operating system design to describe a reusable component for controlling system operation (this terminology was in use as late as with the TOPS-10 system [181, 180]), and later became synonymous with a very simple and primitive operating system.

In most implementations, the reference monitor does not handle access itself; the decision of the reference monitor is implemented separately. This is one of the pragmatic concessions necessary in semi-rigorous reference monitor implementations [251].

A rigorous implementation of the reference monitor concept requires the development of a security kernel. In the words of Schell *"a subset of the hardware and software that was sufficient to provide security even if the remainder of the system* [had been produced] *by an adversary"*.

The actual term "security kernel" was coined by Goodenough who suggested it in analogy to the use of the term in other areas of mathematics [341]; due to the fact that the reference monitor concept is typically implemented using the security kernel primitive, the two terms are frequently used interchangeably.

The security kernel as defined originally was to contain all pertinent functions necessary not only to perform the decisions on references but also to execute those decisions; this in turn implied a necessity for hardware protection mechanisms to ensure that the isolation property is satisfied.

This approach was exemplified in work performed by Ames, Gasser, and Schell [19]; the first actual implementation of a security kernel architecture was performed by MITRE led by Schiller [634, 477] based on the DEC PDP 11/45 and was mainly intended as a proof of concept; as such it already included the use of formal techniques for the specification of the security kernel.

The research and experience of this prototype along with research performed by Schell *et al.* on project GUARDIAN, which was intended to provide a security kernel for the Multics operating system led to a formal specification [635], but ultimately did not lead to a full scale implementation.

Instead, several concepts were later introduced in the Multics Access Isolation Mechanism (AIM); an unrelated effort [755] at Case Western Reserve University led by Walter also attempted to retrofit a security kernel to Multics.

Another early implementation of the reference monitor principle was provided by Kernelized VM, a security-enhanced version of IBM's VM/370 operating system for the S/370 family of mainframe systems developed by Gold *et al.* at SDC.

Unix was also the basis for several implementations. One of these was the Kernelized Secure Operating System (KSOS) which was developed by Ford Aerospace for the DEC PDP 11/70 under the lead of Neumann, Nagle, McCauley, and Drongowski (this was known as KSOS-11) [696, 456, 69] and by Honeywell for the Honeywell Level 6 Machine (somewhat confusingly known also as KSOS-6).

KSOS included significant portions of the operating system within the security kernel (32 system calls compared to the 48 present in Unix Version 6) and therefore can be considered in violation of the reference monitor principle.

However, it was specified in the form of a finite state machine and all system calls were defined in terms of transitions among secure states in the BLP for which proof was obtained using the SPECIAL specification language and using machine-augmented proofs [545].

Another formally verified system that conformed to the reference monitor concept and was the first to be evaluated according to the TCSEC as compliant to the A1 class was the SCOMP system (formerly the Secure Front-End Processor, SFEP) developed at Honeywell Aerospace [659, 660, 221, 97, 746, 747, 748] under the auspices of Silverman, Broadbridge, and Mekota [110].

SCOMP also has its origins in the Air Force GUARDIAN program (see above) and was an implementation of a hardware/software general purpose operating system based on the TCB concept.

The SCOMP trusted operating program (STOP) was a security kernel-based, general purpose operating system that provides a multilevel hierarchical file system, interprocess communication, security administrator functions, and operator commands.

The original design objectives called for the development of a security kernel and an emulation subsystem for existing operating systems, particularly for supporting Unix interface in such way that Unix applications could be used on SCOMP.

However, given experiences in the development of KSOS, this was abandoned and, instead of providing full Unix interfaces, a new interface for SCOMP, the SCOMP Kernel Interface Package (SKIP) was designed that provided basic services for customized systems.

This system was also specified using the Hierarchical Development Method (HDM) developed at SRI and the design verified using the Gypsy Verification Environment prover [18, 263, 463, 262, 12]; this system was later transformed into the (Bull) XTS-200 and XTS-300 systems, albeit at lower assurance ratings.
Another Unix-based security kernel implementation developed at the University of California at Los Angeles (UCLA) was the UCLA Data Secure Unix system (DSU) developed for the DEC PDP 11/45 and 11/70 by Popek *et al.* [559, 558]; the actual security kernel of DSU contained only 2000 lines of code. This system was also in part a precursor of the KSOS system described above.
Besides these research implementations there have been relatively few implementations such as the commercially deployed GEMSOS operating system developed at Gemini Computers [633] by Schell *et al.* for Intel Corporation's iAPX 286 microprocessors which was, among others, used for components of the BLACKER system described in section 3.4.2.
While the realization of the reference monitor principle is claimed for several systems (mostly in the form of semi-rigorous implementations and not always supported by a formal model to support these claims and typically obviously flawed in at least that the completeness property is violated by the type of objects the system considers for mediating access to) the apparent lack of implementations outside the research community has led to discussions to the viability of the mechanism as such [83, 654].
However, all criticism levied against the reference monitor mechanism has focused on aspects of perceived or factual cost and feasibility, not against the mechanism itself.
Another aspect also noted in defense of the reference monitor by Shockley [654] is that the reference monitor is, taken by itself, completely neutral with regard to the policy it is to enforce.
It has, however, been used mainly to enforce mandatory access control mechanisms in the form of the U.S. Department of Defense multilevel security policy which is presumably less than surprising given the source of funding for the research described above.

8.1 Extending the Reference Monitor Mechanism

The mechanism described in [26] assumed that the processing of information would generally occur within the confines of a monolithic computer system that facilitated centralized control. Similarly, [733] provides only minimal considerations for the handling of information exchanged between multiple instances and then assumes that the sensitivity labeling of all instances obeys a single scheme.
It is this aspect that needs extension given the typical configuration of currently deployed information systems. Given the security model and policy of an organization or a set of organizations, it has therefore become necessary to ensure the consistent enforcement of these policies occurs on all components of the computing environment to which they apply. As has been argued before, this enforcement must encompass all resources on a given system and thus occur at the level of non-bypassable operating system mechanisms [424].
The mechanism described here arguably represents a natural extension of the proven reference monitor abstraction that fulfills this requirement for distributed environments by externalizing the actual policy definition and distribution into a redundant network of **external reference monitors** (ERM) and providing only enforcement mechanisms at the level of the actual component systems, controlled by an **externally controlled reference monitor** (ECRM) subordinate to ERMs. In addition, the mechanism described here can serve to provide not only access control but also to exert behavioral restrictions and to provide distributed firewalling and intrusion detection facilities.
For the purpose of enforcing policies, a number of sensors and actuators must exist within the operating system of each PEN instance according to definition 8.1. The set of all enforcement modules represents the Security Policy Enforcement Function (SPEF) in the DGSA terminology.

Definition 8.1 *An **enforcement module** (EM) is a self-contained func-*

tional unit associated with a functional component within an operating system that is capable of

- *detecting any attempted operation within the operating system component,*
- *preventing any such operation,*
- *modifying any such operation,*
- *collecting information within the the operating system component and forwarding it to an ECRM if and only if the ECRM associated with the same node the enforcement module is associated with, or*
- *causing operations or collecting information within the operating system component based on instructions external to the enforcement module if and only if they emanate from the ECRM associated with the same node the enforcement module is associated with.*

An ECRM may perform two types of decision operations as set forth in definition 8.2.

Definition 8.2 *An ECRM operation is referred to as a*

- ***reactive decision** if and only if an enforcement module controlled by the ECRM issues a request for a policy decision.*
- ***regulated decision** if the decision operation is caused by a stimulus or decision request other than an enforcement module.*

Only reactive decisions are commonly associated with reference monitors. The inclusion of regulated decisions reflects the need to model behavior imposed externally without being induced by internal events. While it would be possible to model such events by pseudo-reactive behavior based on events such as timers coupled to policy predicates, this would obfuscate the distinct character of the two decision types.

Reactive decisions are implemented at the PEN instances. As noted in section 4.8, each PEN instance operating system is augmented by several enforcement modules. The instantiation of the modules depends on the characteristics of the operating system to be implemented. However, in order to satisfy the completeness property, a PEN instance must meet the requirements set out in definition 8.3.

Definition 8.3 *A PEN instance must possess enforcement modules placed in such a way that all references of subjects according to definition 5.12 or derivative instances thereof to other subjects or objects according to definition 5.13 and derivatives thereof are recognized by at least one enforcement module, and must ensure that the operation resulting in the reference does not proceed without a policy decision of the externally controlled reference monitor.*

The definitions above do not mandate a specific implementation mechanism for the realization of enforcement modules. Several general strategies can, however, be identified:

New System In rare cases, a new operating system may be implemented which not only may incorporate all required or desired interfaces for the support of security mechanisms, but also incorporate the enforcement module as an intrinsic part of the functional components.

This may also include the provision of application programming interfaces reflecting security policy handling and providing feedback to application programs and subjects as to the effects of a security policy being enforced.

A key benefit from the perspective of the implementation of PEN instances provided by the design of a new system lies in the ability to identify any subject, object, and operation in a homogeneous fashion and to permit the association and correlation between subjects, objects, and operations of different abstraction layers in a direct way for the enforcement modules and ECRMs.

Emulation The existence of application programs may dictate the external application program interfaces of an operating system, and existing device drivers may, in addition, impose limitations on the structure and interface of the operating system core components and kernel.

While particularly the benefits of entity and operation identification within the operating system core and kernel can be retained unless specific interface restrictions preclude such an implementation strategy, the interfaces, and particularly assumptions regarding the system behavior not codified within the application programming interface but rather in the use of such interfaces by existing application programs impose limitations on the realization of certain security policies without incurring incompatibility with existing application programs.

Limitations on the interactions between entities provided for the sake of emulation can be achieved by compartmentalizing individual processes each with the full application programming interfaces dictated by the emulated system into separate processes of the emulator system without the possibility of unmediated information flow between processes present in the emulated system. Any mediation must become well-defined and occur at the level of the ECRM; otherwise an emulator is restricted to the use of full separation between compartmented processes in the implementation of information flow policies. This is exacerbated both by implicit information flow that is frequently inherent in emulated systems such as global resources including process tables and file system structures, and the granularity of calls to the operating system core and kernel that imply the execution of several atomic operations, each of which may have policy-controlled restrictions and implied subordinate operations or interactions with policy handling, to fulfill the original request and whose semantics particularly in the case of fault conditions cannot be reproduced faithfully.

Virtualization Virtualization of operating systems can occur in two directions. One is in the emulation of a full operating system for individual processes and users and the compartmentalization at the level of the base system providing virtual machines in order to provide information flow control. This approach, however, is limited to a narrow and unexpressive set of security policies and models. Its appeal lies mainly in the simplicity of implementation.

An alternate approach is the virtualization of components of an existing operating system by providing the external interfaces of such components and mediating the interactions between the components as well as between components and application processes. This compartmentalization of system modules requires the base system to be sufficiently modular and with well-defined overall and particularly module interface semantics.

Augmentation Finally, in what may be described as a hybrid between the Emulation and Virtualization implementation strategy, an existing system can be augmented in such a way that certain components of the existing system are surrounded or wrapped by a virtualization instance which exerts policy control over the component while exhibiting the behavior and interfaces of the unvirtualized environment to the virtualized component. The facilitation of other policy-controlled behavior in an existing system requires the embedding of new components in a system that exhibit external interfaces similar or identical to existing interfaces while performing operations disjoint from those performed by modules of the given interface signature, i.e. a reverse virtualization.

The main benefit of this approach lies in the ability to retroactively provide the entity and operation identification within the operating system core and kernel and the resulting ability to implement a broad spectrum of security policies. As in the secondary case of Virtualization, detailed knowledge of the system to be augmented is prerequisite, as is a pre-existing overall operating system structure that supports compartmentalization between controlled entities and the operating system.

The system selected for the reference interpretation and implementation pursues the Augmentation approach.

In accordance with the distinction between regulated and reactive decisions, the ECRM mechanism itself can be dissected into two

functional elements, both of which make use of the same decision mechanism described in definition 8.4.
The following definitions are given in natural language in the interest of brevity; an excerpt thereof can be found in section 4.8.

Definition 8.4 *Externally controlled reference monitors must contain a nonempty set of* ***security policy decision functions*** *(SPDF). All SPDF have a policy request as one of their parameters, and a set of well-formed formulae of all applicable policies as the remaining parameters and return the decision in accordance with definition 5.23. All SPDF must fulfill the properties given in definitions 5.1, 5.20, 5.22, and 5.24.*
A decision is generated either if an enforcement module stimulates the ECRM either directly or indirectly in such a way that the request of the originating enforcement module is granted if and only if the result of at least one security policy decision function was not a negative reply tuple, or it is generated if a stimulus internal to the ECRM causes the formulation of policy requests.
Each decision is reached within an arbitrary but finite time period from the issuing of the policy request. If a SPDF does not yield a decision result after the time period, the computation of the SPDF is terminated and the result is set to the negative reply tuple corresponding to the policy request.
A decision must be promulgated to all affected enforcement modules, which must enforce the decision effective immediately.

Definition 8.4 leaves open the possibility that certain SPDF operate only on a subset of available policies. Consistency between the results of multiple SPDF is not required. This ensures that certain computationally inexpensive SPDF can be used alongside computationally complex SPDF. The source of the applicable policy formulae to an ECRM is given in definition 8.5:

Definition 8.5 *All PEN instances contain a* ***policy repository*** *instance. An ECRM within the same PEN instance as a policy repository will obtain all well-formed formulae for inclusion in the parameters of its SPDF. If all SPDF of the ECRM yield negative reply tuples, the policy repository must*

issue queries for all applicable policy well-formed formulae for a given policy request to a nonempty set of external reference monitors.
A policy well-formed formula is applicable to a given policy request if, for any constant X of the policy request and any constant Y of the well-formed formula, the predicate derived_or_self(X, Y) *holds.*

The applicability constraint imposed by definition 8.5 is necessary but not sufficient in that the semantics of predicate parameters for predicates with more than one subject or object must additionally be taken into consideration which can occur by means of additional constraint definitions. Moreover, the introduction of hierarchical policy mechanisms must impose additional applicability constraints. The information on which external reference monitor to contact can be obtained from a hierarchy as discussed in section 8.2.
Regulated decisions require a mechanism for evaluating such policy requests deterministically. Such a mechanism is outlined by definition 8.6.

Definition 8.6 *All PEN instances contain an ordered list of policy requests, each associated with an arbitrary waiting period. At fixed but arbitrary time intervals, each element of the ordered list is examined as to whether the waiting period has expired. If the period has expired and the operation lifetime has not expired, the policy request is submitted as a stimulus according to definition 8.4 to the ECRM.*
If the operation lifetime has expired, the policy request is removed from the ordered list.
Entries to the list are added and removed by explicit instructions from ERM instances to which the PEN instance is subordinate in a policy hierarchy, for which the identity and authenticity of the entry can be verified, and for which the subordination of the PEN can be proven.

Definition 8.7 *All PEN instances contain an* ***audit repository*** *instance. An ECRM within the same PEN instance as a policy repository will transmit all instances of policy requests and all instances of the policy replies generated in response to a given policy request either internally or externally to the audit repository.*

A 2-tuple consisting of a policy request and the set of replies on which a policy decision was based is called a ***policy datum****.*
The audit repository will ensure for each policy datum that

- *the association to the origin of the policy datum to the ECRM can be proven.*
- *the integrity of the policy datum can be proven under the assumption that the integrity of the trusted subsystem from with the datum originates has not been breached between the time of creation of the policy datum and its referral to an external trusted system.*
- *the origin of the policy datum can be proven by verifying the identity and authenticity of all entities providing corroboration for the datum under the assumption that the integrity of none of the corroborating entities has been breached between the time of creation of the policy datum and its referral to an external trusted system.*
- *completeness of the policy data can be proven under the assumption that the integrity of the trusted subsystem from with the datum originates has not been breached.*
- *there exists a relation* **R** *forming a partial order over the set of policy data from a fixed but arbitrary audit repository instance.*
- *confidentiality from unauthorized entities is maintained and that any disclosure occurs only through the ERM controller residing within the same ERM instance.*

Policy controller node instances may be considered to a first approximation as PEN instances without the requirement for enforcement modules. This, together with policy requests to be decided emanating from PEN instances instead of from enforcement modules, require different definitions for the types of operations from the ones given for PEN instances in definition 8.2 as shown in definition 8.8. In most cases, there is no requirement for distinctions between the policy controller node and the external reference monitor it contains, and hence the following discussion and definitions are given in terms of the external reference monitors.

Definition 8.8 *An ERM operation is referred to as a*

- ***reactive decision*** *if an ECRM instance subordinate to the given ERM in a policy hierarchy is the source of the policy request, or if the ERM has access to one or more well-formed formula for which, for any constant X of the policy request, and any constant Y of the well-formed formula, the predicate* derived_or_self(X, Y) *holds.*
- ***regulated decision*** *if the decision operation is caused by a stimulus or decision request other than an ECRM subordinate to the given ERM in a policy hierarchy.*

Definition 8.9 provides the definition of security policy decision functions corresponding to definition 8.4 while definition 8.10 provides the analog to definition 8.5.

Definition 8.9 *External reference monitors contain a nonempty set of SPDF. All SPDF have a policy request as one of their parameters, and a set of well-formed formulae of all applicable policies as the remaining parameters and return the decision in accordance with definition 5.23. All SPDF must fulfill the properties given in definitions 5.1, 5.20, 5.22, and 5.24.*
A decision is generated either if an ECRM stimulates the ERM either directly or indirectly in such a way that the request of the originating enforcement module is granted if and only if the result of at least one security policy decision function was not a negative reply tuple, or it is generated if a stimulus internal to the ERM causes the formulation of policy requests.
Each decision is reached within an arbitrary but finite time period from the issuing of the policy request. If a SPDF does not yield a decision result after the time period, the computation of the SPDF is terminated and the result is set to the negative reply tuple corresponding to the policy request.
A decision must be promulgated to all affected enforcement modules, which must enforce the decision effective immediately.

Definition 8.10 *All PCN instances contain a policy repository instance. An* ***ERM controller*** *within the same PCN instance as a policy repository will obtain all well-formed formulae for inclusion in the parameters of its SPDF. If all SPDF of the ERM yield negative reply tuples, the policy* *repository must issue queries for all applicable policy well-formed formulae for a given policy request to a nonempty set of external reference monitors.*

PCN instances are functionally equivalent to PEN instances in the handling of regulated policy requests with the exception of the source of such requests as described in definition 8.11; definition 8.12 provides the analogon to definition 8.7.

Definition 8.11 *All PCN instances contain an ordered list of policy requests, each associated with an arbitrary waiting period. At fixed but arbitrary time intervals, each element of the ordered list is examined as to whether the waiting period has expired. If the period has expired and the operation lifetime has not expired, the policy request is submitted as a stimulus according to definition 8.9 to the ERM.*
If the operation lifetime has expired, the policy request is removed from the ordered list.
Entries to the list are added and removed by explicit instructions from ERM instances to which the ERM instance is subordinate in a policy hierarchy or by authorized individuals.

Definition 8.12 *All PCN instances contain an audit repository instance. An ERM controller within the same ERM instance as a policy repository will transmit all instances of policy requests and all instances of the policy replies generated in response to a given policy request either internally or externally to the audit repository.*
The audit repository will ensure for each policy datum that

- *the association of the policy datum to the ERM can be proven*
- *the integrity of the policy datum can be proven under the assumption that the integrity of the trusted subsystem from with the datum originates has not been breached or under the assumption that a datum has been notarized by a trusted third party.*
- *the origin of the policy datum can be proven by verifying the identity and authenticity of all entities providing corroboration for the datum under the assumption that the integrity of none of the corroborating entities has been breached between the time of creation of the policy datum and its referral to an external trusted system.*

- *completeness of the policy data can be proven under the assumption that the integrity of the trusted subsystem from with the datum originates has not been breached or under the assumption that a datum has been notarized by a trusted third party.*
- *there exists a relation* **R** *forming a partial order over the set of policy data from a fixed but arbitrary audit repository instance.*
- *confidentiality from unauthorized entities is maintained and that any disclosure occurs only through the ECRM residing within the same ERM instance.*

While definitions 8.12 and 8.6 deal with policy data, the properties are also required for individual elements, hence definition 8.13.

Definition 8.13 *For all policy requests and reply tuples X, the following must hold:*

- *it can be proven for any valid X, that X has not been altered since it was created, transmitted, or stored by the originator*
- *for any valid X, the originator of X can prove non-interactively that he is the source of X.*
- *Only the source ECRM (ERM) and the identified destination ERM or ECRM can observe the plaintext of X.*

8.1.1 Entity Identification and Authentication Mechanism

Definitions 8.7, 8.12, and 8.6 require the determination of entity identification and authentication as well as a proof of the subordination relation; entity identification and authentication is also an implicit requirement of definition 8.13.
While a large number of schemes for for non-interactively identification and authentication of entities and data origin authentication and integrity exist [473], the following discussion assumes without loss of generality that a mechanism based on public key encryption is used.

The rationale for this is that public key encryption provides benefits in not limiting the number of parties communicating through a requirement for predistributed key material and permit the use of off-line trusted third parties, reducing the number of messages required compared to in-line and on-line third parties.

8.2 Policy Hierarchy

One of the possible means of structuring a set of external reference monitors is, as described in section 6.3, in the form of a hierarchy of policy domains where policy domains are the vertices of a tree, i.e. a directed, connected, and acyclical graph with precisely one vertex of in-degree zero.

This is an orthogonal concept to the metapolicy algebra discussed in section 5.7 and is primarily of interest in case of hierarchical delegation of policies. Any delegation mechanism must include precautions against inconsistencies between policies issued at any two vertices of a policy tree.
The hierarchy established by the CA tree is also suitable as a surrogate for a separate entity identity schema. Given the required lattices as per definition 6.5 and 6.6, a subject or object identity vector can thus be considered a path within the policy domain tree, where the individual subject and object entities are uniquely identified by the path. Moreover, any vertex of a subtree dominated in the subject (object) lattice by a vertex X is part of the policy domain of X.

To ensure that, in a hierarchical policy structure, no unauthorized entities can pose as PCN, a verification mechanism for policy information and decisions is required as discussed in the preceding section. This can, without loss of generality, be accomplished through the use of public key certificates.
Using the terminology from [67], this implies that both data and entity identification and authentication is effected through verification of public key certificates whose validity period, revocation status, and other properties are in accordance with the policy set forth by

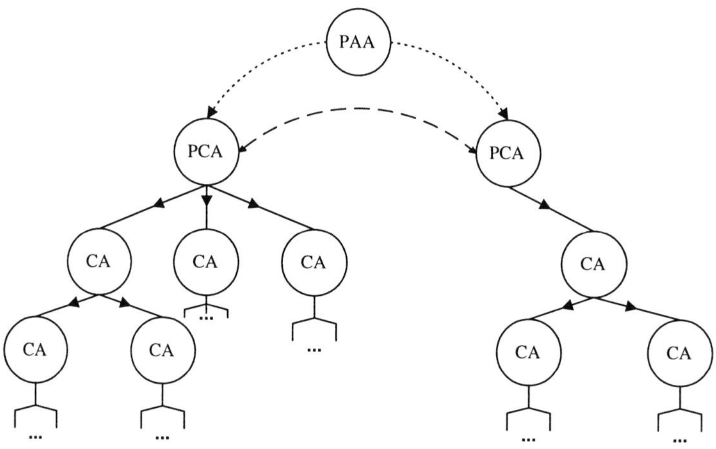

Figure 8.2: Multiple rooted certification authority tree with implicit PAA

the corresponding PAA and all PCAs and CAs required for validating an entity.

Each policy domain can thus be associated with a CA, in the simplest case discussed here again in the form of a tree, permitting each entity within the entire tree to use a single public key for the verification of the rooted certificate chain.

Provided that the PAA policy permits this, multiple rooted tree configurations are also conceivable provided that bridge certificates are used. This scenario is depicted in figure 8.2.

Entity identification can thus be accomplished by mapping the vertices of the policy domain tree (cf. definitions 6.9 and 6.10) onto entities for which CAs provide verification of name-entity association.

For reasons of efficiency, it can be assumed that a terminal vertex in the trust hierarchy (i.e. a PCN or PEN) provides this name-entity association itself. It should be noted that the case of multiple rooted CA trees, the PAA assumes the role of the root of the tree and hence

also the greatest lower bound for the lattices providing subject and object identity.
To further align naming and authority relations among entities, both the policy domain tree and the certification tree can further be aligned with a name service such as the Internet domain name system [481, 120]; this is, however, independent of entity authentication and entry validation of the directory system.
A simple policy authority delegation scheme can restrict the authority to impose policy well-formed formulae (implicitly, this affects applicability) of a PCN to well-formed formulae for which the subject or object identities are dominated by the policy domain tree node. This, however, does not eliminate the problem of invalid formulae being introduced into the set of applicable policy formulae.
The determination of authority must include the verification of the entire certificate chain along the edges of the CA graph to avoid man-in-the-middle attacks by CA nodes with names containing entity identity subelements.

8.3 Secure Coprocessors

Particularly in a configuration consisting of a retrofitted COTS system, the assurance with which the isolation property of the ERM and ECRM TCBs can be claimed is quite low given fault rate estimations and information on the size of the code base operating with supervisor privileges and hence susceptible to tampering or faults resulting in violation of the isolation property.

As was discussed in section 3.5.1, this can be — provided that the corresponding assurance requirements exist — addressed through the use of a secure coprocessor which contains a TCB of limited size assisting in verification and validation and is also secured against tampering [127, 134]; such uses are also discussed in [765, 724, 725, 779].
The design of such a secure coprocessor is nontrivial particularly if cryptographic primitives are realized within the coprocessor since

differential power, timing and other side channel attacks [376, 375, 475, 270, 196, 149, 287, 648, 452, 294, 637, 150, 474, 136, 451, 527, 751, 638] can compromise hardened hardware implementations which must ensure that adequate masking is achieved; otherwise a knowledgeable adversary can gain information on cryptographic keys which can result in the entire secure coprocessor being compromised. An attacker with both knowledge and sufficient funding will also be able to circumvent the tamper-resistance mechanisms [760, 29, 759]; however, the work factor required to circumvent the tamper--resistance mechanisms can be considerable.
The assurance criteria particularly for cryptographic devices claiming tamper resistance are the subject of a Federal standard [669, 671, 511] and evaluation procedures due to their high sensitivity.

Originally developed at IBM T.J. Watson Research laboratories, the IBM 4758 model 2 product is a a high assurance secure coprocessor with cryptographic capabilities (the SIDEARM coprocessor in the LOCK project did not possess cryptographic capabilities) traces its origins to the ABYSS [765, 758] and Citadel architectures [766, 534, 535, 536].
This coprocessor contains an autonomous general purpose computer system consisting of a CPU, volatile and nonvolatile memory, and logic for performing both symmetrical and asymmetrical cryptography as well as physical random number generation based on a thermally stabilized diode and an independent real-time clock within a tamper-resistant and tamper-evident enclosure that can detect and respond to several types of attacks [670, 672, 185, 544, 536, 35].
To ensure that the code running on the coprocessor is authentic and authorized, a multistage boot process is employed starting from a manually verified minimal environment which uses the cryptographic hardware modules to verify the desired properties of the next stage in the boot process until a hierarchy of certified code has been established as an operating environment of the TCB, thus ensuring verifiability, secure delivery, and self-validation requirements[1].

[1]The same rigor must also be applied to all interface definitions, otherwise security properties may be violated [35, 96].

The IBM 4758 model 2 communicates with its host environment via a PCI interface.
Given such a tamper-resistant processing environment the TCB components of the ERM that perform critical functions, i.e. derivation of policy decisions, receiving policy requests and issuing commands to ECRMs, and performing audit data requests and replies can therefore be embedded in such a way that the work factor for an adversary is significantly increased.
Given the availability of custom circuitry for high performance asymmetric encryption, any data exchanged between ERM or ECRM TCBs can be transformed into well-formed messages with integrity, authenticity, and confidentiality, properties and protected against replay attacks. The verification of the properties and subsequent processing can in turn occur within the confines of a tamper-resistant TCB.

8.4 Instrumentation and Instantiation

Given an application domain for policies such as security policies in the context of networks and operating systems described in chapters 5 and a formal derivation of the modeling primitives for a given operating system (i.e. using a model of the operating system as well as the underlying hardware — while such work has indeed been performed on a small scale e.g. by Good *et al.* [264] as well as in part by others [141, 79], the scope of such an effort for a fully functional general-purpose system appears to be significantly beyond the state of the art [436]), the creation of an isomorphous mapping of primitives onto the model of the model of the operating system could be performed within a given formal model.
However, the operating systems considered in this dissertation for pragmatic reasons are not represented by a model embedded whose semantics are embedded into a formal theory; rather, these systems may be both significantly underspecified and even these specifications need not be available for the implementation (as is the case with the system described in section 6.4).

As a result, mappings of primitives as described in section 6.4 (and used in the reference implementation described in chapters 9 through 11) constitute an element of uncertainty within any chain of reasoning over the policies described here.

8.4.1 Instrumentation of Operations

For operations the requisite mappings are, according to definition 8.2, required to be bidirectional. Since, as already noted in the sample interpretation outline in section 6.4 this need not be a homomorphous mapping in the general case, it is frequently necessary to insert several instrumentation points (cf. section 9.1) concurrently which map onto a single predicate as well as optionally constants representing the primitive within a lattice of operation constants analogous to the ones discussed in section 5.1 for subjects and objects (see section 6.4.4.2 for an example of such a nonepimorphous mapping and section 8.4.4 for a description of the hierarchy mechanism).
The simplest case is presented by reactive decisions; here, control flow within an enforcement module is interrupted and transferred to the ECRM. The ECRM is equipped with a platform-dependent mapping mechanism that can map the operation at which the interception took place onto a predicate and retrieve (see sections 8.4.2 and 8.4.3) the matching subject and object constants. As a result of this translation, a hypothesis is formed which can subsequently be presented to the ECRM security policy decision function. If the policy request is granted, control flow returns to the interception point.
Even in the simple reactive decision case, however, several conditions can arise in which additional information must be gathered or retrieved to permit the formulation of the hypothesis prior to submission to the SPDF.
This is mainly due to the fact that the modeled operating system may not provide all requisite information on subject and object entities at the interception point. The mapping mechanism in this case must augment the information immediately provided by the interception

point. This can occur in several ways depending on the instrumented operating system and subsystem from which a hypothesis originates:

- The mapping mechanism can have stored an association of a certain operating system entity with a corresponding element of the formal theory; this can have occurred implicitly in the course of earlier operations and hence hypotheses on the part of other instrumentation points and enforcement modules involving the same operating system entity or as a result of a prior mapping retrieval as described below for the same enforcement module and instrumentation point.
- Retracing of data structures within the instrumented operating system to identify the entity to which an instrumented operation is associated.

  While the parameterization of an instrumentation point may not immediately permit the mapping onto an subject or object entity, it is frequently possible to identify such entities within the operating system (and hence in its mapping) by retracing linked data structures.
- Tagging of operations at semantically higher layers by adding data structures transparently on the part of enforcement modules at such layers permits enforcement modules at semantically lower levels to read out the tag information and forward it to the ECRM mapping mechanism (see section 9.1.2 for an example).
- Finally, pertinent information can also be retrieved actively by the mapping layer by calling on one or more instrumentation points within the same or other enforcement modules (see chapter 9).

  This mechanism must be employed sparingly to limit adverse effects on overall system performance. Depending on the system architecture of the instrumented host operating system, deadlocks may arise and can particularly inhibit multiprocessor performance while data structures are locked for retrieval of information required by the mapping mechanism.

While it would be highly desirable to keep the mapping mechanism

stateless, this is not possible for the general case of reactive policy decisions (besides the performance benefits from repeated mapping as noted above) since, as described in section 8.4, a single modeling primitive in the formal model can be mapped onto by multiple primitives at the level of the instrumented operating system.
This necessitates the storage of partial model operations (including potential partial model operations, this e.g. may arise if a the modeled operation is not completed because a partial operation is canceled because of an error condition) and subsequent purging of the partial operations database after a heuristically determined interval.
The actual hypothesis must be generated once all required information for the model operation is available and prior to completion of the model operation; it should be noted that such behavior can imply the presence of covert channels as success or failure of partial operations may be observed under some instrumented operating systems.
Implicit state is also preserved during the determination of the status of a policy request; however, since the relevant operation is blocked until the granting or denial of the policy request, this state information is retained in the form of control flows which also allows the demultiplexing of policy requests generated by multiple instances of the same instrumentation point or multiple instrumentation points mapping onto a single model operation.

For regulated decisions, several additional constraints arise. The existence of multiple instrumentation points mapped onto a single model operation requires that a regulated decision be promulgated to all instances if instrumentation points affected. As a result, the regulated decision must either be mapped onto idempotent operations for individual instrumentation points or the mapping mechanism must select a single necessary and sufficient instrumentation point for effecting the regulated decision. The former is not only of relevance to avoid multiple consummation of a policy decision but also to avoid race and infinite loop conditions caused by secondary policy requests resulting from promulgating regulated decisions.
In addition, the instrumented operating system as a whole or an en-

forcement module and its associated instrumented operating system module may be in a state that does not permit immediate promulgation and consummation e.g. since resources are blocked or the system is operating at a privilege level that does not permit certain operations to be initiated. To address this issue, the ECRM mapping mechanism must be able to both queue regulated decisions (preserving a partial order over the decisions) and to monitor the instrumented system and components for transitions into states permitting the regulated decisions to be promulgated.

8.4.1.1 Implementation Issues

Control flow for reactive decisions must be blocking regardless of the instrumented operating system; consequently, instrumentation points requiring a policy decision must transfer control to the ECRM and specifically first to the mapping mechanism.

For enforcement modules residing in nonprivileged code this will generally be achieved via a trap or system call-type event whereas at the operating system kernel level an immediate transfer of control within the current execution context is desirable for performance reasons.

Results are communicated on return from this transfer of control flow and must be translated into an error message suitable for the instrumented operation by the instrumentation point in case the policy request is not granted to avoid unspecified conditions for calling processes and modules.

The mapping mechanism for operations will generally consist of a central dispatching point communicating with the enforcement modules via a message passing interface to limit the exposed external interfaces of the trusted computing environment.

As the required behavior for reactive and proactive behavior can only be encoded inefficiently in a generic form (which would presumably result in executing an interpreted intermediate code), there must exist a separate functional component of the mapping mechanism for each model operation to be considered.

This results in a potentially restricted set of model operations supported by a given ECRM (e.g. in case of heterogeneous operation of multiple generations of instrumented operating system enforcement modules) and its associated mapping mechanism which can be communicated to an ERM or ECRM; the effect of such a restricted set of supported operations can be more stringent restrictions on permitted operations since the ECRM may then not be able to provide adequate granularity decisions but conversely simplify the process of generating responses to policy replies.
A secondary result of an implementation as outlined here is that the interpretation of the formal model in which policies are to be applied resides to a significant extent in the implementation of the mapping mechanism. It is therefore imperative to derive the implementation of the mapping mechanism elements in such a way that correspondence to the standard interpretation semantics can be shown or verified; this can e.g. be accomplished by stepwise refinement in the Z notation as discussed briefly in section 4.8, ideally by refinement to code[2].

8.4.2 Instantiation of Subjects

Unlike the operations described in the preceding section, the number of subject entity identities is a priori unbounded since not only is the set of nodes within the model discussed here (see section 4.8) unbounded, subjects may also be created and deleted over time. As per definitions 6.1 through 6.1 and observation 6.1 given in section 6.1, a globally unique representation of each such subject identity exists and can be derived by an appropriate generation rule.
A mapping mechanism can trivially provide a bijection[3] of subject identities within the instrumented operating system onto constants within the formal theory. This bijection must be found as part of the creation of interpretations as exemplified in section 6.4.

[2]Bearing in mind the pragmatic restricions laid out in section 8.4

[3]Generally, this will be a partial bijection since one can assume the left-hand set to be a subset of constructable entity identities.

Conversely, subject type information is bounded and can be determined a priori; as with subject identities, a partial bijection can be defined based on the interpretation for a given instrumented operating system.

8.4.2.1 Implementation Issues

The disambiguation of the subject identity can occur in several steps. First, such disambiguation must permit distinguishing individual nodes; this is accomplished by the node identity established as per section 8.1.1.
Subsequent levels of identification can then be established by assigning classes of subjects within the host operating system a prefix according to definition 6.4 (although this is not necessary since type information is preserved simultaneously, this can significantly improve searches for entities and reduces the complexity of determining membership in the relation subject_subset_identity according to definition 6.2) and a nonce to identify an individual entity. For efficiency reasons, the bijection is realized in the form of a composed relation based on disjoint subdomains.
While the use of nonces introduces a probabilistic element into the disambiguation, this is merely an implementation-specific optimization; deterministic alternatives can rely on high-resolution timestamps in conjunction with in-memory addresses for entites to encode the identity. However, the latter implementation (particularly obtaining persistent time-stamps since entity identities must be preserved over system failures and restarts) introduces a significant performance overhead; while not necessarily a problem in the case of subjects this must be given serious consideration for objects as described in section 8.4.3.

Disambiguation of subject types is not necessary since the mapping bijection can take the subdivision according to definition 6.3 or a representation thereof into account. Unlike identity information, both mapping and its inverse do not require retaining persistent information as the mapping relation itself is sufficient to provide both

at any time. Conceivably, however, situations can arise (e.g. after a catastrophic system failure) in which instrumented host operating systems are reconfigured to provide fewer or more subject types than prior to the reconfiguration. In this case pending policy decisions (both reactive and regulated) may no longer be applicable.
To permit rapid lookup of individual subject identities and, to a lesser extent, of subject types the individual components must be stored in such a way that prefix elements are decomposed and indexed upon to permit prefix searches for the subject_subset_identity and subject_subset_type relations.
The duration for which such database entries must be held cannot be determined a priori since these depend on the set of policies active at any given moment; as a result, heuristics must be used to purge entries selectively for which no corresponding element in the instrumented operating system exists (this applies particularly to remote subject entities).

8.4.3 Instantiation of Objects

The considerations discussed in section 8.4.2 apply equally to object entities, both to object identities and types. However, as the number of objects instantiated as well as turnover of objects at any given time is significantly larger than for subjects, particular care must be given to the complexity of the mapping bijections used as well as to the efficiency of both lookup and purging mechanisms for the mapping databases.

8.4.3.1 Implementation Issues

An optimization step particularly applicable to bulk transfers consisting of identifiable objects such as packets of a packet-based network protocol is to avoid entry and deletion of object identities for individual packets into the object identity mapping database; rather, flows can frequently be identified at the level of an operation in the mapping mechanism and retained temporarily.

This does not obviate the need for the presence of at least one entry in the object identity database as other policy decisions may be required to act upon these objects; however, this requrirement can be satisfied by entering a single proxy object entry into the database access to which will result in triggering a mechanism to fill the database with the remaining database entries.

8.4.4 Axiomatization of Entity and Operation Hierarchies

While policies can be formulated using the basic predicates and ground constants obtained in the mapping relations described in sections 8.4.1 through 8.4.3, this would render such policies highly inefficient and make repeated use of policy decisions impossible.
The structures over entity identities and types defined in definitions 6.1 through 6.1 and 6.5 through 6.8 and elaborated in the preceding sections permits the use of the subject_subset_identity and subject_subset_type as well as the predicates object_subset_identity and object_subset_type in the formulation of policies by providing a set comprehension mechanism. However, as the entities over which the predicate is to range requires an anchoring element, it is frequently desirable (though by no means the only possible structuring mechanism as described in section 6.3) to structure particularly entity identities in a hierarchical fashion to permit the definition of security policies based on policy domains, and indeed the policy domain tree in the sample hierarchy mechanism introduced in definitions 6.9 and 6.10 can be embedded in the same structure as the identities themselves as shown by definitions 6.11 and 6.12, respectively.
However, a similar structure is also desirable for entity types since one can, through formal concept analysis and application of domain-specific modeling, frequently identify hierarchical relations among entity types to which policies are to apply (see chapter 9 for a discussion of the semantics of the layering mechanism for general-purpose operating systems) and is fundamentally also already provided by the structuring of mapping relations described in the preceding sec-

tions. Such a mechanism then permits reasoning over several semantic layers, e.g. by imposing a derivation relation fact between user(S) and process(S).
Similar considerations also apply to operations. This can, in a subordinate case occur intrinsically to a given ECRM in identifying subordinate operations (e.g. the predicate read_file(S, O) in the reference interpretation given in chapter 6 is mapped onto native API functions such as `ZwReadFile`, this in turn can be mapped onto the I/O manager level IRP `IRP_MJ_READ`); such hierarchization are, however, properly considered part of the interpretation itself although they may also be modeled within the policy framework to preserve generality. In the general case, however, the hierarchization of operations must occur at the level of the formal model through the definition of facts as discussed above. However, in the interest of brevity the minimum model excerpt defined in sections 5.1 through 5.2.4 does not contain such instances although definition 5.15 can be extended intuitively.
A structuring mechanism for either entities or operations is thus exemplified by the lattice algebra embedding discussed in sections 6.1 and 6.2 and extensions thereof. It should be noted, however, that while algorithm 5.1 implicitly requires the presence of the structure it is neither necessary that the structure be a lattice nor, strictly, even a partial order over the respective sets.
By identifying designated constants[4] within the range of constants of the set of mapping relations discussed in sections 8.4.1 through 8.4.3 and applying an endomorphism to map sets of constants onto generator constants (as per definition 5.15), reasoning over entity types as well as relations can make use of the algebraic structure thus embedded into the formal theory. It should be noted that the semantics of this embedding can be twofold.
First, in identifying entities that are subordinated (or derivative) within the model to which policies are to be applied (this would e.g. apply to the predicates file(O_j) and data_file(O_i) for types, and for subordinate or derivative entity identities as in the case where a stream of *datagram*(O_j) objects is obtained in the course of a policy

[4] Also referred to as "generator constants" in analogy to common usage in abstract algebra.

derivation from a *file*(O_i) entity when a file is being transmitted to another node via a network file protocol in the general purpose operating system scenario).
Second, the semantics can be part of security policies to be applied; this is mainly of interest if a security policy can generalize certain types of entities and operations (e.g. in case of a security policy that is oriented purely toward access control). In this case, however, the relevant fact entries must be qualified by the security policy domain or corresponding construct within the formal model.

As a result, policies can be expressed at abstract levels within the formal model while the hierarchies described above ensure that these are also applicable to actual hypotheses formulated by the respective mapping mechanisms for both reactive and regulated policies through deriving subordinate instances of policy rules through automated deduction or related operations.
An additional benefit implicit in this mechanism is the possible reuse of policy decisions (as well as, depending on the security policy decision function used, intermediate results; however, here a balance must be struck that prunes extraneous additional results from the set of facts to be considered by the security policy decision functions since performance can be degraded both by excessive lookup times and having to consider a multitude of facts in the derivation and decision process) both at the level of an ERM as well as for ECRMs provided that adequate policy decision lifetimes are defined.

8.4.4.1 Implementation Issues

The suppression of extraneous derivative intermediate results and policy decisions as well as the limitation of constants populating the decision space of security policy decision functions is critical to the performance of the overall policy mechanism.
However, strategies and heuristics to be used are not only highly dependent the general technique used for SPDFs (e.g. resolution, term rewriting), but rather also on specific implementations of each technique. In the case of using a resolution-based automated theo-

rem prover, two techniques provide significant performance benefits. First, pre-selecting relevant entities from the database of entities based on the appropriate entity identity reduces the number of atoms to consider significantly and can be implemented by highly efficient database queries, although one can construct policies and atom universes where certain hypotheses cannot be proven and hence a negative policy reply will be generated. However, as both ECRM and ERM security policy decision functions will generally be forced to operate under both temporal and spatial resource constraints, such degenerate decision instances would most likely terminated due to resource exhaustion given the state space explosion that the theorem prover is confronted with.

Second, a significant optimization in a resolution-based theorem prover (sacrificing the general structuring mechanism for a specific algebraic structure such as the one outlined above) can be realized by utilizing the explicit type information provided in the policy model for sort information that would otherwise be inferred in providing additional inference and reduction rules.

CHAPTER

NINE

Layering of Enforcement Mechanisms

Information operations can be represented at a number of abstraction (semantic) layers starting with humans acting as subject entities directly performing mission-related operating system.
Each such operation at a higher semantic abstraction level may be mapped onto one or several operations at the next lower abstraction layer; this refinement process terminates for operations related to the technical components of the information system (which is what this discussion is solely concerned with) once the operations are mapped onto elementary primitives of the information system.

9.1 Layer Capabilities

The enforcement of security policies requires the mediation of all operations within a PEN that directly or indirectly can affect the policy.
This requires an instrumentation mechanism independent of a specific implementation that provides a structured approach to addressing mediated capabilities and which can be extended arbitrarily, particularly beyond the basic services of an operating system.
The point at which mediation can occur may not provide all necessary information regarding subjects and objects involved or about the characteristics of an operation.

The instrumentation mechanism must therefore not only provide the mediation capabilities for policy enforcement but also for information gathering to the ECRM.
While there exist a number of semantically equivalent components within an operating system, a layered structure is a key component of most modern operating system designs.
This layering approach permits the provision of simple abstract services to higher layers while encapsulating the specific properties of a lower abstraction layer or the capabilities of a given system. Similarly, the OSI reference model [326] provides an layered abstractions for network protocols and data flow that is at least in part honored in most protocols.
Such a structure is furthermore provided by the concept of a set of enforcement modules which provide the instrumentation for both enforcement and information gathering in accordance with the instructions of an ECRM for a given abstraction layer or functional component, ideally corresponding to the structure of the instrumented system.

Each enforcement module provides the mediation for a functional abstraction defined in terms of security policy objects and operations; at the same time the enforcement modules also provide **instrumentation points** that can supply and record information as dictated by an immediate security policy (for auditing or intrusion detection purposes) or as a secondary effect of a security policy if activated or queried by the corresponding ECRM.
Individual instrumentation points are grouped into **instrumentation groups** in the form of a partial order according to the semantics of the entities and operations for which they provide information. In addition, a partial order is defined over enforcement modules, with multiple elements grouped into equivalence classes (cf. sections 5.1 and 6.1).
This provides an analogon to the entity naming scheme and therefore also an unique identification for instrumentation points and groups. Moreover, even if a given specific instrumentation point or group does not exist, the corresponding instrumentation groups at lower

levels as well as instrumentation points at higher abstraction levels are efficiently determinable.
Entity types are thus embedded into the subject and object identity lattices according to definitions 5.2 and 5.6 at the level of the abstraction layer specific enforcement modules and can be extended to individual instrumentation groups.
This is of critical importance for the ability to properly derive both policy requests and replies since the derivation of an dominant entity type within the lattice for a given entity identity can occur efficiently. Similar considerations also apply to the derivation of dominant operations at higher abstraction levels.
The determination of the proper identity of entities, particularly in case of the formulation of policy requests, can require information that is not immediately available within a single instrumentation group or an enforcement module. Therefore, enforcement modules also act as information sources for an ECRM or ERM in reaching policy decisions.
Depending on the structure of a given system to be augmented by the introduction of enforcement modules, a single conceptual enforcement module may also consist of multiple instances or of multiple sub-modules acting as a single module. Such implementation-specific considerations do not affect the entity naming and control flow for a PEN.

9.1.1 Device Interface Layer

The lowest abstraction layer of enforcement modules involve access control to the interfaces such as e.g. network interfaces, serial communication lines, and SCSI interfaces.
Most operating systems provide such services at least for some of these interfaces; these must be placed under the control of the centralized security policy enforcement and effectively act as enforcement modules and instrumentation points.
The level of control is typically at the level of the principal in the system security concept.

This mechanism can be extended by employing a finer granularity control mechanism identifying additional subjects such as application programs; this is elaborated below.

Both application programs and users need not be aware of such an extension since the mechanisms for reporting access violations provided by the host operating system can be used to report violations of such fine-grained security policy violations.

A secondary policy mechanism beyond access control involves services for confidentiality, integrity, and authenticity including end--point authentication. This cannot be used at this most basic layer since the necessary features are not always present or can be tied unambiguously to an entity.

An example of such a situation is a modem attached to a serial port. In this case, the only control mechanism at the interface level is the verification of compliance with the security policy if a set of subjects and a set of objects wishes to exchange data with such a device, resulting in access being simply either granted or denied at this level.

If a protected and identified higher-level protocol is used, however, this can override the basic protection since the necessary entities and operation characteristics are then known.

Finally, as a tertiary mechanism, both access events and subsequent operations on devices (which may require specific characterization in case of specialized devices, otherwise generic predicates will generally be suitable for the device abstraction of common operating systems) can be instrumented in such a way that some or all operations on the device interface level result in an instrumentation event that is then used for auditing and intrusion detection mechanisms.

The distinction between the device interface layer and higher layers, particularly the network protocol layer is becoming less clear with plug-and-play interfaces that imply operations initiated by devices or hosts other than the local node. In such cases mechanisms analogously to those discussed in section 9.1.2 are required.

9.1.2 Network Protocol Layer

Even though there exist some application areas where other network protocols are still used and which need to be dealt with separately, the Internet Protocol is by far the dominant protocol and hence without loss of generality the focus of this discussion. While generally the same mechanisms may be applied to other protocols, this would render the discussion too cumbersome.

The implementation of the Network Protocol Layer Enforcement Module is highly performance sensitive and requires a tight integration of all components to reduce critical code path lengths. However, for the purposes of the description here individual functional components are identified as an in itself layered architecture.

Since it must be assumed that a node may operate in an environment exposed to hostile entities and that the network protocol stack of a retrofitted node's host operating system may not be able to deal with malformed protocol data units (PDU) or sequences of PDUs, a first layer of protection from external nodes (regardless of whether they are part of a policy-controlled environment) is required that performs integrity checking on incoming and outgoing data and ensures that any and all PDUs passed on to the host network stack can be handled by it.

Another role that must be assumed by the Network Protocol Layer is basic packet filtering and circuit-level control. This applies to both incoming and outgoing traffic and must be coordinated with other components within its own layer as well as with other layers. As many policy rules as possible should be formulated in a way that they can be dealt with at the component level for performance reasons; this is elaborated in chapter 10.

In cases where security policies require confidentiality, integrity, and authenticity at the circuit level, this can be assured at this layer as well. The obvious choice of technology for such requirements in the case of the IP family of protocols is IPSec [701] since it is part of IPv6 and provides the necessary features already either by itself or lends itself to straightforward extension; it is also an already widely accepted IETF standard. This also permits the use of such features

when communicating with nodes that are not equipped with the mechanisms described here.
Actual benefits as compared to other security mechanisms besides enforcing a consistent set of security policies across all nodes of a distributed system result from the integration of information derived from other areas of the system.
By identifying the subject performing an operation involving the network layer (which, depending on the host operating system may be either derived from information available in the local layer or from the system call layer) security policy rules can be enforced at a much higher level of abstraction.

Similarly, the identity of the process performing the operation can be traced back with the help of other subsystems and can be related to the identity of an application or system service that is performing the operation.
This permits e.g. the specification of policy rules that identify which subjects may use which application programs to communicate along with additional constraints; in addition, the set of nodes with which communication may be established can be restricted based on various factors including temporal restrictions provided that suitable predicates for representing these are established.

While policy rules can in fact stipulate specific sequences of operations (this may occur at the level of individual instrumentation points), this is presumably of limited interest in practical application scenarios due to the performance sensitivity of this enforcement module layer and the large state space implicitly required, which could also result in vulnerabilities to denial of service type attacks.
Further benefits for the Internet protocol family of network protocols can be achieved by employing a modification to IPSec in terms of the operational model employed, namely positive identification and authentication of peer nodes.

For this to be functional, each node must be identified, typically by means of a public key certificate. The certificate must be issued by a certification authority that is contained in the security policy that

is applicable for decisions on the permissibility of a connection to a peer node.
It should be noted that this modification is merely in the behavior of the node with regard to the establishment and removal of entries in the equivalents to the security policy database (SPD) and security association database (SAD) of the IPSec reference model (cf. section 10.2); interoperability with other non-PENs that are employing IPSec mechanisms and for which security policies permit communication are not affected.

Such a mechanism interacts closely with the general packet and circuit filtering discussed above; in cases where the peer is also a PEN, the certificate should contain the same (node) subject name for the sake of simplifying policy rules; if the IPSec implementation is not embedded within the ECRM, a second key pair for the IPSec mechanism is required since the latter may be compromised by external manipulation of the node.
This also permits the enforcement of policy rules requiring positively identified and authenticated (certified) peer nodes; if a peer node is compromised, the node's certificate can be placed on a certificate revocation list or similar blacklist propagation mechanism, limiting at least the potential exposure of other nodes after the detection and blacklisting of the compromised node.

One disadvantage in requiring positive peer identification and using other features of IPSec is that it makes the integration of proxies, network translation devices, and other intermediate nodes that modify data streams hard or impossible since strict end-to-end semantics are then implicitly enforced.
This is intentional, but may require considerable changes to existing network infrastructures.

A final sublayer in the Network Protocol Layer can then optionally perform operations typically reserved for such intermediate nodes such as transformations of data before reaching an application (e.g. decryption of data that was stored in encrypted form on an external server) or performing malware scanning.

This applies to both incoming and outgoing data streams, but is most relevant for incoming streams. Other than the earlier sublayers, this sublayer is not necessarily fully transparent since it may induce noticeable changes in e.g. the traffic pattern observed by an application.

9.1.3 File System Layer

This layer has two main purposes. One is to provide security mechanisms at the file system level in its own right, the other is as an information source for policy decisions that involve other layers.
Security at the file system level is part of the basic security functionality provided by virtually all general-purpose operating systems. However, there are two deficiencies that need to be remedied. First is the lack of conformance to a consistent, consistently enforced security policy or set of policies. Even in networked environments, such controls are honored only in largely homogeneous areas. Another issue is the need for protecting data once it has left the immediate control of a node's operating system. This can generally occur in two ways.
One is by accessing the storage media locally with an operating system other than the one enforcing the security mechanisms (possibly just another instance of the same operating system, only configured differently), the other occurs when the storage media are exposed. While the latter threat has always existed in case of removable media or exposure of portable systems, this scenario is becoming more urgent as technologies such as network-attached storage and storage-area networks become more prevalent. The key problem in both cases is that the previous assumption that the storage medium and path to the node providing security enforcement is secure is no longer valid.
What is therefore required is a protection mechanism operating at the file system layer which transparently protects whatever storage medium or mechanism is employed even if the operating system providing the protection is not operational at the time of attack.

One solution to this is the use of cryptographic techniques for confidentiality, integrity, and authenticity for data transfers to and from the storage medium under the control of the ECRM. While there are several possible interception points at which this can be accomplished, interception at the file system layer permits the abstraction from specific file systems. This approach also has the potential benefit of preserving the syntax and semantics of lower abstraction layers, thus being transparent to such layers which is of particular relevance in the case of a retrofitted host operating system.
Use of a mandatory cryptographic mechanisms for storage paths also deals with the problem of accessing the storage medium from another operating system or an access method that does not honor security policies.
The transparent cryptographic mechanisms is are not a priori an access control mechanism; rather, it ensures that access to data objects by subjects (e.g. processes) occurs only when mediated by the ECRM and hence in accordance with the applicable security policies. It is therefore conceivable that mandatory transparent cryptographic mechanisms ares used at the enforcement module level while the actual security policy enforced permits unrestricted access to the data objects and does not consider other constraints.

To permit the identification of entities across node boundaries within the distributed system, files (as well as other objects such as network streams) are affixed with a label that is handled by the enforcement module subsystem and is transparent to the remainder of the system (i.e. both lower and upper layer drivers of the host operating system as well as application programs and users are oblivious to the labeling mechanism).
These labels are protected against manipulation by being tied to the content (i.e. unique identifying characteristics) of the object; the actual policy information is contained in either the ERM node providing policy information or is temporarily delegated to the (protected) security subsystem, the ECRM.
To ensure enforcement even if the enforcement module subsystem is inoperative, automatic cryptographic protection for integrity and

authenticity purposes can and should be used on the contents of thus labeled objects, policy elements may further dictate the use of confidentiality mechanisms.

The cryptographic protection mechanism also necessitates the provision of policy information in case an object is transferred to another node to ensure identification of transformations applied and for obtaining key material, algorithmic, and mode of operation identification if required.

There is one category of file objects to which labels must not be affixed; this category consists mainly of files which are required for bootstrapping the entire system and which are accessed prior to the enforcement module being activated.

Even if such files were only labeled and not encrypted or otherwise transformed, this could lead to unpredictable results.

As a general rule, however, each file object must carry an object label even if only to identify the object. Based on the identity of the file object, the ECRM can then determine what — if any — actions must be applied to a given file based on the security policy or security policies applicable to either the object or to an operation.

A simplistic policy element concerning file system security could consist only of enforcing access controls that are independent of the host operating system and enforced consistently for identified objects across an entire distributed system based on the object labels uniquely identifying each file object.

An application program or other process attempting to access such a file would be confronted with an error code matching regular error codes for the operating system to retain compatibility with the failure modes expected by both application programs and users.

For this purpose, as with other decisions regarding operations, the ECRM can — depending on what relevant security policies dictate — either query an authoritative ERM for a decision, possibly including key material, or it can resort to consulting a local cache of policy decisions and rules for which an authoritative ERM has specified a lifetime. The file system layer furthermore implicitly provides access to other important information. This information correlates users and files they are using. It is necessary to distinguish three types of files.

The first type of file is the executable file as seen by the operating system. Such executables, which usually consist of several parts (a main file and a number of dynamically loaded shared objects or dynamically linked libraries), can be identified and matched against security policy rules containing approved applications.

The second type of file is harder to identify when located at the file (operating) system layer and involves all scripting languages, i.e. mechanisms that involve files classified as non-executable by the operating system but executed by an intermediate application program.

This class of applications includes macro languages found in many applications and has been the source of a large number of successful attacks. Here only heuristics and elaborate checks can attempt to identify and protect against malicious code.

The third type of file consists of plain data objects. This information can be combined with other information collected at different layers. In particular, the integration of the file system layer permits the dynamic "sandboxing" of applications.

One example of such sandboxing in case of a MLS-like policy is the dynamic restriction of a process from making certain network connections once it has accessed a data object whose classification label does not match with the classification of a given network peer. The same mechanism obviously also is applicable to operations within the node local file system and can be used to implement a purely local MLS configuration. In most cases, however, caching and common networked file systems will require coordination of policy across node boundaries.

Finally, similar to the topmost sublayer described in section 9.1.2, it is also possible to embed additional transformations into the file system access mechanisms.

One example of such a transform would be a malware scan (which can easily be extended to other types of undesirable files or contents; the main issues here are in determining what level of checking constitutes unacceptable performance impact, and that any such blacklisting is by necessity incomplete even when heuristics are used since one needs to assume that a potential attacker is aware of the heuristics and may be capable of using these to his advantage, e.g. by means of denial of service attacks).

9.1.4 Process Management Layer

The process management layer provides access to and mediation for high level abstractions within a node. Processes represent the highest level of granularity for subjects within the context of a node itself.

A process is associated with the ability to initiate and execute operations; while this may be subdivided into multiple execution threads by implementations, for the purposes of this discussion the process abstraction is sufficient. A process is also associated with a state and set of resources such as a (typically protected and separated [147, 126]) address space.

Control and management mechanisms for processes are a central part of advanced general purpose operating systems, which typically already provide protection and isolation as well as sharing mechanisms for resources as well as scheduling mechanisms to ensure availability for a given objective. Particularly in the case of retrofitted enforcement mechanisms, it is therefore important to both provide complete control over the resources constituting processes and also to retain the operational semantics as far as possible.

For this, the conformance of the initial state with applicable security policies must be ensured; this includes the layout and content of the address space associated with the process and possible shared resources via communication channels to other processes such as shared memory segments.

This then permits control over executable images which can be subject to policies as well as to integrity protection mechanisms that also ensure the authenticity of executable images and ancillary resources. For this purpose (cf. definitions 5.12 and 5.13) applications, which

can consist of multiple processes, are defined as subjects while executable and data files are considered objects in the model described in chapter 5.
Security policy control must also be enforced over process management aspects such as the ability to set the program counter, blocking or unblocking, or particularly changes in scheduling priorities.
A particular problem for the proper enforcement of security policies at a sufficient level of granularity is the widespread use of impersonation or aliasing in common operating systems [624, 625, 626, 627].
While this significantly reduces complexity of operating systems for both node-internal and particularly for networked operations, the granularity obtainable by this mechanism is frequently insufficient.
Instead, source authentication and impersonation transparency are suitable mechanisms for enforcing security policies; the requisite information is typically obtained and transmitted orthogonally to existing mechanisms.
Enforcement mechanisms must also ensure that any and all interprocess communication is in compliance with all security policies.
This includes source identification and authentication mechanisms for all communication channels as well as the identification of recipients at the policy level though not necessarily at the level of individual processes or applications.

9.2 File System Enforcement Mechanisms

The following sections provide a brief outline of the implementation mechanisms used for realizing a file system enforcement module by retrofitting the Microsoft Windows NT operating system family.
Section 9.2.1 describes the relevant subsystems involved while section 9.2.2 is concerned with the implementation itself.

9.2.1 File System I/O Structure in Microsoft Windows NT

As outlined in section 6.4, the Microsoft Windows NT family of operating systems [675, 676] exposes several APIs via environmental subsystems. While these APIs are largely procedural in nature, the internal processing is asynchronous and packet-based in nature.
In this regard, it shares more with OpenVMS [259] than with Unix [265], although one major difference to OpenVMS is that, like Unix System V Release 4 and later derivatives, it has a unified file system cache and virtual memory architecture.
Regardless of which environmental subsystem is used, the I/O operation eventually results in a call to the system service dispatcher in kernel mode. This dispatcher handles the distribution of the operations into the various kernel components. For the discussion of file system mechanisms, only some components are of interest.
Besides the I/O Manager, the Windows Management Instrumentation (WMI), Plug and Play (PnP) Manager, and the Power Manager components (these appear only with Microsoft Windows 2000 and later revisions) are also relevant for device level operations.
The central component for file system operations, however, is the I/O manager. It creates I/O request packets (IRP) from incoming requests[1] and ensures that all drivers for which an IRP is relevant are called with the IRP in the proper sequence. Each IRP sent to a kernel-mode driver represents a pending I/O request to that driver. An IRP will remain outstanding until the recipient of the IRP invokes the `IoCompleteRequest()` service routine for that particular IRP. Invoking `IoCompleteRequest()` on an IRP results in that I/O operation being marked as completed, and the I/O Manager then triggers any post-completion processing that was awaiting completion of the I/O request. Each request must be completed exactly once.
This mechanism lends itself to a layered processing approach in which IRP messages are cascaded across several driver layers (pos-

[1]With the exception of Fast I/O which bypasses this step, the mechanism appears to be loosely patterned after the OpenVMS concept by the same name although contrary to the latter implementation it introduces a significant number of additional code paths that need to be dealt with.

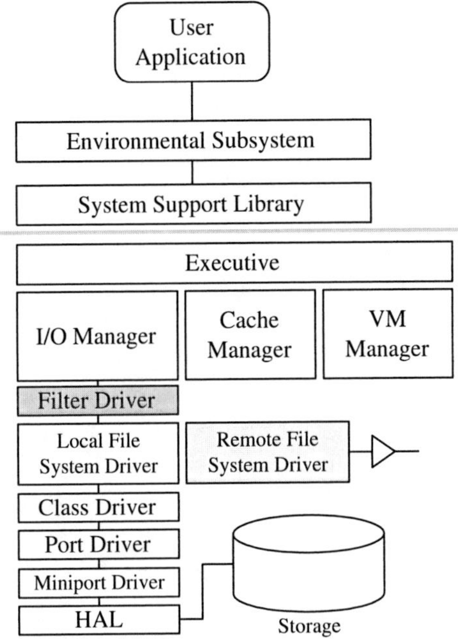

Figure 9.1: Components involved in Windows NT file system I/O

sibly with additional IRP messages created during the course of processing at lower levels).

As a side effect of this architecture, one can alter the functionality of the operating system by interposing additional layers in the driver stack. One example of such an interposition is shown in figure 9.1.

The placement of the filtering layer in figure 9.1 has the advantage of such a module being able to intercept and operate on generic (file-system independent) operations from upper operating system layers. This type of filter is called a file system filter driver in the terminology of Microsoft Windows NT. Most importantly, the depicted interposition layer allows operations on the file level.

Common disk encryption mechanisms typically work by adding special disk drivers or lower level filter drivers; as a result they are dependent on specific hardware or are not able to work on individual files; in addition, they do not support remote file systems.

While handling files individually entails a significantly higher complexity, it is necessary to support the semantics found in the system described here. Interposition at this level is also largely[2] oblivious to the type of file system.
Microsoft Windows NT does not fully adhere to the packet-based I/O model for all types of drivers, though. A special case exists in case of file systems, therefore also for file system filter drivers.

This exception is the Fast I/O mechanism; here the I/O Manager, Cache Manager, and the various file system implementations (if they support this mechanism) interact by means of explicit cross-module calls instead of creating IRP messages.
This performance enhancement adds considerable complexity to the design of any file system filter drivers since additional communication paths must be handled.
While it is possible for a driver (particularly a filter driver – this has the result that lower-level drivers are also not confronted with Fast I/O for a given call) to signal that Fast I/O is not supported with the result that an equivalent request is created in the form of an IRP and sent again by the system service dispatcher, the double performance penalty thus incurred is not justifiable.

Microsoft Windows NT uses the filter driver mechanism — not only for file systems — itself to support additional functionality that is optional or can be made available for different file system types with a single driver.
One example beginning in Microsoft Windows 2000 is the Single Instance Store (SIS) file system filter driver that conserves disk space by removing multiple copies of a file and replacing them with links to a single shared copy in a common directory.

[2]There are some differences in behavior for remote file system redirectors that need to be addressed due to an inconsistency in the operating system design.

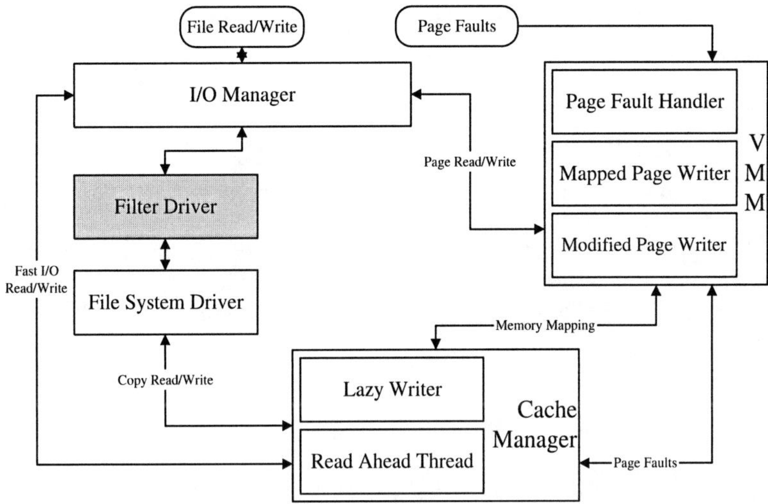

Figure 9.2: Interactions between file system components in Microsoft Windows NT

Another application example of a file system filter drivers is a malware scanner; again, this type of application requires access to file system semantics.

Some care needs to be taken in case a file system filter driver modifies data on underlying file systems since the unified cache and VM architecture results in Fast I/O requests bypassing the regular file system (filter) and accessing the cache directly as shown in figure 9.2. Failing to update all relevant access paths could thus lead to inconsistencies.

Some additional details can be found in [493, 587], although regrettably most of the internal interfaces that need to be supported are largely undocumented.

It should, however, be noted that apart from issues arising from defects removed and occasionally introduced and some minor additions, the internal file system APIs of this platform have remained relatively stable despite major changes in the portions of the system visible to most users and developers.

9.2.2 Implementation of the File System Enforcement Module for Microsoft Windows NT

This section discusses several aspects of implementing the transparent file system security mechanism by means of inserting a file system filter driver into the operating systems of the Microsoft Windows NT platform.

9.2.2.1 Structure and Bootstrapping

Each kernel mode driver must provide a `DriverEntry()` function that is called by the I/O manager on driver load.
This function performs initialization operations of the driver such as reading parameters from registry settings, allocating data structures, `DeviceObject`, and `SymbolicLink` objects, and initializing the call table (`MajorFunction` table).
The `MajorFunction` table is a list of dispatch points supported by the driver. Each I/O request is packed into an IRP by the I/O manager and contains all information describing the request including the desired operation (the `MajorFunction`). The driver may set an entry in this table for each `MajorFunction` it wishes to process with its own dispatch functions.
The enforcement module driver calls on the ECRM for informing it about activation of instrumentation points as well as to inquire about policy rules for permitting operations at instrumentation points to proceed.
While the ECRM interface is logically a separate kernel module, it can be accessed directly by means of inter-module calls instead of using an IRP-based mechanism; this eliminates most issues regarding reentrancy and permitted IRP levels for calls and is also beneficial in terms of achievable performance.

Therefore, no operations related to policy determination or e.g. cryptographic operations on file objects are visible outside of the ECRM.

If such operations are delegated to a tamper-resistant cryptographic coprocessor, this ensures that the policy decisions and key material used for enforcing such decisions cannot be modified unless considerable effort is expended compared to a software-only variant.

The filter driver mechanism must be logically located above the driver modules whose behavior it wishes to change, augment, or replace.

It is also possible for such a filter driver to create new calls to (among others) such lower layer drivers as may be the case if some more elaborate information or modification to the file system are required.

The filter driver described here is located above the file system and is therefore able to work on and identify individual files and does not deal with individual device types.

File system filter drivers operating below the file system driver level (but above storage class drivers) are somewhat misnamed as they are only capable of working on amorphous data blocks without file system semantics in the calls reaching such layers as file operations are broken up into block-based operations by the respective file system drivers.

The ability to operate on files allows the implementation to gather information on the entities wishing to perform the respective operation and forward this information to the ECRM for further processing. If necessary, each individual file can therefore be treated differently if possesses an identifying feature and the security policy or security policies to be applied in such a case dictate this behavior.

To ensure that security policies are enforced uniformly, all file systems on a node must be intercepted and brought under the control of the enforcement module.

This is achieved by registering a callback function with the I/O manager which is called whenever a file system is loaded. This ensures that the filter driver can attach itself to all file systems, even those that are loaded dynamically after booting. Dynamic loading of file systems can, for example, occur when removable media are loaded.

An implication of this is that the filter driver must be loaded prior to all file systems.
This can be achieved by assigning it either to the `Filter` driver group or associating the necessary tag value with it in the registry settings for the driver load sequence.
The only file system for which such a filter driver is not notified of a load event is the `Raw` file system (permitting access to the raw device without any file system semantics interpretation). In this case the filter driver must attach itself explicitly to this file system.
Another exception from the notification mechanism that must be dealt with explicitly is the LAN Manager redirector used for accessing network shares. This, however, appears to have been an oversight by the developers of Microsoft Windows NT since this behavior is no longer observable since Microsoft Windows 2000.
Once the notification callback is called, the filter driver can attach to the file system or file system recognizer, respectively, and is then able to intercept the file system control requests (with the minor functions `LoadFS` and `MountVolume`) and attach itself to mounted volumes. Once it is attached to a mounted volume, the filter driver can intercept all necessary I/O requests.

9.2.2.2 Considerations for Modified Read and Write Behavior

As a consequence of the unified virtual memory and file system architecture in the Microsoft Windows NT operating system family, it is not sufficient to modify only the behavior of read (`IRP_MJ_READ`) and write (`IRP_MJ_WRITE`) operations.
Doing so would lead to a partially encrypted (or otherwise modified by the filter driver) cache since in addition to simple read and write requests, memory mapping operations (e.g. used for mapping executable files into memory) would go a different route from ordinary operations.
In case of smart read-ahead by the file system or of explicit mapping, there would also arise a possible inconsistency. Paging requests therefore must also be handled explicitly.

9.2.2.3 File Size Considerations

The actual file sizes on the file system that is intercepted by the filter driver and the file sizes reported to the upper levels of the system and eventually to the user-mode API may differ from one another. There are two reasons for this. One is that it is necessary to maintain an in-line object label identifying the data object (file) for administrative purposes. This could be handled more elegantly for file systems supporting multiple concurrent data streams per object such as NTFS, but since even within the confines of the Microsoft Windows NT operating system family several file systems (such as removable media, e.g. floppy disks using the FAT or CD-RW media using the UDF file system) and network file systems (e.g. NFS file servers) do not necessarily offer this feature, the more cumbersome in-line mechanism must be used to ensure interoperability and complete coverage.

The second reason is that in case an encryption mechanism or similar transform is applied to a data object, the algorithms used may dictate padding to a certain multiple of bytes; the end of the padding must also be stored in-line.

Neither the object label nor the padding data size change may be exposed to the upper levels of the driver and executive architecture. As a result, the file system filter driver must adjust the file size reported by the underlying file systems in IRP messages such as `IRP_MJ_QUERY_INFORMATION` possibly used for obtaining size as well as `IRP_MJ_SET_INFORMATION`, which could be used to adjust file length information, and the similar `IRP_MJ_DIRECTORY_CONTROL` and `IRP_MJ_QUERY_DIRECTORY` requests.

The padding and object label information is stored at the end of the file stored on the lower-level file system. This avoids complications with memory-mapped files and continuous offset adjustments that would be necessary if the information were to be stored as a header since otherwise page-sized requests from upper levels of the system would in fact straddle page boundaries and thus incur a significant performance penalty. To the file systems below the filter driver, however, the data looks like the contents of an ordinary file.

Another issue the filter driver must deal with is the locking of files by applications since it may have to modify the file as a result of actions by upper layers that would not require file modification without the presence of the security mechanisms. To be able to access such files, a so-called locking key must be known. Therefore, `IRP_MJ_LOCK_CONTROL` requests also must be intercepted.

9.2.2.4 Information Gathering and Processing

The processing of `IRP_MJ_CREATE` requests is of critical importance as an instrumentation point. This IRP is issued when a file is accessed for the first time (i.e. not just for file creation) by an upper level function, so it is possible to perform a number of bookkeeping tasks at the same time as the opening.

For this purpose, the filter driver must itself issue IRP messages to the subordinate file systems, as well as communicate with other layers and modules within the security system. Subordinate requests are usually necessary to read in the object label (if present) at the end of the file.

Based on the object label, the object is classified within the internal bookkeeping mechanisms together with other relevant information regarding the file (e.g. files or memory maps shared with other processes) and further processing can be determined based on the applicable security policies as determined by the ECRM (e.g. whether transparent en-/decryption on subsequent read/write requests is required).

The information on the file stored in this step is referenced in any further processing of the file since gathering the necessary information would — even disregarding performance issues — not be possible due to restrictions on subordinate IRP messages especially during paging operations.

Another request that must be dealt with is `IRP_MJ_CLEANUP`. This request typically precedes the closing of a file. In case the object label itself must be changed or if the size of a file has changed, the new object label gets written during the processing of this IRP.

The IRP_MJ_CLOSE request must also be processed since the internal bookkeeping data structures associated with the file must be released.
This information gathering mechanism can be triggered implicitly by opening files and then querying the filter driver from the central policy enforcement mechanism; in addition, explicit functions such as the calculation of signatures over applications can also be triggered at the kernel level without interactions with any user level components.

9.2.2.5 Selective Use of Enforcement Mechanisms

Even though the enforcement mechanisms are loaded very early during the boot process, there are some files in Microsoft Windows NT which must be processed prior to the driver becoming active (e.g. NTLDR.EXE, NTDETECT.COM, and PAGEFILE.SYS). This means that such files must not have an object label and also must not be encrypted. However, other processing, such as signature verification which does not involve intrusive changes to the files is still possible.
Another file class that may not have an object label and generally should not be encrypted is paging files. The latter is mainly due to performance reasons; depending on the threat model it might be acceptable to wipe the paging files on each — orderly — system shutdown.
In addition, security policies might stipulate that certain files, directories, or even entire volumes (volumes are a subcase of directories) are not subjected to labeling and encryption.
For this exclusion mechanism to work, the full names of the files must be kept at hand; since the Microsoft Windows NT operating system stores the device identity separately from the path and in a different format inside the file object, the filter driver must maintain a translation table to minimize overhead during lookup comparisons.
An additional complication arises in conjunction with removable media; simply querying the root directory without a medium being present would result in an error condition. Instead, one can resort to an undocumented function for querying the properties of symbolic

links; this is a valid practice since the "drive letters" in Microsoft Windows NT are in fact implemented as symbolic links.

9.3 Network Enforcement Mechanisms

The following sections provide a brief outline of the implementation mechanisms used for realizing a network system enforcement module (itself consisting of a family of interrelated sub-modules) by retrofitting the Microsoft Windows NT operating system family. Section 9.3.1 describes the relevant subsystems involved while section 9.3.2 is concerned with the implementation itself.

9.3.1 Network Subsystem Structure in Microsoft Windows NT

Unlike the file system mechanisms described in section 9.2, the networking mechanisms provided by the Microsoft Windows NT operating system family do not share a common abstraction for network communication.

Therefore, in addition to multiple environmental subsystems providing different access mechanisms to network communication subsystems, there exist several conceptually different networking application programming interfaces, namely

- WinSock
- Named Pipes
- Mailslots
- Remote Procedure Call
- NetBIOS
- Telephony

Other services such as DCOM [186] or the .NET framework [700, 388] may be layered on top of these interfaces; while some of these interfaces have their own security and encryption mechanisms (such as RPC), others rely on the connection being assumed as secure and simply enforce access controls (e.g. named pipes and mail slots which are implemented as file systems and can use the access control mechanisms for file systems, see [675, 676]).

Of these mechanisms, the telephony interfaces (TAPI) are in an unique class based on the mechanism used by user level programs to communicate with kernel-level components.

The TAPI user level component (`TAPISRV.DLL`) provides access to a number of TAPI service providers (TSP); while most of these map to networking subsystems discussed later in this section, this also includes direct access to device drivers for modem devices (which can themselves be used to establish arbitrary network connections including interfacing to other network protocols).

This particular component therefore requires specific enforcement mechanism support (e.g. in the form of device-level enforcement for modem-type devices) to avoid the introduction of unenforceable information flow paths.

In the general case, the network architecture of the Microsoft Windows NT family consists of a number of layers. At the lowest level is the physical device, access to the device is regulated by the hardware abstraction layer (HAL). Device drivers are realized as NDIS (Network Driver Interface Specification) modules consisting of the generic NDIS library and the device-specific NDIS miniport drivers; the library fully encapsulates the miniport drivers.

Accessing the NDIS library is the TDI (Transport Driver Interface) mechanism. This itself consists of transports (or protocol drivers), supporting the various transport mechanisms such as NetBEUI and TCP/IP, and TDI clients which provide services for sockets and NetBIOS calls. None of these modules can be called directly from applications since they are protected kernel mode interfaces.

Upper-level APIs such as NetBIOS and Windows Sockets are implemented at the user level. Figure 9.3 on page 258 schematically shows the components involved in the use of one such interface, namely the Windows Sockets interface.

The Windows Sockets API (or WinSock) is modeled after the original BSD Unix sockets API [464] and has undergone significant revisions

under various platforms before arriving in its current form [23, 24]. It is available for both the NT-based and DOS-based operating systems from Microsoft Corporation.
The Windows Sockets API is itself composed of several modules. From an application's perspective the sockets API consists of the exposed API DLL; this DLL communicates with the SPI (Service Provider Interface) layer.
This layer is controlled by the transport service provider DLL which in turn calls on a number of transport helper DLLs and namespace helper DLLs to perform its operations.

Moreover, the transport service provider DLL forwards the thus generated calls to the System Support Library DLL that represents the interface to the abovementioned kernel components.
Since the Microsoft Windows NT design is predicated on a file system model and represents sockets as file handles, a translation mechanism is required. This service is performed by an Ancillary Function Driver (AFD).
Of particular interest in this is the ability to stack several of the transport helper DLLs so as to provide additional services at each level (there is no layering mechanism for namespace helper DLLs). WinSock here distinguishes between "base protocols" and "layered protocols".

The former are protocols capable of performing actual communication with a remote endpoint, the latter must rely on base protocols for actual communication and only provide added value.
Provided that all elements of such a stack are conforming to the interface specifications set forth in [23, 24], it is possible to implement several stacked layers of such layered protocols.

At an abstraction level below the user level API mechanisms, the protocol driver layer accepts requests from API-level mechanisms and translates these into respective network protocol elements.
The number and type of protocol drivers are variable among nodes and may include but are not limited to TCP/IP, NetBEUI, IPX/SPX (provided in a single protocol driver instance), and AppleTalk.

Typically, each protocol driver supports all protocols of a protocol family (e.g. IP, ARP, RARP, ICMP, IGMP, UDP, and TCP in case of the TCP/IP protocol driver, `TCPIP.SYS`).

All protocol drivers communicate with API-level components (as well as other components such as the previously noted Windows Sockets ancillary function driver) using part of the TDI; which is specified in the form of IRP classes.
For connection-oriented protocols, `TdiDispatchCreate` creates a file object (also referred to as an address object) by through the use of an `IRP_MJ_CREATE` IRP which represents the node-local connection endpoint. This subsequently must be associated with an opened file object representing an address, referred to as an connection object.

Depending on the initiating node, subsequent IRP messages must then transition the connection object into listening or connecting state, which is then transitioned into an accepting state on the part of the listening node, which occurs using the IRP creation mechanisms `TdiDispatchDeviceControl`, `TdiDispatchFastDeviceControl`, and `TdiDispatchInternalDeviceControl`, respectively.
After a connection object has been discarded, `TdiDispatchClose` is used to discard the address object after `TdiDispatchCleanup` has ensured that no pending IRP messages exist for the address object; connectionless protocols omit the listening and connecting phase of this control flow.

TDI also permits the use of callback mechanisms and the intermediate caching of network protocol data units for efficiency purposes; this requires the registration of events with TDI client interfaces. Typically, this results in TDI clients generating `TDI_SEND` IRP messages and reacting to `TDI_EVENT_CHAINED_RECEIVE` and the `TDI_EVENT_RECEIVE*` family of IRP messages.

For communication with the device drivers controlling the network interface adapters, the TDI protocol drivers communicate by way of a library encapsulating device-specific properties.
This library, NDIS, provides a procedural interface for the TDI as well as for the actual device drivers (miniport drivers), which com-

municate to the remainder of the operating system only through the NDIS library.

Internally, however, the Microsoft Windows NT implementation of NDIS itself uses IRP-based messaging for control flow. The NDIS library provides services for both connectionless (e.g. IP) and connection-oriented (e.g. ATM) protocols as well as a number of other services [676].

NDIS also provides several other security-relevant services that need to be addressed, such as the ability to forward datagrams from one network interface to another without processing by the remaining operating system network protocol stack or the offloading of certain network processing (specifically TCP/IP-related operations) to the network interface device and hence the NDIS level.

9.3.2 Implementation of the Network Enforcement Modules for Microsoft Windows NT

The provision of the semantics appropriate for a network enforcement layer requires the insertion of instrumentation at least at two of the protocol layers described in the preceding section, namely at the NDIS and TDI layers.

9.3.2.1 NDIS Instrumentation

With the exception of parts of TAPI discussed in section 9.3.1, all network traffic within the Microsoft Windows NT family is transmitted by way of NDIS devices, regardless of the API and protocol used; it is also possible for a user level process to directly communicate with the NDIS layer (again, TAPI is an example of this behavior).

It is therefore imperative for the provision of the required interpretation semantics to intercept and instrument the processing at the NDIS level. For this purpose, several implementation alternatives exist, two of which are of sufficient generality for the purposes discussed here.

One possible approach is the use of an NDIS Intermediate Driver, which permits the interpositioning of code between miniport drivers and the remainder of the NDIS library.

While appealing and providing a well-defined interface for interposition, this approach does not provide the most general mechanism since NDISWAN miniport drivers are not supported in the NDIS revision (version 5.0 and 5.1, respectively) used by Microsoft Windows 2000, XP, and 2003.

This would require the mandatory use of backwards-compatible NDIS version 4.0 mechanisms, which for obvious reasons is highly undesirable given the improvements and features added in NDIS version 5.*x*.

The alternative to intermediate drivers providing the most general coverage of mechanisms supported is in the manual interception of control flows destined for and within NDIS.

Since it is possible that the configuration of both protocols and network interfaces may change at any time during runtime (analogous to the file system handling discussed in section 9.2.2.1), a general mechanism is required that supports not only bootstrapping mechanisms but also provides monitoring and dynamic interception of such configuration changes.

For this purpose, an NDIS layer enforcement sub-module can be loaded and started prior to the initialization of the network subsystem; the configuration mechanisms is analogous to that discussed in section 9.2.2.1.

Unlike the mechanism described in section 9.2.2.1, however, the interception cannot be effected by registering with the I/O manager and redirecting the flow of IRP messages, but must occur directly by redirecting function entry points to the enforcement sub-module itself and subsequent transfer of control flow to the NDIS library once the required operations have been performed on the part of the enforcement sub-module.

To ensure that policies can be enforced uniformly, all network interfaces on a node must be intercepted and brought under the control of the enforcement sub-module.

This occurs by intercepting the NDIS functions `OpenAdapter` and `CloseAdapter` and tracking the activation and deactivation of any (virtual) network interface; the actual interception mechanism relies on modifying the addresses contained in the export table of the module providing the NDIS library upon loading of the NDIS module.

Similarly, to be able to track information and control flows — particularly for callback mechanisms — the enforcement sub-module must retain information on which protocol drivers are registered with (and hence may access) the NDIS layer. This is accomplished by intercepting the `NdisRegisterProtocol` functions for registration and, correspondingly for unloading and deregistration, the `NdisMRegisterUnloadHandler` and `NdisDeregisterProtocol` functions.

The information thus obtained permits the correlation of information and subsequent coordination with instrumentation provided by enforcement sub-modules at the protocol driver level discussed in section 9.3.2.2.

While NDIS is the proper location to capture all control and data flows pertaining to network traffic and therefore also to perform protocol-specific operations, the information available at the level of the NDIS library (and hence the interception mechanism) are severely limited.

At the level of the NDIS library, it is not directly possible to identify the subject (i.e. process) a data flow is associated with since the data flow from a process directed towards NDIS is translated into IRP messages at the kernel level, thus obliterating the information on the subject.

Conversely, data flows directed towards subjects are not associated with processes directly, but only with protocol drivers.

It is therefore necessary (as described in section 9.3.2.2) to correlate information regarding the subject association with a data flow by coordinating the information available at the NDIS level with information from higher abstraction levels.

A similar problem exists with regard to the payload of the individual data flows processed by NDIS. At the NDIS level, only data already

processed into protocol data units (PDU) are presented, and NDIS is expected to operate opaquely on the data provided in either direction. In this case, the information as to which protocol is associated with a PDU is obtained indirectly through the information gathered on registration of protocols.
A list of known protocol drivers must be maintained (this can occur through known identifying characteristics within the protocol driver or indirectly through the file system enforcement mechanism providing an unique fingerprint for a given protocol driver by way of the ECRM), and protocol-specific operations must be invoked on the PDU based on the information thus obtained.
One such mechanism (for the TCP/IP family of protocols) is discussed in section 10.2; for the purposes of the NDIS enforcement sub-module this processing is opaque.
However, since PDU may be constrained either by the respective protocol or by the network interface, it is not always possible to transform PDU in place. Instead, a given PDU (regardless of inbound or outbound processing) may result in several PDU after processing by the protocol-specific enforcement sub-module and, moreover, the protocol-specific enforcement sub-module can withhold the processed PDU (and hence process additional PDU from the same data flow) prior to emitting one or more PDU for further processing by the NDIS layer.

The necessary information for identifying subjects, objects, and operations are transmitted by the TDI sub-module as discussed in section 9.3.2.2.
Individually, the instrumentation provided by the NDIS layer enforcement sub-module can monitor the activation and deactivation of protocols and adapters as well as monitor in- and outbound data flows, including the elimination of inbound traffic as well as outbound[3].
Another operation that can be performed by the NDIS layer without interoperation with other sub-layers is data flow normalization, i.e.

[3]The `NdisCancelSendPackets` and `NdisGeneratePartialCancelId` mechanisms are, while not strictly necessary for this purpose, supported only from NDIS 5.1 onward.

providing well-defined temporal characteristics for all or selected data streams such as inter-PDU time intervals. This, however, requires potentially large buffers in the absence of flow control mechanisms that can be applied transparently to the communicating parties proper.

9.3.2.2 TDI Instrumentation

While the enforcement mechanism itself is located at the NDIS level as described in the preceding section, the implementation of the Microsoft Windows NT operating system family necessitates the addition of a further enforcement sub-module at the protocol driver level.
The need for this additional sub-module stems from the lack of information regarding the association of subjects (and potentially of operations) as well as of objects of higher abstraction levels from which a given object or PDU is derived at the NDIS layer.
However, as noted before, there are potentially multiple protocol drivers active within a given node, each of which requiring specific actions for deriving the requisite information for reaching policy decisions by the ECRM in conjunction with other sub-module information. For the purposes of this dissertation, the discussion concentrates without loss of generality on the TCP/IP protocol driver.
Interception of the protocol driver occurs analogous to the mechanism described for the NDIS library in section 9.3.2.1; entry points are dynamically redirected on initialization of the protocol driver and forwarded after processing. As with the NDIS layer, this facilitates dynamic addition and removal of protocol drivers at runtime provided that the proper enforcement sub-module for a given TDI protocol driver is available.

The main operation performed at the TDI enforcement sub-module is the collection of information regarding subjects, objects, and data flows (as noted in section 9.3.1, the latter information is available implicitly through the observation of calls to the TDI); subject (i.e. process information that can be correlated with other subject information at the ECRM) information is implicitly available through the calling mechanism.

In case of an outbound data flow, the information thus gathered must be made available to the NDIS sub-module to permit proper processing.

While it would be conceivable to transmit this information out of band or to store it at the ECRM itself, both possible alternatives would require not only considerable storage, but also imply complex storage management since the processing order is not necessarily the same for data flows at the TDI and NDIS layer, and special cases such as canceled data flows would need to be taken into account to avoid stale storage.

To avoid these problems as well as performance issues arising from extraneous communication between sub-modules (typically in the form of IOCTL messages that require considerable processing overhead), data flows can be annotated in-band with the requisite information. The NDIS sub-module can extract this information[4] and continue processing as described in the preceding section.

Similarly, inbound data flows can be reverse-associated with the information regarding subjects, objects, and data flows. This requires one instance of communication between the NDIS and TDI sub-modules for each flow (in the worst case of connectionless protocols, this is once for each PDU, although heuristics and information from other sub-modules such as the one described in section 9.3.2.3 can be established to identify virtual flows based on addressing information in the more general case of connectionless protocols).

9.3.2.3 Windows Sockets Instrumentation

As an example of an user level enforcement sub-module, additional application-level instrumentation can be realized, providing a mechanism for precise control over API behavior.

[4] There exists a mechanism for this purpose in the `NDIS_PACKET_STACK` structure introduced in NDIS 5.1; prior NDIS versions require the allocation of a new, larger packet for the integration of this data.

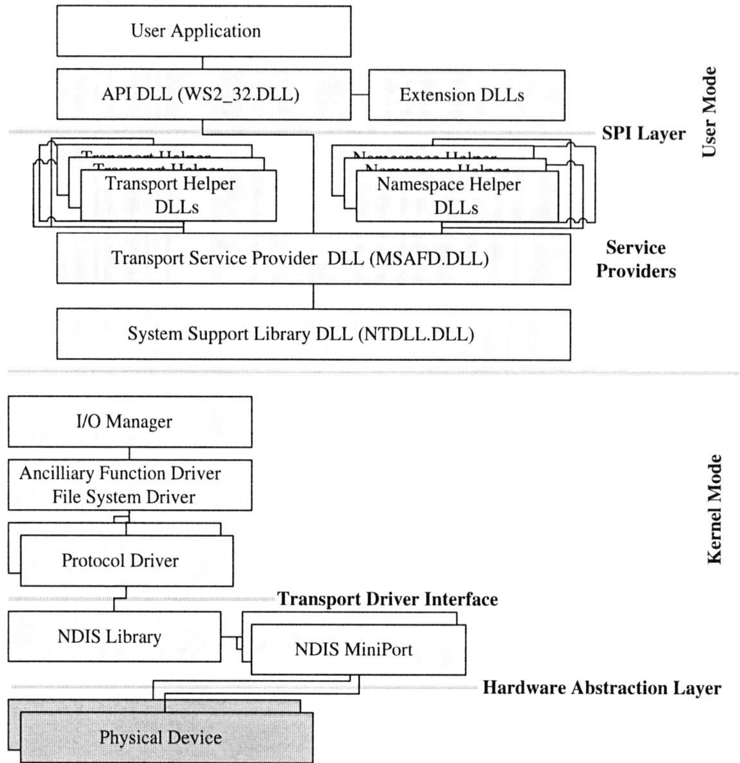

Figure 9.3: Components involved in WinSock networking mechanisms in Microsoft Windows NT

In case of the Windows Sockets API, there exists a mechanism for additional interposition in the form of a WinSock Layered Service Provider (LSP) which provides full access to the processing environment at the user level, which is particularly relevant for information gathering instrumentation.

At the same time, the fact that such a sub-module is located at the user level implies that it is largely unsuitable for actual enforcement since a hostile user level component or application can compromise such a sub-module, although integrity protection mechanisms can limit this vulnerability.

A Layered Service Provider must announce itself to the Windows Sockets API by registering a new Provider Catalog Entry; this catalog entry can be considered the new layer.

The ordering of processing within the Windows Sockets API subsystem in the presence of multiple layered service providers, additional Service Provider Chain Catalog Entries have to be created. In the example case discussed here, this is performed for both TCP and UDP protocols.

To ensure that the a given Layered Service Provider cannot be registered twice, each of the Provider Catalog Entries is assigned a unique catalog number (GUID). This mechanism as well as the deregistration is part of the provider and can only be called by a thread with administrative privileges.

The mechanism described here is of interest mainly as a speed optimization mechanism for information gathering and does not replace the sub-modules discussed previously.

The mechanisms discussed here require the presence of a system call interface for the support of overlapped I/O, which was not properly implemented until the release of Microsoft Windows NT 4.0 Service Pack 4.

An example of the use of this mechanism is the in-band analysis of application-level encryption such as the SSL/TLS protocol.

Unlike the NDIS and TLI layers, the Windows Sockets API is presented with a pre-assembled stream which can be intercepted, analyzed, and forwarded efficiently and which is not constricted in the way that recursions in the network stack are not possible due to interrupt privilege level conflicts.

This permits the inclusion of application-specific protocols and data used therein in the definition and enforcement of security policies to which otherwise only computationally complex and storage-intensive analysis would provide insight which are likely to be considered unacceptable in the performance and service timeliness degradation thus introduced.

CHAPTER

TEN

Distributed Firewalling and Intrusion Detection

10.1 Environmental Analysis

The policy-based mechanisms at the network protocol layer described in section 9.1.2 can be employed to exemplify the capabilities inherent in the security architecture and its layered enforcement mechanisms. The capabilities described here are commonly associated with firewalling mechanisms in that they permit the specification of permitted network traffic at a very fine level of granularity, but extend significantly beyond this as will be described below.

As described in section 3.4.3, this is necessary due to the changes in the use of computer systems that have rendered topological network protection schemes largely obsolete and require protection mechanisms that can handle cryptographically or otherwise tunneled traffic and advanced protocol semantics.

The network security architecture assumed by the systems described in [484, 129, 636] stipulated networks segregated into internal levels of various trust levels as well an external, untrustworthy networks. Protection of the individual layers occurs using firewalls or guard mechanisms (see section 3.4.1). These mechanisms rely on several assumptions, typically tacit, for their validity:

- The internal (protected) network is trustworthy.
- There is a meaningful distinction between internal and external networks.
- Attacks are initiated from external hosts directed at internal networks.
- The syntax and semantics of the protocols used are known.
- Code passing through a firewall is executed with the consent of a user.

While all of these assumptions were true at some point and are reflected in the design of firewalling mechanisms, most of them gradually lost their validity at some point in time; any new mechanism for countering the current threat environment may no longer make use of them. This section highlights some of the threats that need to be countered by security policies in the network area. It is not, however, a complete risk analysis and merely intended to illustrate the raison d'être of fine-grained distributed policy enforcement.

10.1.1 Mobile Devices and Remote Access

The topology of a network can no longer be used as the linchpin of its security [794]. Bellovin discussed this mainly in terms of performance [66], but network structures including mobile devices and remote access solutions render this if not obsolete then at least secondary. It must be assumed that a mobile device may, possibly using a transitive and therefore unanticipated network connection, contact sensitive network nodes. A similar scenario results if a user has established a remote access connection to a protected network and also has for some reason or accidentally established another network connection, possibly opening a routable connection to the Internet bypassing all protection and auditing mechanisms in the process.

10.1.2 Mobile Code

Many current applications include scripting mechanisms of varying levels of expressiveness; in addition, mechanisms such as Sun's Java or Microsoft's ActiveX, COM, COM+, DCOM, and .NET, permit the transmission and execution of full applications. While most of these mechanisms contain security mechanisms, the track record in using these is not encouraging if only because deactivating e.g. macro language mechanisms would render some documents unusable and exposure of sensitive services can be provided outside of a benign environment (e.g. components originally intended for protected internal use only) inadvertently.

A threat assessment including systems based on such technologies must therefore assume that malicious code can reach a sensitive network node.

Trojan horses of various levels of intricacy are another threat that can be classified as mobile code; given that the execution of such code is frequently caused by a social factor, technical means probably cannot fully prevent such infiltration of sensitive nodes but merely limit their impact.

10.1.3 Tunneling Mechanisms

A number of application programs can transmit data and instructions across transmission channels designed for different purposes. In some cases, this is done surreptitiously by application programs, in other cases the tunneling occurs deliberately.

The most notable recent example for such a mechanism is the SOAP [101] and mechanisms built on it such as .NET, following similar earlier developments in the CORBA area.

In both cases exposed application interfaces are called without intervening security mechanisms. The entire responsibility for security is placed on the developers of the thus exposed applications and their protocols and the administrators of such applications and components once these are deployed.

Even advanced application-level firewalls can not necessarily detect such mechanisms since the tunneled data conforms to the syntax of the "host" protocol except by resorting to ad-hoc detection schemes.

10.1.4 Application Complexity

Even ostensibly simple contemporary client-side WWW applications typically involves a complex set of protocols and applications ranging from advanced HTML parsers and display engines — which themselves may be vulnerable — to JavaScript / ECMAScript and may involve implicit operations and code execution on a client. Very little of this is perceived consciously by a user or can even be controlled at any meaningful level of granularity. This is further exacerbated in case additional services are integrated.

Distinguishing deliberate actions on the part of the user from implicit operations due to application behavior or malicious operations resulting from mobile code (see section 10.1.2) represents a considerable difficulty.

A related problem occurs when trying to specify permitted application behavior for intrusion detection mechanisms. Full specifications of generic COTS or even reasonably complex custom applications are very hard to obtain. Linking permitted operations on the part of either a user or of the application to observed behavior by the application will therefore frequently have to resort to heuristics or imprecise specifications.

In doing so one is confronted with missing significant events or having to accept a considerable false-positive frequency [374].

10.2 Network Security Mechanisms

The embedding of enforcement mechanisms into individual PENs, particularly at the network protocol and device interface layers, represents the foundation of the network security mechanisms found in the architecture discussed here.

As discussed in section 9.1.2, network security mechanisms, particularly when retrofitted to existing systems of limited assurance, must accomplish several objectives that are associated with different layers in the OSI reference model and implicitly require varying amounts of contextual information for this purpose.

The mechanisms for enforcing security policies at the network level (e.g. as described in section 9.3.2.1 for the Microsoft Windows NT family of operating systems) can, at the implementation level, be considered as multiple interlocking coroutines where control flows are initiated on receipt of network data flows regardless of direction, an arrangement that is colloquially referred to as a *bump in the stack* mechanism.

This implementation itself is largely dictated by the restrictions on interrupt-driven processing found in most operating systems for asynchronous type I/O processing.

The remainder of the networking mechanisms of an operating systems can remain completely unaware of the presence of the enforcement modules and, while dynamic changes in the data flow may occur, these are indistinguishable for both the remainder of the PEN and — unless explicitly required by policy — for entities communicating with a PEN.

The mechanisms to be used vary with the type of network protocol used; the main distinction required is between point-to-point protocols and routable protocols.

The former require a fully transparent interposition and potentially the retention of state information, e.g. if additional information such as subject and object identity and type labels need to be transmitted in-band or information flows need to be modified directly, changing size information.

In this case (e.g. for basic HDLC transmissions) information on frame numbers or retransmission windows for error conditions that cannot be efficiently off-loaded outside the enforcement module must be retained.

For routable protocols, particularly packet-based protocols, the state information to be retained can be minimized effectively by having the enforcement module operate as a router (depending on the protocol, this can also be an anonymous router or bridge) interposed between all communications although the retention of state information in the general case cannot be avoided completely (e.g. in the case of fragmentation of datagrams mandated by a protocol).

As noted in section 9.1.2, while adequate performance can be obtained only from a highly integrated mechanism, a discussion of the individual characteristics can conceptually separate the functional components.

At the most primitive level, the network enforcement module can restrict the expressiveness of data link and network layer protocols within the ISO OSI reference model.
This can serve multiple purposes, beginning with the protection of COTS operating system network stacks that cannot handle PDUs that are either malformed or legitimate but undesirable in that they may cause malfunctions by the COTS operating system network stack.
Examples of such behavior include the setting of flags or the use of certain options in communication protocols. As a result, only well-known (or, in the case of opaque systems, at a minimum tested) PDU formats are transmitted to and from the protected PEN.
Similar mechanisms can also be employed in the behavioral characteristics of protocols such as dynamic negotiation of capabilities. Here, the enforcement module can either prevent the exhibition (and in some cases exposure) of such capabilities or dynamically augment such behavior.

Such behavior is not only desirable for the protection of brittle or flawed network stacks but also for concealment and masquerading otherwise characteristic behavior due to peculiarities of a given implementation in interpreting a specification or outright implementation flaws thereof.
One behavioral characteristic that can be adjusted at the level of individual PDUs are the temporal characteristics exhibited; this is relevant both for denying an adversary information from traffic analysis and for limiting the bandwidth of covert channels.

The amount of state information required to be retained for this purpose depends, among others, on the variability that is permitted, whether flow control can be exerted or all PDUs have to be buffered, and whether k-th order derivative variations are also to be considered (e.g. rate of increase).

While such mechanisms can be realized statically, it is desirable to permit control over the modifications and constraints imposed by means of policy mechanisms.

This is particularly the case for non-trivial characteristics that may serve protocol optimization; policy enforcement can interfere significantly with the performance characteristics of a PEN node (e.g. fragmentation or the windowing and selective acknowledgment mechanisms in the TCP protocol [335, 336, 197, 447]).

Dynamic policies can also take individual PEN capabilities into account to limit the adverse effects of such policy requirements.

Depending on the network protocol employed, several other properties such as protection of confidentiality, integrity, and authenticity can be enforced at the PDU level. In case of the Internet Protocol used without loss of generality here for illustration purposes, this is the case with the IPSec mechanisms [701]; functionally similar end-to-end security mechanisms for the data link and also network layer can be employed in the case of other network protocols.

As with information on entities, the enforcement of using IPSec mechanisms requires additional communication between nodes.

However, by interposing the enforcement module as an intermediate non-bypassable routing mechanism, this can be accomplished transparently.

Independent of the operating system the mechanism is based on (e.g. as described in section 9.3.2.1), egress PDUs can be transparently modified in such a way as to fulfill policy confidentiality, integrity, and authenticity requirements and augmented with requisite information on entities and operations (while this is feasible at the PDU level, it is desirable for efficiency reasons to identify data flows and perform the transmission of entity and operation information only once per flow) [367, 368, 369].

The first step in processing the PDU in terms of the IPSec reference architecture [367] is to query the SPD to find which policy must be applied on a given packet.

The selector is constructed from the traffic informations found in the packet, namely the source and destination IP address, the transport protocol, and the source and destination ports when not being constrained to the IP protocol itself.

The SPD acts as an intermediate caching mechanism in lieu of direct queries to the ECRM and retains validity for short time intervals only. If no entry in the SPD exists, the ECRM must be queried directly with the decision result transferred implicitly into the SPD for further use.

The policy thus obtained can specify (among other operations and constraints) three different classes of behavior for the egress PDU, discard, relay, or further processing using IPsec.

If the PDU must be discarded or relayed, IPsec processing ends, although additional policy-derived operations (e.g. instrumentation point activation resulting in audit events and changes in the policy fact set). If the packet must be processed IPsec then there must exist an SA or SA bundle for the given traffic.

If that SA exists, then it is retrieved from the SAD. If no entry exists in the SAD, a new SA or SA bundle must be determined or negotiated, which can occur using the IKE protocol suite (see below).

The addition of data implicitly may require fragmentation of PDUs that exceed the MTU or PMTU for a given path which, while transparent to the source of the data flow, can not only impede the performance of the data transmission but may require the retention of additional state information and interaction with higher protocol layers to ensure proper transmission.

Analogous to the treatment of egress PDUs, ingress PDUs can be intercepted, analyzed in accordance with both the requirements of the IPSec mechanism itself (e.g. discarding PDUs whose required integrity protection has been violated) and additional policy requirements (e.g. auditing or the recording of information into the policy decision fact database). For PDUs belonging to data flows for which policy requires protection and therefore all ingress PDUs, the SAD is queried to retrieve the SA used to protect the PDU.

The IP destination address, the IPsec protocol and the SPI are used to query the SAD. If no SA exists then the packet is discarded and additional operations based on active security policies may apply. If, however, a matching SA is found, then the IPsec processing defined by the SA is applied and selectors are verified for matching those defined by the SA; if this is not the case the packet is discarded and additional operations based on active security policies may apply.

These two steps are iterated until no further IPSec processing is required; subsequently it can be verified that an entry policy matching the packet selector exists and that the policy thus specified has indeed been enforced.

Since it may not be possible to restrict communication to tunnel mode for all data flows, intermediate fragment reassembly at the level of the enforcement module can be required for these PDUs.

While this technically violates the requirements set forth in [49] when considering the enforcement module as a router, it is transparent to the remainder of the network stack and external entities. Similarly, the labels inserted into the data flows are transparently removed prior to forwarding to entities outside the enforcement module.

The information required to determine the permitted data flows and the required transformations on the data flows to comply with all applicable policies require positive identification of the remote node or nodes for the finest granularity policies to take effect.

For PEN and PCN nodes this is accomplished via the AH mechanism providing connectionless integrity and data origin authentication as well as replay protection, which provides end-to-end protection.

In terms of an IPSec implementation, this implies that AH with the replay protection option must always be part of an SA bundle associated with a data flow (regardless of the direction of the flow).

As a result, it can be established at the individual PDU level of an IP data flow whether a given communicating peer node is a PEN (or PCN). This positive identification and authentication is also possible with NOPEs that provide appropriate information and can be used to base efficient decisions on admissibility of data flows without reassembling data flows and analysis at higher protocol levels. Similarly, the audit information gathered from this type of data flow can also be based on unambiguous identification and authentication under the assumption that the trust and proof mechanisms employed have not been compromised. Between PEN and PCN as well as

PEN and PEN nodes, this yields a trusted channel (distinct from the ERM/ECRM and ERM/ERM channels).
For nodes identified as PENs or PCNs, data flows can, based on the establishment of the trusted channel, contain label data which provides information on entity identities and types. The upper bound on the communication complexity thus imposed is the transmission of single PDU flows that exceed MTU or PMTU restrictions; in this case the number of PDUs is doubled. For data flows which are identified as such, this overhead asymptotically decreases to zero.
The key exchange mechanisms required for the AH (as well as ESP) mechanism are irrelevant for the purposes of this discussion. However, unless pre-shared keys and SAs or SA bundles are used, it may be necessary to use a key exchange mechanism such as IKE [120, 550, 448, 290, 525]; in this case the trust domains of the host operating system (which may itself employ IPSec mechanisms if not aware of or configured to take the presence of the underlying security mechanisms into account) and the EM can be different, resulting in two separate key negotiation mechanisms (and, presumably, different keys for this purpose). The behavior of the IKE protocol is itself directly influenced by the general policy mechanism and can require the IKE component of the network enforcement sub-module to query the ECRM.
These mechanisms can, moreover, be used regardless of whether the remainder of the network protocol stack itself uses a mechanism such as IPSec although naïve implementations may incur a corresponding performance penalty. Furthermore, as noted in section 9.1.2, it also provides interoperability with NOPE nodes that do support IPSec mechanisms and assists in identifying and authenticating nodes. However, particularly policies concerned with information flows may curtail communication and data flows as no information on transitive information flows outside the policy-controlled domain can be given reliably.

At the transport level, virtual circuits (as may occur e.g. in case of the ATM communication protocols, used synonymously here with connections more common in TCP/IP parlance) permit that additional controls and information gathering can be applied based on policy requirements.

For this purpose, state information pertaining to the individual virtual circuits must be retained; unlike in the case of individual PDUs, this requires that the behavior of another protocol stack is mirrored or simulated unless the enforcement module is integrated with the network stack itself (i.e. typically the host operating system stack will need to be simulated).

This requirement for simulation of another finite state machine with an a priori unknown number of states and conjectured state transitions is problematic first in that the enforcement module must perform the simulation of each virtual circuit based on the protocol used (which must be known beforehand) and base any and all instrumentation and hence policy decisions on the protocol model [666].
This may result in an unnecessarily restrictive behavior as some state transitions of the simulated virtual circuit may not be permitted although they may be legitimate, otherwise the risk of desynchronization between the model and the subject of the model exists.

Second, the global state of a given virtual circuit may also depend on the local states and behavior (e.g. fault conditions encountered) of a number of unobservable nodes; depending on the protocol used (notably TCP), this requires that heuristics are to be used to determine some state transitions, particularly regarding the closure of given virtual circuits.
For both problems the use of a model for capturing and, where necessary, simulating behavior permits efficient and effective policy control and gathering of instrumentation information by mapping a number of states indistinguishable in the model used into equivalence classes.
The use of heuristics for determining or inferring state transitions may lead to the occurrence of transitions or lack thereof, however, it is trivial to ensure that any such application of heuristics satisfies safety conditions and stochastic fairness conditions if so required.

Both techniques are well known from network firewalling [120] and have been studied and deployed intensively.

Instrumentation points or groups as well as concomitant policy decision instances can therefore arise at this level within the model of virtual circuits or abstractions thereof that are independent of node-specific considerations as a side effect.

As with the data link and network layers, this mechanism is largely independent of whether the communicating parties are PEN/PCN or NOPE nodes, although the behavioral control and optimizations regarding the transmission of entity labels can occur only in case of PEN/PCNs.

The transparent interposition found at the transport layer can furthermore be extended to session and application layers, as is e.g. the case with network file systems exemplified in section 9.2.2.

While some generic approaches can be identified, most such interposition mechanisms are application-specific and are constrained similarly in deriving legitimate model state changes from observations that need not provide adequate information on the state transitions in the application program and protocol.

It may well be more effective, therefore, to integrate enforcement modules at the application layer level into the application itself and use mechanisms within the trusted environment of the security architecture to monitor this application program and all its components for potential manipulation that may impede the operational characteristics of the application-specific enforcement module.

However, a general mechanism applicable to the session layer exists in the transparent interception of transport- or session-level end-to-end cryptographic mechanisms, particularly of TLS/SSL and SSH [178, 120] in case of the Internet Protocol.

By dynamic interposition and hence being able to intercept material destined for application programs in plaintext, it becomes not only possible to perform policy-based operations based on the plaintext of the intercepted sessions itself but also of higher level protocols encapsulated in the protocol information itself, most notably of HTTP [210, 573].

Provided that enforcement modules for such higher level protocols exist, additional processing can occur at higher semantic levels.
Moreover, as a number of protocols use not only TLS/SSL or SSH but also HTTP (and particularly combinations thereof) as a transport protocol in itself, interception at this level can provide valuable information as well as policy enforcement, e.g. for protocols such as the SOAP [279, 111, 37, 120] and additional protocols and applications such as the .NET framework [700, 575, 388].
Given that the interception of application level protocols may require arbitrary length data streams for analysis and that both latency and state memory size considerations counsel against performing such reassembly and ordering below the transport level; an example of a layering mechanism that can perform such interceptions effectively was given in section 9.3.2.3.
The interception is, as in all cases, not limited to access or flow control and similar semantically invariant operations on the data flow (e.g. temporal flow control and auditing) but rather may also include the modification of both ingress and egress data streams if policies dictate so [566].

10.3 Assessing Intrusion Detection

The considerable popularity gained by intrusion detection [22, 43, 120] both in research and in the field together with continued reliance on systems of dubious reliability, survivability, security, and assurance can be a cause for concern in that the perceived benefits from such an approach may preclude necessary adjustments in the defensive posture by other means. In fact, it appears that the use of COTS systems even in sensitive areas together with a defense consisting mainly of firewalling and ID systems threatens to eclipse much of the progress made earlier regarding sound models of security and verifiable implementations [631].
The role of intrusion detection in the framework of the architecture described here is as one of the lines of defense which each node is equipped with. It is best considered as part of a defense in depth

strategy although it is conceivable that it is used without the benefit of policy enforcement as discussed in section 10.6.

10.4 Effects of Component-Level Instrumentation on Intrusion Detection

Fine granularity observations of system behavior have been shown to yield results particularly valuable for anomaly detection even without the use of computationally complex processing [757].
The policy enforcement requires mediation of all operations which can be performed on entities of the policy; one may therefore assume that the same subjects, objects, and operations are also the ones relevant to auditing and intrusion detection since otherwise a contradiction would be obtained.
The obvious drawback of such instrumentation is the volume of observations that must be handled; naïve centralized collection of all unprocessed observations from network nodes is not feasible in real time and impractical even for forensic purposes.
As a direct consequence processing at the source of the observations is required; this occurs under the control of the ECRM which is also the controlling entity for ID-related activities. This processing can consist of several elements which may also be used in conjunction:

1. The ECRM can modulate sensor output and level of detail based on existing and deduced policy rules; this is obviously more efficient than filtering sensor data post facto.

   While a certain performance loss is incurred due to the presence of the instrumentation point and determination of sensor activation, this penalty is significantly lower than postprocessing.

   Since the processing steps are controlled by the ECRM (or ERMs, respectively) the modulation can occur dynamically as required.

2. The collection of observations or arbitrary policy decisions based on such observations can induce the collection of derivative information from the same or other modules and components required for additional processing and decisions.

3. Based on active policies, one or more ERMs can be contacted. The information transmitted may include observation data as well as preprocessed data and derived observations.

   The method for transmission of such information is also subject to policy control and may stipulate transmission of some data with minimum delay while other data can be collected at the node and transmitted at a later point in time in a burst together with other data.

The efficacy for distributed systems requires the addition of another premise as the instrumentation described so far is insufficient for providing a common model of observations.

However, since the security policy mechanism described in chapter 5 provides the means for expressing security policies in terms of the equivalence class predicates regardless of the underlying interpretation (i.e. host operating system or application), provided such an interpretation exists, one can derive a corollary to this, namely that observations from nodes can now also be represented in the form of such equivalence classes.
While of no particular relevance in the case of observations from an individual node where all observations are within the domain of a single interpretation, this gains importance once observations from multiple nodes with heterogeneous interpretations must be taken into consideration.

Unless there is a common model for observations (at least at a certain abstraction level, see section 10.5) within a heterogeneous system, the individual observations are incommensurate and contribute to the overall complexity of an intrusion model.

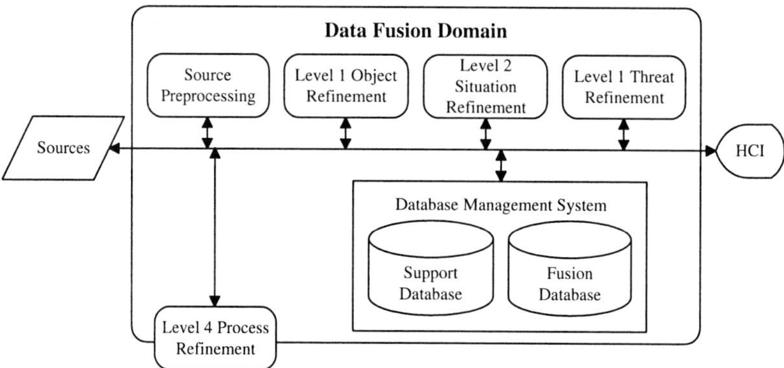

Figure 10.1: The Joint Directors of Laboratories (JDL) process model for data fusion

10.5 The Data Fusion Model for Intrusion Detection

Data fusion [767] as defined in [682] is "*the process of combining data or information to estimate or predict entity states*". The model depicted in figure 10.1 consists of several levels, shown adapted from the model presented in [767]:

Level 1: Object Refinement Aimed at combining sensor data to obtain reliable and accurate estimates of an entity's identity and properties.

Level 2: Situation Refinement Dynamically attempts to develop a description of current relationships among entities and events in the context of their environment.

Level 3: Threat Refinement At this level the current situation is projected into the future to draw inferences about enemy threats, friend and foe vulnerabilities, and opportunities for operations.

Level 4: Process Refinement A meta-process that monitors the overall data fusion process to assess and improve real-time performance.

While primarily developed for real-time defense applications involving physical targets (which have some properties not applicable to IDS targets such as a kinematics envelope permitting to rule out certain sensor data), the paradigm clearly has merit for modeling advanced intrusion detection mechanisms [53].
Most current IDS operate on raw data material such as network traffic or operating system audit trails; typically the latter is not specific to intrusion detection but rather intended to meet auditing requirements such as the CAPP [328, 500] (formerly TCSEC C2 [733]) and does not match the requirements of the IDS in volume and granularity.
This also counteracts the approach of using C2 auditing as the baseline for unifying heterogeneous systems as apparently became evident during the development of DIDS [673].

10.5.1 Level 0

Source Preprocessing (also referred to as Level 0) in this model occurs at the level of individual PENs. and must in itself be separated into three processing sublevels.
At the lower sublevel (0A), each enforcement module is equipped with a number of instrumentation points, i.e. locations in code where a number of items can be collected.
These include the fact that the instrumentation point was reached in the flow of execution as well as information available at the particular point such as the process and thread IDs as well as the account under which the executing process is running. These instrumentation points represent the source for raw data, no homogenization occurs at this stage.
Each of the instrumentation point is uniquely identified and can therefore be activated as necessary.

In cases where the auxiliary data collected must be configurable (e.g. in the case of collecting network traffic), this must be modeled by several separate instrumentation points.
To reflect this as well as to permit the creation of classes of sensor activation, instrumentation points can be coalesced into instrumentation groups; the sets thus formed may intersect.

At the middle sublevel (0B) the data collected by the instrumentation points is processed by annotating the lower sublevel data with the type and identity constants wherever these are known.
The rationale for the insertion of this sublevel is that, while some audit or intrusion detection mechanisms and policies may require the use of raw instrumentation data, significant parts of these contain ephemeral references to subjects and objects or are not unique.

At the upper sublevel (0C) the data from one or more instrumentation points is translated and collated into predicates of the formal theory defining operations in the terms of security policies. The type of translation required depends on the system on which it occurs.
In some cases a bijective mapping between predicates and specific system calls or operations exists; in others a sequence of calls or operations must fulfill certain criteria such as the presence of certain parameters in all elements of the sequence.
This requires that the ECRM retains a history of operations and the entities involved in the operations from which a sequence of operations can be determined which can e.g. be done efficiently using colored Petri nets (CP-nets) [342, 343], also permitting the introduction of temporal constraints on the matching.

The semantics of CP-nets permit the representation of fully concurrent operations and offer a well understood mechanism for confluent preconditions that can be modeled and the derivation of CP-nets from well-formed formulae in formal can be done automatically and verified rigorously.
The relative cost of retaining the operation history is further reduced by the fact that the inverse to the mapping which must be performed for the Level 0 ID processing is required for security policy enforce-

ment, the overhead for retaining this information for intrusion detection is therefore negligible.
Unless annotated level 0B data is used, this MDF layer provides a data reduction mechanism and a normalized set of features independent of the host operating system of the sensor platform.

10.5.2 Level 1

The goal of determining the identity of an entity, particularly of subjects, associated with level 1 data fusion in the JDL model is already achieved by the normalization in level 0B for entities falling under the direct control of the policy enforcement mechanisms. Entities outside this domain of control cannot be implicitly identified positively.
For such entities it becomes necessary to perform level 1 data fusion if events related to more than one activity are to be correlated. This level establishes hypotheses based on observations normalized by level 0 processing as well as existing hypotheses that a certain set of observations represents an distinct individual entity; this is also referred to as a "track".
Such a track may consist of a cluster of operations centered around a certain object or set of objects as established by an anomaly detection mechanism from which a heuristic may determine that a common factor (i.e. the hypothesized subject) is the cause for the clustering occurrence.
Another technique that can be used for level 1 fusion may be termed an inverse intrusion signature mechanism. While one is typically interested in the fact that an intrusion may have occurred, the general mechanism of identifying characteristic operations can also be used in such a way that the pattern of observations give rise to the hypothesis that this signature is associated with a single causative entity.
A major difficulty for level 1 fusion is the ease with which masquerading can occur, both inadvertently or deliberately as a decoy. One example for such masquerading is the use of readily available

network scanning tools which will generally use address spoofing or use intermediate compromised hosts to mount reconnaissance operations.

The pattern of observations will then contain a large amount of correlated features although there may be several independent subjects involved.

Given the large number of such reconnaissance operations and subjects, this appears present a particularly urgent need for taking additional characteristics from different instrumentation points or sensors into account in firming up the identity hypothesis for a subject.

In the example mentioned here such characteristics would include temporal patterns such as apparent network latency.

The relevance of determining the identity of a subject presumed to be involved in hostile actions is twofold.

One is in situations where offensive information operations against hostile subjects may be both justified and feasible; here the primary interest is in ascertaining the identity with the least degree of uncertainty possible.

It should be noted that this is only feasible if the enforcement portion of the security architecture is employed in such a way as to permit positive identification and authentication of remote entities due to the potential for mistakenly responding to e.g. forged network addresses.

This permits the a priori classification of a large portion of entities into FFN categories. Entities which have been identified positively cannot, however, be statically classified as friendly or even neutral due to the possibility of these entities being subverted.

The remaining necessity arises when one wishes to perform higher level (2 and 3) data fusion; here the entity identification is a necessary prerequisite.

Level 1 fusion is feasible relying on the sensor suite of an individual node; while in some cases information stemming from other nodes may be desirable, the benefits of such cross-node fusion at level 1 is as yet conjectural and would constitute a significant increase in complexity.

10.5.3 Level 2

During level 2 fusion a number of tracks or hypothetical entities are aggregated into what is commonly referred to as a situation.
The representation of such situations can be accomplished using the primitives established earlier in that relations are modeled in the first order theory. These can be:

- Aggregation relations which are modeled as predicates establishing equivalence classes over entities.
- Temporal relations which cannot be expressed naturally in the context of a first order predicate logic but either using predicates modeling temporal relations (but not permitting inference on these; this becomes relevant in level 3 fusion) or using temporal (linear) logic.

  The latter is rather unproblematic since temporal or indeed most modal logic systems — with some exceptions such as propositional dynamic logic that are presumably of limited interest in this context — can be efficiently translated into first order logic [523, 639, 321].
- Causal relations which are expressed as connectives in the formal theory that need not be constrained to implication and equivalence.
- Similarity relations may be considered a special case of aggregation relations which are qualified additionally by a similarity metric that in turn is dependent on the identity or the equivalence class membership of the entities involved.

In most cases a probability metric is required for the relations; while this metric can be modeled as a partial function over the elements of the relations, such an approach is clearly unappealing.
Instead one can follow [683] and obtain a mathematically sound method for inferring relational probability metrics, namely the use of a Bayesian network technique [284] to combine

- estimates of entity states X_i based on a set of observations Z_i in a level 1 hypothesis and
- estimates of entity states X_i based on a set of relations R_i among nodes of a level 2 hypothesis.

The distribution of discrete states x_d for X given its assignment to the given node in a level 2 hypothesis ζ will be determined by the data from each of the sources.
With $\mathrm{p}_{\mathrm{L1}}(x_d)$ denoting the probability currently assigned to the discrete state x_d by level 1 data fusion of observations associated with node X and $\Lambda(x_d)$ denoting the data communicated to X from the hypotheses related to Y in a level 2 association hypothesis, one obtains equation 10.1

$$\mathrm{p}_{\mathrm{L2}}(x_d, \varsigma) = \frac{\mathrm{p}_{\mathrm{L1}}(x_d)\Lambda_\varsigma(x_d)}{\sum_{x_d} \mathrm{p}_{\mathrm{L1}}(x_d)\Lambda_\varsigma(x_d)} \tag{10.1}$$

where the factors $\Lambda_\varsigma(x_d)$ represent the product of the data from each of the nodes Y communicating with a given X as shown in equation 10.2 and are interpreted as relational states among entities.

$$\Lambda_\varsigma(x_d) = \prod_{\langle X,Y\rangle \in \zeta} \Lambda_Y(x_d) \tag{10.2}$$

Ordered pairs of entities are hypothesized as having a relational state $r_i(X, Y)$; a hypothesis Y may be present in several, also contradictory, relations representing hypotheses relative to X with a probability distribution $\mathrm{p}[r_i, (X, Y)]$.
While only binary relations are discussed here, n-ary relations can obviously be represented by nesting binary relations.
Updating a hypothesis Y contributes information for evaluating the probability of each state x of a possibly related entity X. As with attributive states relational states r can be decomposed into discrete and continuous components r_d and r_c as appropriate. The contextual evidence is given by equation 10.3.

$$
\begin{aligned}
\Lambda_Y(x_d) &= \sum_{y_d} p_{L1}(y_d) p[y_d \mid x_d] && (10.3) \\
&= \int \sum_{y_d} p_{L1}(y_d) p[y_d \mid r x_d] p[r \mid x_d] dr && (10.4) \\
&= \sum_{y_d} p_{L1}(y_d) \\
&\quad \sum_{r_d} p[y_d \mid r_d, x_d] p[r_d \mid x_d] \int p[r_c \mid r_d, x_d] dr && (10.5)
\end{aligned}
$$

Given the level 2 hypothesis that an entity corresponding to X_i stands in a specific relation to another hypothesized entity corresponding to a hypothesis X_j, inferences can be drawn.

Somewhat problematic in this is the proper assignment of probabilities to individual hypotheses which can aggregate into significant uncertainty given the high dimensionality of the state space. We assume that this significantly curtails the efficacy of inferences that can be drawn automatically compared to fusion scenarios based on physical observations.

Alternatives, particularly in the field of logic involve the use of non-monotonic logic [238, 504, 183]; this also permits the use of epistemic logic under the same framework. Even when dealing with decidable aspects of non-monotonic logics as is typically the case in knowledge representation the problems of highly nondeterministic reasoning and complexity remain [122].

The specification of knowledge producing actions such as observations and belief revision within the framework of multi-modal logic has e.g. been the subject of research [742, 523] as has been the case for fuzzy logic [786, 430]; both approaches can be applied here.

10.5.4 Level 3

Level 3 fusion is mainly concerned with predicting the state space established by the hypotheses at level 2 at a point in time in the future and is intended for the estimation of impact of courses of action such as the likelihood of an outcome state and cost metrics for such actions.

The primary area of interest here is presumably the interactive evaluation of scenarios; given the high degree of uncertainty introduced at level 2 it is not immediately apparent how such a mechanism could e.g. be employed for automated reactive behavior without incurring significant risks of aberrant behavior.

10.6 Operational Models for Intrusion Detection

A number of scenarios are possible when combining the aspects of policy-controlled enforcement and intrusion detection; these are outlined below. In the following discussion, all scenarios include mechanisms introduced by previous scenarios:

Passive Detection This situation is equivalent to the deployment scenario generally used for IDS, namely intrusion detection decoupled from policy enforcement. The node is fully exposed to all threats from undesirable behavior of insiders as well as to external attacks.

Here the benefits of having extensive instrumentation at the various abstraction layers can be assumed to be roughly equivalent to that provided by [757] and [488] in that they provide a more detailed view of host-based system behavior than what can be expected from native instrumentation typically intended for TCSEC C2 or later equivalents.

For network-based components, the immediate benefit is in the ability to access network PDUs that are subject to end-to-end encapsulation or encryption and in distributing the processing load, addressing bandwidth limitations at central detectors in the process.

In this scenario a policy can be distributed ad hoc and ensuring equivalent behavior across heterogeneous platforms including possible preprocessing before reporting to designated processing and fusion nodes.

Augmented Detection In this scenario policy enforcement is also decoupled from intrusion detection. Hence, the threat environment is as described in the previous scenario.

However, a dynamic element is introduced at two levels. First, local policy rules can, in response to observations from local sensors, derive additional rules for activation of sensors. This implicitly includes the derivation of intrusion scenarios using the deduction system (which amounts to the use of the deduction engine as a production system) or a secondary anomaly detector capable of generating rules as output. All such rules are implicitly time-bounded using the lifetime mechanism described earlier.

The second dynamic element consists of the ECRM nodes sending sensor data or derived hypotheses to ERM nodes which then react to these notifications from one or more nodes. Again, the reaction on the part of the ERM can be induced by the deduction system or an anomaly detector. Policy elements for intrusion detection such as an increase in output volume for certain sensors on specific nodes can thus be obtained and propagated to the ECRM nodes.

It should be noted that the extent to which the sensor activation, output modulation, and subsequent processing by both ERM and ECRM takes place must be weighed carefully since degradation or denial of service can otherwise occur.

Enforcement Augmentation In this enhanced scenario, security policy enforcement is in place and is merely augmented by intrusion detection capabilities; this represents the intended environment for the architecture described in this paper. In the presence of policy enforcement, some premises for the previous scenario are — depending on the policies — no longer applicable.

Policies which enforce that only legitimate operations are performed under a given security model and clearly sets of operations and the circumstances under which these may be performed can a priori eliminate a significant amount of behavior that would otherwise need to be analyzed for signs of intrusion as well as behavior which, while legitimate under the security policy, is abnormal.

Some of this ambiguous area for a given set of policies can probably be covered by deduction-based detection; most, however, will need to be analyzed using anomaly detection. Attempts at violating the security policies need to be analyzed in conjunction with legitimate behavior preceding it or concomitant to it; it is mainly in the elimination of some behavior subsequent to initial breaches of policy that would otherwise need to be analyzed forensically.

Attack detection (i.e. operations performed by unauthorized subjects not under the control of policy mechanisms) represent a category which still needs to be addressed. However, given the volume of sensor data and particularly of observations which require further processing, the distributed processing or preprocessing becomes highly valuable even though the amount and quality of sensor data is otherwise reduced.

Fully Reactive While the drawbacks inherent in this scenario are severe and will presumably preclude it from consideration in most cases, it is included here nonetheless in the interest of symmetry and completeness.

This scenario adds definitions to the policy rule sets for rewriting policies to restrict operations normally within the purview of certain subjects or concerning sensitive objects (e.g. limiting access to objects with a given classification on a node for which attacks followed by anomalous behavior of internal subjects has been observed).

Even more so than in the Augmented Detection and Enforcement Augmentation scenarios with reactive sensor and processing behavior this creates severe risks of denial of service for legitimate subjects.

Finally, it is also conceivable that the Fully Reactive scenario is extended through the policy-based activation of active information operations; since this would only be relevant in situations where external attacks are detected this implies that in all likelihood the identity of the adversary cannot be positively verified. This, however, almost ensures that a skilled adversary will be able to use predictable reactive systems against the defender.

CHAPTER

ELEVEN

Digital Watermarking as an Enforcement Mechanism

Labeling human-readable output with sensitivity labels is a requirement for implementations conforming to classes in the TCSEC B division [733] and its successors (e.g. the Common Criteria LSPP [328, 501]). Since the creation of the requirements, however, developments particularly in the area of printing hard copies have made it difficult to embed such labels since line-oriented printing devices have been supplanted by page-oriented devices with a variety of page-description languages which do not lend themselves as easily to the required postprocessing steps.

The need for adequate labeling is also prevalent in civilian application areas; while it is generally not feasible to restrict or prohibit the creation of hard copy as that would be considered an undue burden on the work flow, the risks associated with uncontrolled and careless handling of sensitive printed material are considerable.

This risk not only includes leaving sensitive material unguarded in insecure locations but rather extends beyond actual use in that such material is discarded intact without adequate prior shredding, burning, or similar destruction techniques.

In other areas, such as multimedia data — e.g. audio and video data — the issue of applying sensitivity labels must also be addressed even though the relevant standards do not as yet address these issues.

Another issue that needs to be addressed is the automatic and reliable creation of auditable records for the creation of hard copies of sensitive material.

Printing accounting systems currently in use can only provide records of the creation of hard copies and at best provide an application and document name in such records. The latter are not mandatory and are easily circumvented or fabricated.

In addition to sensitivity labels, it can be generally desirable — possibly even for materials considered publicly available — to have an automatable capability for identifying a document unambiguously even if only a fragment of a single page of such a document is found.

While it is possible to add identification markings to individual pages, conventional mechanisms such as bar codes are highly localized and cannot be used e.g. in case only a figure or a photograph of a labeled document is reproduced elsewhere and needs to be identified.

By providing a mechanism to embed not only the identity of the document but also that of the user associated with the process initiating the creation of the hard copy, a location-independent audit record is provided which can also be back-linked with other audit events to provide insight into the distribution and handling of sensitive material.

The characteristics outlined above can be implemented by means of an enforcement module implementing a filtering mechanism that is interposed into the operating system in such a way that any and all print output must pass through the filtering mechanism in accordance with the reference monitor model.

Compared to application-level labeling mechanisms this is desirable for several reasons; the first one being that interception at the operating system level removes the possibility for tampering with the labeling facility by unprivileged users.

The second important reason is that embedding at the operating system level while retaining the external application programming interfaces implies that any application program using these interfaces will run unmodified and unaware of any additional processing.

In terms of the overall architecture the print subsystem represents merely a high level abstraction layer enforcement module that can be controlled by the same policy mechanisms used throughout the system.
The second component required for fulfilling the requirements as discussed above is digital watermarking [153, 34]. Documents typically consist of a number of different media types such as text, charts or similar vector-based material, and raster images where the composition will vary from page to page of the printed output.

Each of these media types requires specific handling to ensure that the requirements for robustness and visibility are met; the naïve approach of converting the output to a raster format and applying a raster marking technique is both inefficient and produces visible artifacts. Since the desideratum is that the technique must be able to recover the label even from a fragment of a document, this implies that all of the media types present on a printed document should be handled.

11.1 Digital Watermarking Requirements for a Labeling System

The following requirements must be met by digital watermarking algorithms useful for labeling with sensitivity and identification as primary application areas and hard copy as the principal output medium:

1. The watermarks must be robust against D/A — A/D processing.

   This implies that the algorithm must embed the marking in the carrier signal; it also introduces a robustness requirement against jitter.

2. The watermarks must be robust against cropping.

   A fragment of the original human-readable output should suffice to recover all or part of the marking.

3. The watermarks must be robust against affine transforms.

   Some processing steps applied to a hard copy (e.g. automatic paper format resizing applied by some printers) can introduce global affine transforms. This is opposed to local affine transforms such as those produced by irregularities and imperfections in photocopying which are subsumed under requirement (1).

4. Payload sizes must be ≥ 64 bit.

   The payload must be sufficiently large to carry information such as the classification and categories, identification of origin, timestamps, the identity of the subject initiating the human--readable output, and the owner of the document.

   While all this information can be stored in database facilities, it is necessary to have a sufficiently large domain from which entries can be selected. Depending on the need for blind detection, more than one payload entry may be necessary.

5. Blind detection schemes must be used.

   Since the original signal typically does not exist in this application scenario, the watermark must be recoverable without taking recourse to a comparison with a reference signal.

6. Visibility of the watermark should be low to imperceptible.

   This requirement is due to ensure that the signal degradation introduced by the marking does not adversely affect the usability of the human-readable output.

These requirements are kept qualitative since the precise metrics for which level of attack a marking must survive (e.g. whether 50% or 90% of the original may be cropped in case of requirement 2) depend on the application area.
In case of print output, there are several types of objects that can be identified for marking:

- Raster images
- Vector graphics
- Individual lines of text
- Text paragraphs

However, vector-based are typically too sensitive for marking except under well-defined circumstances [119]; in addition, blind detection of such markings in the face of modifications such as cropping and affine transforms is difficult at best without semantic knowledge.
Watermarking techniques for raster images are well established [793]; depending on the algorithm used a certain minimum size for raster images must be used. For individual lines of text, several algorithms also exist [107], the same is true for text paragraphs [425].

11.2 Identification Markings

As noted above there are several purposes for watermarking hard copies. The embedding of initiator identification requires that the subject initiating printing can be identified; this will generally be a process abstraction of the operating system which itself can be connected to a human user (alternatively, a system process may initiate such a process and then represents the highest abstraction level entity which may be associated with the printout). The payload for the marking process is then the highest abstraction level subject that can be found.
If possible (such as is e.g. the case with the mechanism described in [793]) the watermarks should be embedded using a secret key; the use of multiple secret keys (i.e. ideally one per node which performs the embedding) is desirable to limit the exposure of compromised key material. The obvious drawback in such a case is that the number of hypothesis tests becomes linear in the number of nodes which are suspected of being the source of an embedding. As described here, all marking techniques use secret keys for embedding and to ensure minimum exposure a per-node key scheme is used.

While this mechanism can be applied for isolated systems, the consistent labeling of subjects and objects in distributed environments provided by the overall architecture permits the mapping of labels to entities that would presumably not be able to provide full identification or at least be somewhat inefficient.
However, digital watermarks have only limited payload size while the labels provided by the architectural framework may a priori be of arbitrary length. Therefore a compact representation capable of imposing a fixed upped bound on the size of the payload must be used.
Such a compact representation is achieved by applying a cryptographic hash function to the vectors and embedding the hash value in lieu of the full subject type and identity vectors.
From the properties of ideal cryptographic hash functions [427] one can easily see that selecting any arbitrary but fixed sequence of bits from the output bits generated by such a function the individual bits selected retain the properties of the entire function, particularly the avalanche effect.
Assuming that m is the number of unique vectors, an ideal cryptographic hash function, and a payload size of k bits, the number of compact representations is $N = 2^k$; the ratio of the number of compact representations without collisions to the total number of vectors is $\prod_{i=0}^{m-1}\left(1-\frac{i}{N}\right)$. Since $1-x \leq e^{-x}$ (for $x > 0$) one obtains

$$\prod_{i=0}^{m-1}\left(1-\frac{i}{N}\right) \leq \prod_{i=0}^{m-1} e^{-\frac{i}{N}} = e^{-\sum_{i=0}^{m-1}\frac{i}{N}} = e^{-\frac{m(m-1)}{2N}}$$

Thus, the probability of a collision after inserting $\sqrt{2N}$ keys is $1-\frac{1}{e}$. For a 32 bit payload this implies a collision once for every 90,000 subjects; for a 64 bit payload one collision in 6 billion subjects can be expected, which may be considered acceptable — hence the requirement for a 64 bit payload.
Obviously, these calculations do not take into account that the payload may not be extracted faithfully; if this is to be avoided, the payload size must be increased further to accommodate error detec-

tion or correcting codes, possibly causing visible degradation of the output.
For timestamps which might also be inserted into the payload, the requirements for payload are considerably smaller; one minute resolution can be considered adequate given the presumed purpose of establishing probabilities of culpability; 26 bits are sufficient to encode an entire century at this resolution.

11.3 Sensitivity Labels

Since only page-oriented printing systems need to be considered here, there is considerable freedom in the placement, type, and size of the visible sensitivity labels.
By default, this means that a textual representation of the overall sensitivity of the output based either on the properties of identifiable objects present on the page to be printed or on the properties of all objects accessed by the subject (process) prior to submitting the printing request will be placed on the logical top and bottom of the output depending on the assumed orientation of the page. Alternatively, the overall sensitivity for all individual pages in a job can be derived from the properties of all objects accessed by the subject (process) prior to submitting the print job.

11.4 Implementation

The implementation of the labeling system consists of three components. First, a mechanism is required that intercepts all data destined for printing, modifies this data, and subsequently continues processing in the printing subsystem with the modified data.
This filtering mechanism is heavily dependent on the host operating system it is to be integrated in. In cases where several paths exist that can result in a printout, any path that is not filtered by this mechanism must be disabled.

The filtering component must also create detailed audit logs of operations involving the printing system, namely the time and date of a printout and the identity of the subject (i.e. in most cases the user) submitting the print job.
Additional audit information such as the objects the subject accessed prior to submitting the print jobs depend on the presence of additional security components.
In most cases this functionality can, again, be added to the host operating system without requiring source access to the operating system or other components at the application level.

Specifically, in case of the Microsoft Windows NT family of operating systems, the printing architecture is a layered one. An overview of the Windows printing architecture is shown in figure 11.1.
Here, user mode components are not permitted direct access to hardware[1] and must use the GDI32 (Graphics Device Interface) library to issue commands for displaying or printing data. The GDI32 operates in the user mode domain and communicates the requests to its counterpart in the kernel mode domain, the GDI engine.

The GDI engine is itself separated into several components. For translating the GDI commands into printer-specific commands, special printer drivers are required.
The key feature required for operation is that the printer drivers do not communicate directly with the hardware, rather they are only attached to the GDI Engine as shown in figure 11.1 and return the commands generated in reply to the generic DDI (Display Device Interface) requests to the GDI Engine which in turn sends the data to the I/O Manager for actual issuing of the printer commands.

There are several generic (class) drivers (e.g. for raster and Adobe PostScript capable printers) supplied by the operating system, support for specific devices in this framework requires only a mini printer driver which is instantiated by the class driver and communicates only with this class driver.

[1] with the exception of Direct Draw for video output facilities; these have to be disabled for configurations where the reference monitor property is to apply

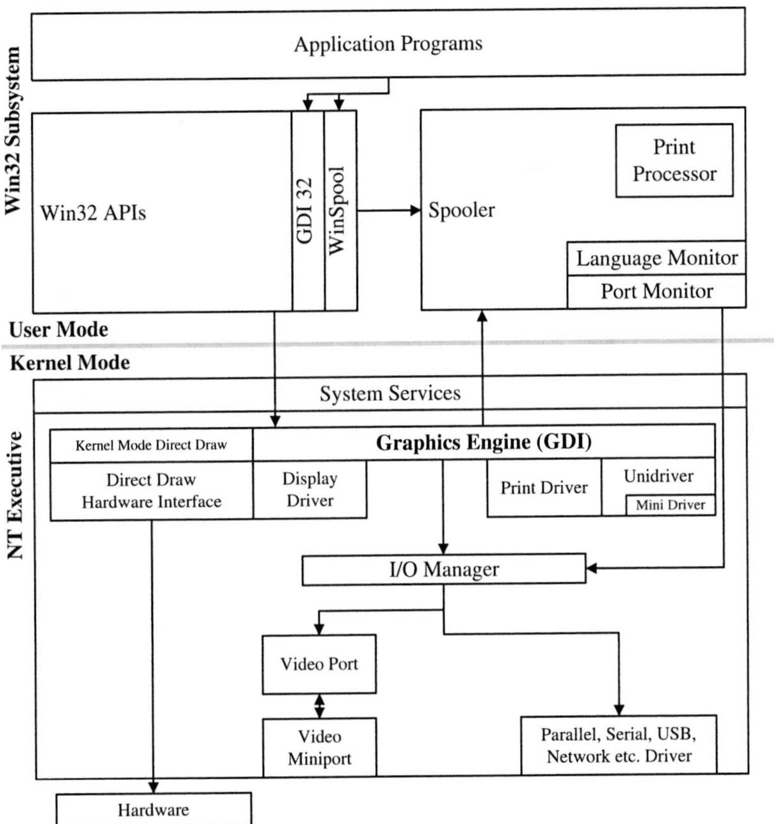

Figure 11.1: Microsoft Windows NT Printing Architecture

In addition, device manufacturers can also supply self-contained drivers that do not rely on class drivers. All printer drivers are implemented as dynamically linked libraries (DLL).

11.4.1 Filtering Mechanism

To ensure that both system-supplied and custom drivers are properly intercepted, the filtering mechanism consists of a printer "wrapper" driver DLL (PWD) which supplies the same entry points as a normal printer driver and replaces the original DLL with itself while renaming the original DLL (see figure 11.2). The original DLLs are moved to a location inaccessible to the GDI engine under a temporary unique name.

The GDI engine will subsequently instantiate and call the wrapper DLL instead of the original driver since it is indistinguishable in both name and the entry points supplied once a request is detected.

The wrapper can now either directly forward calls not relevant to the security functionality to the original library, invoking the same entry points it has itself been called with along with the parameters obtained from the caller, or it can intercept the calls and subject these to further processing whose results are eventually also forwarded to the renamed original driver.

This processing is to be performed in the Labeling and Marking Subsystem (LMS), the second major component in this scenario.

At the start of a print job, the access control subsystem can be invoked which can in turn verify that all prerequisite rules for printing in this instance have been fulfilled by submitting the request for decision to the ECRM. For the remainder of the print job, the LMS is invoked by the PWD for each page.

Processing in the case of the Microsoft Windows NT printing system consists of modifying the sequence of DDI calls for each page. The PWD must buffer the calls for each page in a meta-format and submit the DDI stream to the LMS (doing so for each call would also be possible, but would incur a sizeable performance penalty due to the doubling of call overhead).

The restriction to pages as the unit for marking is derived from the semantics of print processing under Microsoft Windows NT; final processing for labeling and marking is possible only once a page has been finished by the print client since up to that point modifications to the entire page are still possible.

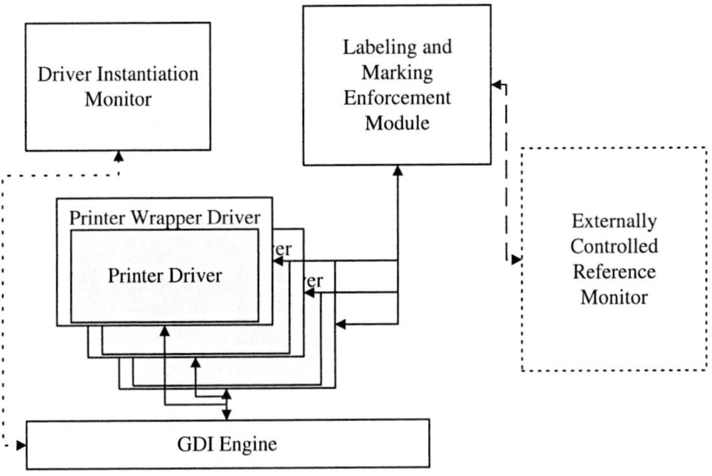

Figure 11.2: Labeling and Marking Components

As a result, some latency is introduced into the print process.

11.4.2 Driver Instantiation Monitor

The printers available to a system can change during runtime (e.g. if a USB printer is attached for which a driver exists in the system's driver database).
It is therefore not possible to establish the PWDs at boot time or even during the initial configuration of the labeling enforcement module; restricting printing to disallow such plug-and-play operations would — while easier to control on the part of the enforcement mechanism — impose an unnecessary burden on users.
Instead, the GDI subsystem must be monitored for events leading to the instantiation of a new printer driver.
This can be accomplished by monitoring the Windows registry keys (i.e. the configuration manager) containing information on registered printers.

This information is maintained at the HKEY_LOCAL_MACHINE\SYSTEM\CurrentControlSet\Control\Print) registry hive location for attempted changes and redirecting the references on the fly.
The reason for this is that the printing subsystem is not a proper device driver but rather derives from legacy implementations of prior operating system platforms. As a result, the standard course of action for such cases, namely the observation of instantiation of new Device executive objects with certain characteristics is not suitable. As modification of system module code is not desirable, observing the effects of new printer registration via the registry is the only viable option.

11.4.3 Labeling and Marking Mechanism

The LMS can add human-readable sensitivity labels (or any other graphical output) to the page by simply adding DDI commands returned to the PWD for the given page. In case of the default markings, this is accomplished by overlaying a box with the background (paper) color at the logical top and bottom of each page in which the sensitivity label is placed. As this step occurs immediately before the finishing of each logical page, there is no possibility of the labels becoming obscured by the results of other DDI operations.
Digital watermarking is somewhat more involved than labeling since it requires that the input stream be modified instead of simple augmentation in the case of labeling.
However, raster images of usable size are treated as objects by the DDI and can thus be used directly as input for raster image watermarking. Since the marking does not affect the external features of the bitmaps relevant for placement of the objects in relation to other objects on a page (size, color resolution etc.), this permits direct replacement of original with marked bitmaps regardless of the marking algorithm used.
The same holds true for individual lines of text; DDI operates on lines at a time, although some application programs deliberately issue GDI calls for individual words. Depending on the size of the

line received, the text watermarking module must therefore decide whether line-based algorithms are appropriate or whether word- or character-based algorithms must be used alone. To ensure that the overall layout is not perturbed, the external dimensions of such a marked block must not change. This requirement is lifted for paragraphs identified by the LMS as such and to which line-shifting watermarking techniques are applied to the extent that the external dimensions of the paragraph must not change.

One of the reasons for requiring robustness against affine transforms, including interpolation effects, is that print output (particularly raster images) will frequently be modified to accommodate different pixel aspect ratios.

To facilitate the integration of new and improved algorithms, the LMS is designed to process the DDI commands in a modular fashion. The inbound DDI commands are categorized into three groups (vector, text, raster) and routed to the appropriate modules. Since the input and output in each case is a DDI command stream, the partial results can then be integrated again into a page representation. The main requirement for making this possible is that changes to the size and boundaries of the blocks processed are avoided.

CHAPTER
TWELVE

Conclusions

The present thesis has described an architecture for modeling distributed systems and formulating, analyzing, and enforcing arbitrary security policies within these systems along with mechanisms for the enforcement of such policies in a reference system based on the same modeling mechanisms.
Moreover, selected aspects of the implementation of such modeling and enforcement mechanisms, also without loss of generality for the reference system, were presented.
This chapter provides a reprise and conclusion by presenting an overview of the contributions in section 12.1 and discussing both possible directions for future research and limitations — in part external to the research reported in this dissertation — to the efficacy thereof in section 12.2.

12.1 Summary of Contributions

The underlying assumptions and hypotheses for this dissertation were formulated in chapter 4. Based on these hypotheses, an architecture has been described that permits the formulation, analysis, and enforcement of arbitrary security policies in distributed, heterogeneous systems.

- The core abstractions of resources and entities provided by a broad range of general purpose operating system were formulated by way of conceptual analysis and exemplified for the case of a reference operating system family. Said abstractions were modeled in a first order theory and the conceptual relations between the entities represented by algebraic (lattice) structures, which turn are embedded in the first order theory. This permits reasoning over the structure imposed on concepts and identities within the mechanisms of a formal theory.

- It has been shown that security models and policies can furthermore be efficiently expressed within the same formal theory; this property extends to all policies expressible in deterministic computer systems by virtue of being Turing-complete. Based on the algebraic structure imposed on entities and operations, the policies can be expressed at high semantic abstraction levels while lower abstraction levels can be derived and identified by lattice algebraic operations.

- Security policies or algebraically conjoined sets of policies under constraints imposed by the abovementioned lattice structure can, moreover, be employed for real-time automated reasoning over hypotheses expressing either the intent to perform an arbitrary operation, in which case the proof of a hypothesis indicates that the given operation is permitted and may proceed or the requirement of executing an arbitrary operation, in which case obtaining proof is equivalent the existence of the requirement.

- The separation of enforcement mechanisms from the locus of specification and decision (albeit with the possibility for temporary delegation for efficiency reasons) for security policies particularly in general purpose operating systems by extension of the reference monitor mechanism has been shown. Implicit in the use of a formal theory for modeling of operating systems and security policies is the capability of providing a common view of all applicable security policies on every node on which the enforcement mechanism is realized regardless of the specific characteristics of the node itself. Moreover, by providing layered instrumentation mechanisms within both operating systems and application programs, decisions regarding security policies can be obtained efficiently.

- The feasibility of realizing enforcement mechanisms retroactively in an opaque commercial operating system without privileged access has been demonstrated in the case of a reference operating system family described in this dissertation as well as in a second, significantly different operating system not described in this dissertation.

- Network security mechanisms based on the common modeling mechanism and implicit identification of entities and operations have been developed; this yields not only improved static defensive capabilities compared to traditional firewalling and intrusion detection mechanisms but also increased dynamic defenses based on the improved situational awareness provided by multisensor data fusion from the system-independent instrumentation of all nodes controlled by a set of security policies.

  Similarly, entity-based file system security mechanisms for have been developed that permit the identification and application of policies based on file entities including transformations such as transparent encryption at the file entity level.

12.2 Future Research

While this dissertation has concentrated on only the simplest and best understood logic for modeling systems and security policies in the form of $\mathcal{PC}$, this imposes restrictions on the ease of expression – though not of expressibility – achievable particularly with regard to sequences of operations or possible operations.

As indicated by earlier and ongoing research (cf. sections 7.2.2 and 3.1), modal logic, and in particular temporal logic is an attractive theory in which both the underlying systems and the policies acting upon these models can be expressed (see section 7.2.2). However, consideration of such an approach implies two major problems.
First, the set of traces required for reasoning is a priori unbounded and may include not only partial traces from multiple nodes but arbitrary interleavings of these traces.
Second, the theories thus obtained are – in the most general case – not semi-decidable [241]. It would therefore be of interest to determine the extent to which modal and nonmonotonic reasoning elements [462] (e.g. circumscription or default logic) can be employed by restricting the domain to which it is applied [646, 245, 739, 716, 59].
This is of particular interest in the area of multisensor data fusion for intrusion detection (cf. sections 10.5, 10.5.3, and 10.5.4) where probabilistic entailment arises frequently.

Additional issues of interest for future research arise from the requirements for availability and timeliness that are inherently part of information assurance that have not been considered in this dissertation.
This particularly includes the efficient definition and evaluation of policy requirements that cross node boundaries and introduce temporal reasoning elements as well as considerations for cases where faults arise in nodes (cf. section 7.3) and includes self-referential aspects such as timeliness constraints on regulated policies.
Also, within a given framework for formulating policies such as $\mathcal{PC}$, several issues have not been addressed within the confines of this dissertation. One such issue is the formulation of dynamic delegation and revocation mechanisms in security policies (which are assumed to be handled by forward-resolving predicates in this dissertation) that are highly dynamic.
Similarly, the issue of initial policy design itself has not been addressed; here, the perceived complexity of policies can presumably be reduced through the judicious use of human-computer interaction (HCI) techniques that use heuristics for pre-formulating policy

elements for human assembly. While theorem 3.1 precludes the automated development and verification of security policies, heuristics can also be used to generate potential counterexamples for policy consistency and similar undesirable properties and, in conjunction with HCI techniques, be used to explore policies prior to deployment.

Another subject of further investigation is in obtaining quantitative results on the efficacy of using policy mechanisms for the management of complex heterogeneous systems, particularly compared to reactive systems and multiple, separately configured and managed actuator elements. Similar analysis is also required for determining the effectiveness of reactive policies based on zero-hour vulnerability detections for ameliorating vulnerabilities prior to availability of corrective measures.

12.2.1 Pragmatic Limits to Achievable Assurance

As noted in sections 1.2 and 4.1, the overall assurance of a security architecture is severely limited if, based on a low assurance system, information assurance components are added [756, 257, 387, 38, 492, 415, 30].

While the discussion in this dissertation has shown that indeed the mechanisms are sufficiently general to permit integration into arbitrary, specifically high assurance systems, and, moreover, such high basic assurance systems have been built in the past [220, 633, 762], a pragmatic question arises as to whether such systems may again be implemented for general purpose computers and – more importantly – for widespread use in civilian, non-regulated environments in the future, although this issue is disjoint from the viability of the concepts and mechanisms described in this dissertation.

Unfortunately, the answer to this critical question appears to be less in the realm of mathematics as applied to the problem domain – although arguments pertaining upper limits on feasibility based on complexity and computational theory can certainly be raised – than that of economics and game theory.

The high assurance systems referenced elsewhere in this dissertation were built on behalf of defense organizations with clearly defined security objectives. However, the development of general purpose computer systems have since become primarily driven by consumer and commercial applications.
This change in the principal application domain, however, has significant implications. It has been observed that – given the quantitative data available, particularly with regard to a lack of security breaches and losses commensurate to predictions from the computer security industry – expenditures and efforts on the part of commercial users have to be considered adequate [27, 267].
Moreover, as shown by Akerlof [11] and by subsequent game-theoretic research [168], suppliers of general purpose operating systems not only have a significant incentive to be first and, by extension the dominant supplier (which could be considered an instance of "Metcalfe's Law"), but are furthermore constrained in their incentives by customers' inability to distinguish between effective and ineffective offerings.

While the previous arguments apply to both commercial and free software, commercial vendors additionally have thus far generally not been subjected to liability issues that would otherwise justify expenditures on due diligence.
This results in the observations on current levels of assurance made, among others, by Karger and Schell on several occasions [631, 360]. Moreover, as has been demonstrated by von Neumann and Morgenstern [750] based on the theory of partial differential equations, it is not mathematically possible to maximize for two or more variables at the same time.
One obvious solution to this problem is the restating of information assurance objectives and the consequences of a lack thereof is terms of economic variables [644, 84], a research subject that has gained significant attention.
Blakley argues that information assurance and security must be monetized in such a way that losses and effectiveness can be quantified, insurances can determine premiums and other financial incentives

such as securities can be established along with liability for losses stemming from insufficient security or other information assurance lapses [84].

However, this argument is specious in at least that the cost of such lapses is not necessarily incurred by the causative agent. Hardin [288] has shown that no mathematical or technical solution exists in such a situation, and Varian [744] has provided an analysis based on Nash equilibria[1] [494] that in the total effort scenario, there is too little effort expended at the Nash equilibrium. Moreover, the probability of success in the globally optimal solution is always lower in case of weakest link, which can be presumed to be a realistic scenario at least for some vulnerabilities based on the stipulation of negligence or non-monetary motivations, than in the case of total effort. Furthermore, in contrast to the argument put forth by Blakley, strict liability is insufficient to achieve a global optimum [112].

Based on these existing results, it appears that unless liability in conjunction with fines and other, also non-monetary, disincentives are established, efforts expended on information assurance and security are unlikely to improve.

Hence, as already noted briefly in section 4.2, the reliability and hence also assurance obtained for software is, in the absence of a rigorous formal specification and proof of correctness, largely limited to what can be accomplished by testing – both by suppliers and indirectly by users of such software – as shown by results on reliability growth models [121, 80]. While, as noted by Hoare [309], the defects and defect rates common to non-rigorous development methods have not had the crippling cumulative effect predicted several decades ago, large software systems developed using such non-rigorous development methods – which for commonly used software can be assumed not only for the overall system, but moreover also for assurance-critical components, the ostensibly trusted code base – amenable to statistical modeling and analysis nevertheless exhibit a number of highly undesirable properties.

[1] Although one may argue that the assumption of completely rational players may, given the apparently innate human inability to assess risk, not be entirely suitable.

Assuming a random distribution of tests equiprobable to detect defects with j as the number of tests performed thus far and an arbitrary defect with index i, and assuming a probability that the defect i is undetected after j steps is $e^{-p_i j}$, Brady *et al.* have shown that, based on reliability growth models and statistical mechanics [105, 28], after removal of n defects and long periods of statistical testing, the probability of a defect converges to a polynomial distribution shown in equation 12.1 based on an argument that defects in software can be modeled by a Boltzmann distribution and thermodynamic equilibria [124, 228]; this model is consistent with empirical data for at least one large system [10].

$$p = \sum_{i=n+1}^{\infty} e^{-p_i j} \approx \frac{k}{j} \tag{12.1}$$

The integration constant k derived in equation 12.1, however, follows immediately from the ab initio quality of the software under test. Moreover, Brady *et al.* argue based on Poisson survival statistics that any such testing will remove only the minimum number of defects required to succeed in the number of tests applied. However, given the preceding economical arguments regarding incentives for suppliers resulting in large k and the frequency with which new or modified systems (and hence new test statistics with $t = 0$ at least for parts of such systems), the likelihood of large truly high assurance systems emerging from such circumstances appears to be minute.
Unless a definite mandatory requirement imposing both liability and criminal culpability constraints exists (which would presumably have to be effected through transnational governmental action) [200, 295], the above discussion may lead to the conclusion that wholly satisfactory information assurance in general purpose, particularly civilian, systems is a rather elusive goal.

GLOSSARY

CA A Certification Authority certifies the public keys belonging to individual end entities within its domain of control in a manner consistent with the cognizant PCA and PAA policies and requirements and can revoke end entity certificates in cases of non-compliance.

CCITSE The Common Criteria for Information Technology Security Evaluation occasionally (and somewhat incorrectly) referred to as the Harmonized Criteria, is a multinational effort to write a successor to the TCSEC, ITSEC and other earlier security evaluation criteria. This effort was originally sponsored by the United States, United Kingdom, Germany, France, Canada, and the Netherlands (the sponsoring organisations of the CTCPEC, FC, TCSEC and ITSEC). The CCEB and its successor, the CCIB, developed several versions of the criteria in cooperation with ISO WG 3; the body of version 2.1 was simultaneously adopted as ISO/IEC 15408-1 in December 1999.

CIPSO The Common Internet Protocols Security Option is an IP options field containing a sensitivity label; this field is used for label-based routing. CIPSO does not require negotation of label semantics.

DCOM The Microsoft Distributed Component Object Model is a set of concepts and program interfaces in which application objects can request services from server objects on other computers in a network, which may consist of various transport mechanisms. DCOM is based on the Component Object Model (COM) for intra-node object communication.

DISA The United States Defense Information Systems Agency, renamed to reflect its expanded mission more properly from the original DCA.

HDLC The ISO High Level Data Link Control protocol standard corresponds to Layer 2 (the Data Link Layer) of the ISO OSI reference architecture. It is responsible for the error-free movement of data between network nodes. There exist two distinct HDLC implementations: HDLC NRM (see SDLC) and HDLC Link Access Procedure Balanced (LAPB).

ITSEC The Information Technology Security Evaluation Criteria represent the harmonized European criteria for security evaluations, initially sponsored by France, Germany, the Netherlands and the United Kingdom and was adopted in 1991. It has since been superseded by the CCITSE.

LAPB The Link Access Procedure Balanced is an implementation of the HDLC standard. LAPB is a bit-oriented synchronous protocol that provides complete data transparency in a full-duplex point-to-point operation and mainly used in the ITU X.25 standard.

NCSC The United States National Computer Security Center was originally formed as the Computer Security Center under the authority of the NSA in 1981 in accordance with DoD directive 5215.1. Its mission is to maintain [...] *technical standards and criteria for the security evaluation of trusted computer systems that can be incorporated into the Department of Defense component life-cycle management process* [...]

NDIS The Network Device Interface Specification was developed by Microsoft and 3Com. Novell offers a similar device driver for NetWare called Open Data-Link Interface (ODI). NDIS specifies top-edge interfaces for network addressing and protocol stacks, device drivers interacting directly with the communication device typically as a bottom-edge access mechanism for the MAC sublayer of OSI sublayer 2, as well as for a Protocol Manager which provides bindings between the layers.

NetBEUI The NetBios Enhanced User Interface is a revision of the NetBIOS protocol originally designed by IBM for the IBM LAN Manager; later extensions to the protocol were initiated by both Microsoft and Novell; unlike NetBIOS, NetBEUI refers only to the protocol itself, not the associated application programming interfaces. NetBEUI is a non-routable protocol.

NetBIOS The Network Basic Input/Output System, originally designed by IBM and later extended by Microsoft refers to both an application programming interface and a network protocol; it is, however, commonly used only in referring to the transport-layer mechanism which may be encapsulated in other protocols such as NetBEUI or TCP/IP.

NIST United States National Institute of Standards and Technology, renamed to reflect its expanded mission more properly from the original NBS.

ORA A Organizational Registration Authority is an entity acting as an intermediary between a CA and an end entity. Its sole purpose is to vouch for the identity and domain affiliation of the end entity and to register that end entity with its CA.

OSI The Open Systems Interconnection is a reference model for the transmission of information between any two points in a telecommunication network. Its purpose is to guide product implementors so that their products will consistently work with other products. The reference model defines seven layers of functions that take place at each end of a communication: physical, data link, network, transport, session, presentation, and application layer. OSI was officially adopted as an international standard by the ISO. Currently, it is Recommendation X.200 of the ITU-TS.

PAA The Policy Approving Authority creates the overall requirements for operation of the entire public key infrastructure. It certifies public keys belonging to individual PCAs and can revoke PCA certificates in cases of non-compliance.

PCA A Policy Certification Authority establishes policy and requirements for all CA and end entities within its domain of control. PCAs certify public keys belonging to individual CAs and can revoke CA certificates in cases of non-compliance.

POSIX A group of standards, the "Portable Operating System Interfaces" defined by the IEEE working group P1003 including kernel and system call interfaces and semantics.

SDC The System Development Corporation evolved out of the System Development Division of the RAND Corporation. It was established in 1955. SDC was acquired by Burroughs Corporation in 1980.

SDLC The Synchronous Data Link Control is a variation of the HDLC standard and also known as HDLC NRM (Normal Response Mode) which was introduced by IBM Corporation. Unlike LAPB, SDLC is not a peer to peer protocol. An SDLC network is made up of a primary station that controls all communications, and one or more secondary stations. Where multiple secondaries are connected to a single primary, this is known as a multipoint or multidrop network.

TCSEC The Trusted Computer System Evaluation Criteria was first issued in August of 1983 and was subsequently reissued as a DoD standard in 1985. The TCSEC standard specified requirements for systems in terms of degrees of trust with increasing levels of trust at each level. Each level in this hierarchy builds upon the previous one by adding security features and assurance. The TCSEC have been superseded by the Common Criteria for Information Technology Security Evaluation.

UDF The Universal Disk Format is a CD-ROM and DVD file system standard developed as a means of ensuring consistency among data written to various optical media, by facilitating both data interchange and the implementation of the ISO/IEC 13346 standard. UDF is required for DVD-ROMs, and is used by DVD to contain MPEG audio/video streams. Originally developed as a replacement for the file system specifications in the original CD-ROM standard, ISO 9660, UDF is used by CD-R and CD-RW in a process called packet writing that makes CD writing more efficient in terms of the time and disk space required.

X.25 The ITU standard protocol known as X.25 encompasses the first three layers of the ISO OSI reference model. The Physical Layer is covered by several standards such as V.35, RS232, and X.21. The Data Link Layer is LAPB, an implementation of the ISO HDLC standard. The Network Layer is covered by the X.25 Packet Layer Protocol (PLP). An X.25 packet makes up the data field of an HDLC frame. Additional flow control and windowing are provided for each Logical Channel at the X.25 level.

LIST OF ACRONYMS

ACC .... Access Control Center.

ACL........... Access Control List.

AFD........... Ancillary Function Driver.

AH............ Authentication Header.

API............ Application Programming Interface.

ARPA .... U.S. Advanced Research Projects Agency.

ATM .... Asynchronous Transfer Mode.

BFE .... BLACKER Front End.

BLP .... Bell-LaPadula Model.

BSD........... Berkeley Software Distribution.

BST .... Basic Security Theorem.

CAPP .... Controlled Access Protection Profile.

CCEB .... Common Criteria Editorial Board.

CCIB.......... Common Criteria Implementation Board.

CDRL......... Contract Data Requirements List.

CMU .... Carnegie Mellon University.

CMW .... Compartmented Mode Workstation.

CMWEC ...... Compartmented Mode Workstation Evaluation Criteria.

CMWREQ..... Compartmented Mode Workstation Requirements.

CORBA .... Common Request Broker Architecture.

COTS .... Commercial Off-the-Shelf.

CRADA .... Cooperative Research and Development Agreement.

CSP .... Communicating Sequential Processes.

DAC .... Discretionary Access Control.

DCA .... U.S. Defense Communications Agency.

DCE........... Data Communications Equipment / Distributed Computing Environment.

DDI .... Display Device Interface.

DDN .... United States Defense Data Network.

DEC........... Digital Equipment Corporation.

DGSA .... Defense Goal Security Architecture.

DIA .... United States Defense Intelligence Agency.

DLL........... Dynamically Linked Library.

DoD........... United States Department of Defense.

DODIIS .... U.S. Department of Defense Intelligence Information System.

DOI .... Domain of Interpretation.

DSU .... UCLA Data Secure Unix.

DTOS .... Distributed Trusted Operating System.

EAL .... Evaluation Assurance Level.

ECO .... Engineering Change Order.

ECRM .... Externally Controlled Reference Monitor.

ERM .... External Reference Monitor.

FAT .... File Allocation Table.

FCB .... File Control Block.

FFN .... Friend / Foe / Neutral.

GDI .... Graphics Device Interface.

GFAC .... Generalized Framework for Access Control.

GUID .... Globally Unique Identifier.

HAL .... Hardware Abstraction Layer.

HCI .... Human-Computer Interaction.

HRU .... Harrison-Ruzzo-Ullman Model.

HTTP .... Hypertext Transfer Protocol.

IA .... Information Assurance.

IBM .... International Business Machines Corporation.

IDS .... Intrusion Detection System.

IEEE .... The Institute of Electrical and Electronics Engineers.

IETF........... Internet Engineering Task Force.

IKE............ Internet Key Exchange.

IO............. Information Operations.

IP .... Internet Protocol.

IPC............ Inter-Process Communication.

IPSec.......... Internet Protocol Security Protocol.

IPSO .... IP Security Option.

IPv6........... Internet Protocol Version 6.

IPX............ Internetwork Packet Exchange.

IrDA.......... Infrared Data Association.

IRP............ I/O Request Packet.

IRQ .... Interrupt Request.

IS .... Information Systems.

ISAKMP....... Internet Security Association and Key Management Protocol.

ISO............ International Standards Organization.

ITU .... International Telecommunications Union.

ITU-TS .... ITU Telecom Standardization.

JTA............ Joint Technical Architecture.

KDC .... Key Distribution Center.

KSOS.......... Kernelized Secure Operating System.

LPC........... Local Procedure Call.

LSM........... Linux Security Module.

LSP........... Layered Service Provider.

LSPP.......... Labeled Security Protection Profile.

MAC.......... Mandatory Access Control / Media Access Control.

MDF.......... Multisensor Data Fusion.

MLS........... Multilevel Security.

MMS.......... Military Message System.

MTU.......... Maximum Transmission Unit.

NBS........... United States National Bureau of Standards.

NFS........... Network File System.

NIC........... Network Interface Card.

NOPE......... Non-Policy-Enforcing Entity.

NRL........... United States Naval Research Laboratories.

NSA........... United States National Security Agency.

NTCB......... Network Trusted Computing Base.

NTFS.......... Microsoft New Technology File System.

ODI........... Open Data-Link Interface.

PCI............ Peripheral Component Interconnect.

PCN........... Policy Controller Node.

PDU........... Protocol Data Unit.

PE............. Portable Executable.

PEB........... Process Environment Block.

PEN........... Policy Enforcing Node.

PLP........... Packet Layer Protocol.

PMTU......... Path Maximum Transmission Unit.

PnP........... Plug and Play.

PP............. Protection Profile.

PWD.......... Printer Wrapper Driver DLL.

RACF......... Resource Access Control Facility.

RADC......... United States Air Force Rome Air Development Center.

RBAC......... Role-Based Access Control.

RIPSO......... Revised IP Security Option.

RNS........... Root Name Server.

SAD........... Security Association Database.

SAMP......... Security Attribute Modulation Protocol.

SATMP........ Security Attribute Token Mapping Protocol.

SBI............ Special Background Investigation.

SCC........... Secure Computing Corporation.

SCI............ Sensitive Compartmented Information.

SCSI........... Small Computer System Interface.

SDIO .... Strategic Defense Initiative Organization.

SID .... Security Identifier.

SIDEARM ..... System Independent Domain Enforcing Assuring Reference Monitor.

SOAP .... Simple Object Access Protocol.

SPD .... Security Policy Database.

SPDF .... Security Policy Decision Function.

SPEF .... Security Policy Enforcement Function.

SPI .... Service Provider Interface / Security Parameter Index.

SPX .... Sequence Packet Exchange.

SSH .... Secure Shell.

SSL .... Secure Sockets Layer.

TAFIM .... Technical Architecture Framework for Information Management.

TAPI .... Telephony Application Programming Interface.

TCB .... Trusted Computing Base.

TDI .... Transport Driver Interface.

TIS .... Trusted Information Systems.

TLA .... Temporal Logic of Action.

TLS .... Transport Layer Security.

TNI .... Trusted Network Interpretation.

TSIG .... Trusted Systems Interoperability Group.

TSP .... TAPI Service Provider.

UCLA .... University of California at Los Angeles.

USB .... Universal Serial Bus.

VFS .... Virtual File System.

VNODE .... Virtual File System Node.

WMI .... Windows Management Instrumentation.

WWMCCS .... Worldwide Military Command and Control System.

BIBLIOGRAPHY

[1] Abadi, M., Burrows, M., Lampson, B., and Plotkin, G. A Calculus for Access Control in Distributed Systems. *ACM Transactions on Programming Languages and Systems 2*, 1 (Jan. 1993), 706–734.

[2] Abadi, M., and Lamport, L. The Existence of Refinement Mappings. *Theoretical Computer Science 82*, 2 (1991), 253–284.

[3] Abadi, M., and Lamport, L. Composing Specifications. *ACM Transactions on Programming Languages and Systems 15*, 1 (Jan. 1993), 73–132.

[4] Abadi, M., and Lamport, L. Conjoining Specifications. Tech. Rep. 118, Digital Equipment Corporation, Systems Research Center, Dec. 1993.

[5] Abadi, M., Lamport, L., and Wolper, P. Realizable and Unrealizable Specifications of Reactive Systems. In *Proceedings of the 16th International Colloquium on Automata, Languages, and Programming* (Stresa, Italy, July 1989), G. Ausiello et al., Eds., vol. 372 of *Lecture Notes in Computer Science*, Springer Verlag, pp. 1–17.

[6] Abrams, M. D., Amoroso, E. G., LaPadula, L. J., Lunt, T. F., and Williams, J. G. Report of an Integrity Research Study Group. *Computers & Security Journal 12*, 7 (Nov. 1993), 679–689.

[7] Abrams, M. D., Eggers, K. W., LaPadula, L. J., and Olson, I. M. A Generalized Framework for Access Control: An Informal Description. In *Proceedings 13th NIST-NCSC National Computer Security Conference* (Washington D.C., USA, Oct. 1990), pp. 135–143.

[8] Accetta, M., Baron, R., Bolosky, W., Golub, D. B., Rashid, R., Tevanian, A., and Young, M. Mach: A New Kernel Foundation for Unix Development. In *Proceedings of the Summer 1986 USENIX Conference* (Seattle, WA, USA, June 1986), USENIX, Ed., USENIX, pp. 93–112.

[9] Adamowicz, Z., and Zbierski, P. *Logic of Mathematics.* Wiley Series in Pure and Applied Mathematics. John Wiley & Sons, Inc., New York, NY, USA, 1997.

[10] Adams, E. N. Optimizing Preventive Maintenance of Software Products. *IBM Journal of Research and Development 28*, 1 (Jan. 1984), 2–14.

[11] Akerlof, G. A. The Market for "Lemons": Quality Uncertainty and the Market Mechanism. *The Quarterly Journal of Economics 84*, 3 (Aug. 1970), 488–500.

[12] Akers, R. L., Hartman, B. A., Smith, L. M., Taylor, M. C., and Young, W. D. *Gypsy Verification Environment User's Manual.* Austin, TX, USA, June 1990.

[13] Alexandrov, A., Ibel, M., Schauser, K., and Scheiman, C. Extending the Operating System at the User Level. In *Proceedings of the 1997 USENIX Annual Technical Conference* (Berkeley, CA, USA, June 1997), USENIX Association, Ed., USENIX, pp. 77–90.

[14] Allen, P. G. A Comparison of Non-interference and Non-deducibility using CSP. In *Proceedings of the IEEE Computer Security Foundations Workshop IV* (Franconia, NH, USA, June 1991), IEEE Computer Society, pp. 43–54.

[15] Alpern, B., and Schneider, F. B. Defining Liveness. *Information Processing Letters 21*, 4 (Oct. 1985), 181–185.

[16] Alpern, B., and Schneider, F. B. Recognizing Safety and Liveness. *Distributed Computing* 2, 3 (1987), 117–126.

[17] Alpern, B., and Schneider, F. B. Verifying Temporal Properties without Temporal Logic. *ACM Transactions on Programming Languages and Systems 11*, 1 (Jan. 1989), 147–167.

[18] Ambler, A. L., Good, D. I., Browne, J. C., Burger, W. F., Cohen, R. C., Hoch, C. G., and Wells, R. E. GYPSY: A Language for Specification and Implementation of Verifiable Programs. *ACM SIGPLAN Notices 12*, 3 (Mar. 1977), 1–10. Proceedings of the ACM Conference Language Design for Reliable Software.

[19] Ames, Jr., S. R., Gasser, M., and Schell, R. R. Security Kernel Design and Implementation. *IEEE Computer 16*, 7 (July 1983), 14–22. Reprinted in [722].

[20] Ames, Jr., S. R., and Oestreicher, D. R. Design of a Message Processing System for a Multilevel Secure Environment. In *Proceedings of the National Computer Conference* (Anaheim, CA, USA, Nov. 1978), vol. 47, AFIPS, AFIPS Press, pp. 765–771.

[21] Amoroso, E. G. A Policy Model for Denial of Service. In *Proceedings of the IEEE Computer Security Foundations Workshop III* (Franconia, NH, USA, June 1990), IEEE Computer Society, pp. 110–114.

[22] Amoroso, E. G. *Intrusion Detection: An Introduction to Internet Surveillance, Correlation, Trace Back, Traps, and Response*. Intrusion.Net Books, Sparta, NJ, USA, 1999.

[23] Andersen, D. B. Windows Sockets 2 Application Provider Interface. Tech. rep., Intel Corp., May 1997. Version 2.2.1.

[24] Andersen, D. B. Windows Sockets 2 Service Provider Interface. Tech. rep., Intel Corp., May 1997. Version 2.2.1.

[25] ANDERSON, A. R. Independent Axiom Schemata for S5. *Journal of Symbolic Logic 21* (1956), 255–256.

[26] ANDERSON, J. P. Computer Security Technology Planning Study. Tech. Rep. ESD-TR-73-51, Air Force Electronic Systems Division (AFSC), L. G. Hanscom Field, Bedford, MA, USA, Oct. 1972. AD-758 206, ESD/AFSC. (Also available as Vol. I, DITCAD-758206. Vol. II, DITCAD-772806).

[27] ANDERSON, R. Maybe we spend too much? In *Workshop on Economics and Information Security at the University of California, Berkeley* (Berkeley, CA, USA, May 2002). Electronic proceedings only.

[28] ANDERSON, R. Security in Open versus Closed Systems - The Dance of Boltzmann, Coase and Moore. In *Proceedings of Open Source Software: Economics, Law and Policy* (Toulouse, France, June 2002).

[29] ANDERSON, R., AND KUHN, M. Tamper Resistance - a Cautionary Note. In *Proceedings of the 2nd USENIX Workshop on Electronic Commerce* (Oakland, CA, USA, Nov. 1996), USENIX, Ed., USENIX, pp. 1–11.

[30] ANDERSON III, E. A Demonstration of the Subversion Threat: Facing a Critical Responsibility in the Defense of Cyberspace. Master's thesis, Naval Postgraduate School, Monterey, CA, USA, 2002.

[31] APPEL, A. W., AND FELTEN, E. W. Proof-Carrying Authentication. In *Proceedings of the 6th ACM Conference on Computer and Communications Security* (Kent Ridge Digital Labs, Singapore, Nov. 1999), ACM Press, pp. 52–62.

[32] APT, K. R. Ten Years of Hoare's Logic: A Survey — Part I. *ACM Transactions on Programming Languages and Systems 3*, 4 (Apr. 1981), 431–483.

[33] APT, K. R. Ten Years of Hoare's Logic: A Survey — Part II: Nondeterminism. *Journal of Theoretical Computer Science 28* (1984), 83–109.

[34] ARNOLD, M., SCHMUCKER, M., AND WOLTHUSEN, S. D. *Techniques and Applications of Digital Watermarking and Content Protection.* The Artech House Computer Security Series. Artech House, Norwood, MA, USA, 2003.

[35] ARNOLD, T. Personal Communication. Design of a minimal software interface between the IBM 4758 model 2 and a kernel environment for Microsoft Windows 2000 and Sun Solaris 8., Nov. 2000.

[36] ASHCROFT, E. A., CLINT, M., AND HOARE, C. A. R. Remarks on "Program Proving: Jumps and Functions" by M. Clint and C.A.R. Hoare. *Acta Informatica 6* (1976), 317–318.

[37] ATKINSON, R., DELLA-LIBERA, G., HADA, S., HONDO, M., HALLAM-BAKER, P., KLEIN, J., LAMACCHIA, B., LEACH, P., MANFERDELLI, J., MARUYAMA, H., NADALIN, A., NAGARATNAM, N., PRAFULLCHANDRA, H., SHEWCHUK, J., AND SIMON, D. Web Services Security (WS-Security), Apr. 2002. International Business Machines Corporation, Microsoft Corporation und VeriSign Inc.

[38] ATTANASIO, C. R., MARKSTEIN, P. W., AND PHILLIPS, R. J. Penetrating an Operating System: A Study of VM/370 Integrity. *IBM Systems Journal 15*, 1 (1976), 102–116.

[39] AURA, T. Distributed Access Rights Management with Delegation Certificates. In *Secure Internet Programming: Security Issues for Mobile and Distributed Objects* (New York, NY, USA, Oct. 1999), vol. 1603 of *Lecture Notes in Computer Science*, Springer Verlag, pp. 211–235.

[40] Auslander, M. A., Larkin, D. C., and Scherr, A. L. The Evolution of the MVS Operating System. *IBM Journal of Research and Development 25*, 5 (Sept. 1981), 471–482.

[41] Avitabile, M. An Examination of Requirements for Metapolicies in Policy-Based Management. Master's thesis, Institut für Informatik, Munich Technical University, Munich, Germany, Nov. 1998.

[42] Baader, F., and Nipkow, T. *Term Rewriting and All That*. Cambridge University Press, Cambridge, UK, 1998.

[43] Bace, R. *Intrusion Detection*. Macmillan Technical Publishing, Indianapolis, IN, USA, 2000.

[44] Bach, M. J. *The Design of the UNIX Operating System*. Prentice-Hall, Englewood Cliffs, NJ, USA, 1986.

[45] Baeten, J. C. M., and Weijland, W. P. *Process Algebra*, vol. 18 of *Cambridge Tracts In Theoretical Computer Science*. Cambridge University Press, Cambridge, UK, 1990.

[46] Bai, Y., and Varadarajan, V. A Logic for State Transformations in Authorization Policies. In *Proceedings of the 10th IEEE Computer Security Foundations Workshop (CSFW-10)* (Rockport, MA, USA, June 1997), IEEE Computer Society, pp. 173–182.

[47] Baier, C., and Kwiatkowska, M. On Topological Hierarchies of Temporal Properties. *Fundamenta Informaticae 41*, 3 (2000), 259–294.

[48] Baker, D. B. Fortresses Built Upon Sand. In *Proceedings of the Workshop on New Security Paradigms* (Great Langdale, UK, Sept. 1997), ACM Press, pp. 148–153.

[49] Baker, F. RFC 1812: Requirements for IP Version 4 Routers, June 1995. Obsoletes RFC1716, RFC1009.

[50] BALDWIN, R. W. Naming and Grouping Privileges to Simplify Security Management in Large Databases. In *Proceedings of the 1990 IEEE Symposium on Security and Privacy (SOSP '90)* (Oakland, CA, USA, May 1990), IEEE Computer Society, pp. 116–132.

[51] BALL, E., FELDMAN, J., LOW, J., RASHID, R., AND ROVNER, P. RIG: Rochester's Intelligent Gateway: System Overview. *IEEE Transactions on Software Engineering 2*, 4 (Dec. 1976), 321–328.

[52] BARDEN, R., STEPNEY, S., AND COOPER, D. *Z in Practice*. BCS Practicioner Series. Prentice Hall International, Hempstead, UK, 1994.

[53] BASS, T. Intrusion Detection Systems and Multisensor Data Fusion. *Communications of the Association for Computing Machinery 43*, 4 (Apr. 2000), 99–105.

[54] BAUER, L., SCHNEIDER, M. A., AND FELTEN, E. W. A General and Flexible Access-Control System for the Web. In *Proceedings of the 11th USENIX Security Symposium* (San Francisco, CA, USA, Aug. 2002), USENIX, Ed.

[55] BAUMANN, P., AND LERMER, K. A Framework for the Specification of Reactive and Concurrent Systems in Z. In *Proceedings of the 15th Conference on Foundations of Software Technology and Theoretical Computer Science* (Bangalore, India, Dec. 1995), P. S. Thiagarajan, Ed., vol. 1026 of *Lecture Notes in Computer Science*, Springer Verlag, pp. 62–79.

[56] BAUMANN, P., AND LERMER, K. Developing Reactive Systems with Z. Tech. Rep. 96-35, Software Verification Research Centre, Department of Computer Science, University of Queensland, Queensland, Australia, Dec. 1996.

[57] BAUMANN, P., AND LERMER, K. Specifying Parallel and Distributed Real-Time Systems in Z. In *Proceedings of the 4th International Workshop on Parallel and Distributed Real-Time Systems*

(Honolulu, HI, USA, Apr. 1996), IEEE Computer Society Press, pp. 216–222.

[58] Baumgartner, P., and Zhang, H., Eds. *Proceedings of FTP 2000: International Workshop on First-Order Theorem Proving* (St. Andrews, UK, July 2000), no. 5/2000 in Technical Report, Universität Koblenz-Landau, Germany.

[59] Beame, P. W., and Buss, S. R., Eds. *Proof Complexity and Feasible Arithmetics*, vol. 39 of *DIMACS Series in Discrete Mathematics and Theoretical Computer Science*. American Mathematical Society, 1998.

[60] Beebe, N. H. F. A Bibliography of Publications about the *AT&T Plan 9* Distributed Operating System. Tech. rep., Center for Scientific Computing, Department of Mathematics, University of Utah, Salt Lake City, UT, USA, May 1996.

[61] Bell, D. Modeling the "Multipolicy Machine". In *Proceedings of the Workshop on New Security Paradigms* (Little Compton, RI, USA, Sept. 1994), ACM Press, pp. 2–9.

[62] Bell, D. E., and LaPadula, L. J. Secure Computer Systems: Mathematical Foundations and Model. Tech. rep., The MITRE Corporation, Bedford, MA, USA, Mar. 1973. Number ESD-TR-74-244. Performed for the Electronic Systems Division, Air Force Systems Command, United States Air Force.

[63] Bell, D. E., and LaPadula, L. J. Secure Computer Systems. Tech. rep., The MITRE Corporation, Bedford, MA, USA, Oct. 1974. Number ESD-TR-73-278. This report was published in between 1973 and 1974 in three volumes and represent an expanded treatment of [62]. Performed for the Electronic Systems Division, Air Force Systems Command, United States Air Force.

[64] Bell, D. E., and LaPadula, L. J. Secure Computer Systems: Unified Exposition and Multics Interpretation. Tech. rep., The

MITRE Corporation, Bedford, MA, USA, Mar. 1976. Number ESD-TR-75-306. Performed for the Electronic Systems Division, Air Force Systems Command, United States Air Force.

[65] Bellovin, S. M. Towards a Commercial IP Security Option. In *Commercial IPSO Workshop, INTEROP '89* (San Jose, CA, USA, Oct. 1989).

[66] Bellovin, S. M. distributed firewalls. *;login: the USENIX Association newsletter 19*, Special Issue on Security (Nov. 1999), 39–47.

[67] Berkovits, S., Chokhani, S., Furlong, J. A., Geiter, J. A., and Guild, J. C. Public Key Infrastructure Study: Final Report. Tech. rep., The MITRE Corporation, McLean, VA, USA, Mar. 1994. Contract Number 50SBNB1C6732. Performed for the National Institute of Standards and Technology.

[68] Bernard, A., and Lee, P. Temporal Logic for Proof-Carrying Code. In *Proceedings of the 18th International Conference on Automated Deduction (CADE-18)* (Copenhagen, Denmark, July 2002), A. Vorkonkov, Ed., vol. 2392 of *Lecture Notes in Computer Science*, Springer Verlag, pp. 31–46.

[69] Berson, T. A., and Barksdale, Jr., G. L. KSOS — Development Methodology for a Secure Operating System. In *Proceedings of the National Computer Conference* (New York, NY, USA, June 1979), vol. 48, AFIPS, AFIPS Press, pp. 365–371.

[70] Bertino, E., Bettini, C., Ferrari, E., and Samarati, P. A Temporal Access Control Mechanism for Database Systems. *IEEE Transactions on Knowledge and Data Engineering 8*, 1 (Feb. 1996), 67–80.

[71] Bertino, E., Buccafurri, F., Ferrari, E., and Rullo, P. A Logical Framework for Reasoning on Data Access Control Policies. In *Proceedings of the 12th IEEE Computer Security Foundations*

Workshop (CSFW-12) (Mordano, Italy, June 1999), IEEE Computer Society, pp. 175–189.

[72] Bertino, E., Catania, B., Ferrari, E., and Perlasca, P. A Logical Framework for Reasoning about Access Control Models. In *Proceedings of the Sixth ACM Symposium on Access Control Models and Technologies* (Chantilly, VA, USA, May 2001), ACM Press, pp. 41–52.

[73] Bertino, E., Jajodia, S., and Samarati, P. Supporting Multiple Access Control Policies in Database Systems. In *Proceedings of the 1996 IEEE Symposium on Security and Privacy (SOSP '96)* (Oakland, CA, USA, May 1996), IEEE Computer Society, pp. 94–107.

[74] Bhattacharya, P. B., Jain, S. K., and Nagpaul, S. R. *Basic Abstract Algebra*. Cambridge University Press, Cambridge, UK, 1986.

[75] Biba, K. J. Integrity Considerations for Secure Computer Systems. Tech. rep., The MITRE Corporation, Bedford, MA, USA, Apr. 1977. Number ESD-TR-77-372. Performed for the Electronic Systems Division, Air Force Systems Command, United States Air Force.

[76] Bieber, P. An Epistemic and Deontic Logic for Reasoning about Computer Security. In *Proceedings of European Symposium on Research on Computer Security* (Toulouse, France, June 1990).

[77] Bieber, P., and Cuppens, F. Computer Security Policies and Deontic Logic. In *Proceedings of the Workshop on Deontic Logic and Computer Science* (Amsterdam, The Netherlands, Mar. 1991).

[78] Birkhoff, G. *Lattice Theory*, 3rd ed., vol. 25 of *Colloquium Publications*. American Mathematical Society, Providence, RI, USA, 1967.

[79] Birtwistle, G., and Subrahmanyam, P. A., Eds. *VLSI Specification, Verification and Synthesis*, vol. 35 of *The Kluwer International Series in Engineering and Computer Science*. Kluwer Academic Publishers, Dordrecht, The Netherlands, 1988.

[80] Bishop, P., and Bloomfield, R. A Conservative Theory for Long-term Reliability-growth Prediction of Software. *IEEE Transactions on Reliability 45*, 4 (Dec. 1996), 550–560.

[81] Bittler, K. J. A Policy-Independent Secure X Server. Master's thesis, Portland State University, Portland, OR, USA, 1997.

[82] Black, D. L., Golub, D. B., Julin, D. P., Rashid, R. F., Draves, R. P., Dean, R. W., Forin, A., Barrera, J., Tokadu, H., Malan, G., and Bohman, D. Microkernel Operating System Architecture and Mach. In *Proceedings of the USENIX Workshop on Micro-Kernel and Other Kernel Architectures* (Berkeley, CA, USA, Apr. 1992), USENIX, Ed., USENIX, pp. 11–30.

[83] Blakley, B., and Kienzle, D. M. Some Weaknesses of the TCB Model. In *Proceedings of the 1997 IEEE Symposium on Security and Privacy (SOSP '97)* (Oakland, CA, USA, May 1997), IEEE Computer Society, pp. 3–5. Position paper for the panel discussion: Is the Trusted Computing Base Concept Fundamentally Flawed?

[84] Blakley, R. The Measure of Information Security is Dollars. In *Workshop on Economics and Information Security at the University of California, Berkeley* (Berkeley, CA, USA, May 2002). Electronic proceedings only.

[85] Blaze, M. A Cryptographic Filesystem for Unix. In *Proceedings of the 1st ACM Conference on Computer and Communications Security* (Fairfax, VA, USA, Nov. 1993), ACM Press, pp. 9–16.

[86] Blaze, M., Feigenbaum, J., Ioannidis, J., and Keromytis, A. The Role of Trust Management in Distributed Systems. In *Secure Internet Programming: Security Issues for Mobile and Distributed*

Objects (New York, NY, USA, Oct. 1999), vol. 1603 of *Lecture Notes in Computer Science*, Springer Verlag, pp. 185–210.

[87] Blaze, M., Feigenbaum, J., and Keromytis, A. KeyNote: Trust Management for Public-Key Infrastructure. In *Proceedings of the 1998 Cambridge Security Protocols International Workshop* (Cambridge, UK, Dec. 1999), vol. 1550 of *Lecture Notes in Computer Science*, Springer Verlag, pp. 59–63.

[88] Blaze, M., Feigenbaum, J., and Lacy, J. Decentralized Trust Management. In *Proceedings of the 1996 IEEE Symposium on Security and Privacy (SOSP '96)* (Oakland, CA, USA, May 1996), IEEE Computer Society, pp. 164–173.

[89] Blotchky, S., Lynch, K., and Lipner, S. SE / VMS: Implementing Mandatory Security in VAX / VMS. In *Proceedings of the 9th DoD / NBS National Computer Security Conference* (Gaithersburg, MD, USA, Oct. 1986), pp. 47–54.

[90] Boebert, W. E., and Kain, R. Y. A Practical Alternative to Hierarchical Integrity Policies. In *Proceedings of the 8th DoD / NBS National Computer Security Conference* (Gaithersburg, MD, USA, Oct. 1985), pp. 18–27.

[91] Boebert, W. E., Kain, R. Y., and Young, W. D. Secure Computing: The Secure Ada Target Approach. *Scientific Honeyweller 6*, 2 (July 1985), 10–17. Reprinted in [723].

[92] Boebert, W. E., Young, W. D., Kain, R. Y., and Hansohn, S. A. Secure ADA Target: Issues, System Design, and Verification. In *Proceedings of the 1985 IEEE Symposium on Security and Privacy (SOSP '85)* (Oakland, CA, USA, Apr. 1985), IEEE Computer Society, pp. 176–183.

[93] Bomberger, A. C., Frantz, W. S., Hardy, A. C., Hardy, N., Landau, C. R., and Shapiro, J. S. The KeyKOS Nanokernel

Architecture. In *Proceedings of the USENIX Workshop on Micro-Kernels and Other Kernel Architectures* (Seattle, WA, USA, Apr. 1992), USENIX, Ed., USENIX, pp. 95–112.

[94] Bonatti, P., De Capitani di Vimecrati, S., and Samarati, P. A Modular Approach to Composing Access Control Policies. In *Proceedings of the 7th ACM Conference on Computer and Communications Security (CCS 2000)* (Athens, Greece, Nov. 2000), ACM SIGSAC, ACM Press, pp. 164–173.

[95] Bonatti, P., De Capitani di Vimecrati, S., and Samarati, P. An Algebra for Composing Access Control Policies. *ACM Transactions on Information and System Security 5*, 1 (Feb. 2002), 1–35.

[96] Bond, M. Attacks on Cryptoprocessor Transaction Sets. In *Cryptographic Hardware and Embedded Systems – CHES 2001. Proceedings of the Third International Workshop* (Paris, France, May 2001), Ç. Koç, D. Naccache, and C. Paar, Eds., vol. 2162 of *Lecture Notes in Computer Science*, Springer Verlag, pp. 220–234.

[97] Bonneau, C. H. SCOMP Security Kernel Verification Report. Tech. rep., Honeywell Information Systems, McLean, VA, USA, Apr. 1984.

[98] Book, R. V., and Otto, F. *String-Rewriting Systems*. Texts and Monographs in Computer Science. Springer Verlag, Heidelberg, Germany, 1993.

[99] Bourbaki, N. *General Topology, Volume 1*. Elements of Mathematics. Springer Verlag, Heidelberg, Germany, 1998. Contains chapters 1–4 of *Topologie Générale*.

[100] Bourbaki, N. *General Topology, Volume 2*. Elements of Mathematics. Springer Verlag, Heidelberg, Germany, 1998. Contains chapters 5–10 of *Topologie Générale*.

[101] Box, D., Ehnebuske, D., Kakivaya, G., Layman, A., Mendelsohn, N., Nielsen, H. F., Thatte, S., and Winer, D. Simple Object Access Protocol (SOAP) 1.1 . Tech. rep., W3C, May 2000. Status: W3C Note.

[102] Boyer, R., and Moore, J. S. *A Computational Logic*. Academic Press, San Diego, CA, USA, 1979.

[103] Boyer, R., and Moore, J. S. A Verification Condition Generator for Fortran. Tech. Rep. CSL-103, Computer Science Laboratory, SRI International, Menlo Park, CA, USA, June 1980.

[104] Brachman, R. J., and Levesque, H. J. *Readings in Knowledge Representation*. Morgan Kaufmann Publishers, Los Altos, CA, USA, 1985.

[105] Brady, R. M., Anderson, R. J., and Ball, R. C. Murphy s law, the fitness of evolving species, and the limits of software reliability. Tech. Rep. 476, Cambridge University, Cambridge, UK, Computer Laboratory, 1999.

[106] Branstad, M., Tajalli, H., Mayer, F., and Dalva, D. Access Mediation in a Message-Passing Kernel. In *Proceedings of the 1989 IEEE Symposium on Security and Privacy (SOSP '89)* (Oakland, CA, USA, Apr. 1989), IEEE Computer Society, pp. 66–72.

[107] Brassil, J., and O'Gorman, L. Watermarking Document Images with Bounding Box Expansion. In *Proceedings of Information Hiding: First International Workshop* (Cambridge, UK, May 1996), R. Anderson, Ed., vol. 1174 of *Lecture Notes in Computer Science*, Springer Verlag, pp. 227–235.

[108] Brewer, D. F. C., and Nash, M. J. The Chinese Wall Security Policy. In *Proceedings of the 1989 IEEE Symposium on Security and Privacy (SOSP '89)* (Oakland, CA, USA, Apr. 1989), IEEE Computer Society, pp. 206–214.

[109] Brinch Hansen, P. The Evolution of Operating Systems. In *Classic Operating Systems: From Batch Processing to Distributed Systems*, P. Brinch Hansen, Ed. Springer Verlag, Heidelberg, Germany, 2001.

[110] Broadbridge, R., and Mekota, J. Secure Communications Processor Specification. Tech. rep., Honeywell Information Systems, McLean, VA, USA, June 1976. Number ESD-TR-76-351. Performed for the Electronic Systems Division, Air Force Systems Command, United States Air Force.

[111] Brown, A., Fox, B., Hada, S., LaMacchia, B., and Maruyama, H. SOAP Security Extensions: Digital Signature, Feb. 2001. W3C Note.

[112] Brown, J. Toward an Economic Theory of Liability. *Journal of Legal Studies II*, 1 (Jan. 1973), 323–350.

[113] Büchi, J. R. Weak Second-Order Arithmetic and Finite Automata. *Zeitschrift für mathematische Logik und Grundlagen der Mathematik 6* (1960), 66–92.

[114] Büchi, J. R. On a Decision Method in Restricted Second Order Arithmetic. In *Proceedings of the International Congress on Logic, Method, and Philosophy of Science* (Stanford, CA, USA, 1962), E. Nagel, Ed., Stanford University Press, pp. 1–12.

[115] Büchi, J. R. The monadic theory of ω_1. In *Decidable Theories II*, J. R. Büchi and D. Siefkes, Eds., Lecture Notes in Mathematics. Springer Verlag, Heidelberg, Germany, 1973, pp. 1–127.

[116] Bundy, A. *The Computer Modelling of Mathematical Reasoning*. Academic Press, San Diego, CA, USA, 1983.

[117] Burrows, M., Abadi, M., and Needham, R. M. A Logic of Authentication. *ACM Transactions on Computer Systems 8*, 1 (Feb. 1990), 18–36.

[118] Burtescu, S., Corell, T., Langer, L., Mulliner, C., Rademer, E., Wald, C., Zeidler, A., and Wolthusen, S. D. *COSEDA – Comprehensive Security for Distributed Architectures – Architecture and Functional Specification*. Darmstadt, Germany, Nov. 2002. Internal use only.

[119] Busch, C., Rademer, E., Schmucker, M., and Wolthusen, S. Concepts for a Watermarking Technique for Music Scores. In *Proceedings of the Visual 2000/3rd International Conference on Visual Computing, Mexico City, Mexico* (Sept. 2000).

[120] Busch, C., and Wolthusen, S. D. *Netzwerksicherheit*. Spektrum Akademischer Verlag, Heidelberg, Germany, 2002.

[121] Butler, R. W., and Finelli, G. B. The Infeasibility of Quantifying the Reliability of Life-Critical Real-Time Software. *IEEE Transactions on Software Engineering 19*, 1 (Jan. 1993), 3–12.

[122] Cadoli, M., and Schaerf, M. A Survey of Complexity Results for Non-Monotonic Logics. *Journal of Logic Programming 17* (1993), 127–160.

[123] Caferra, R., and Salzer, G., Eds. *Proceedings of FTP 1998: International Workshop on First-Order Theorem Proving* (Vienna, Austria, Nov. 1998), no. E1852-GS-981 in Technical Report, Technische Universität Wien, Austria.

[124] Callen, H. B. *Thermodynamics and an Introduction to Thermostatistics*, 2nd ed. John Wiley & Sons, Inc., New York, NY, USA, 1985.

[125] Castro, M. *Practical Byzantine Fault Tolerance*. PhD thesis, Laboratory for Computer Science, Massachusetts Institute of Technology, Cambridge, MA, USA, 2001. MIT-LCS-TR-817.

[126] Chase, J. S., Levy, H. M., Feeley, M. J., and Lazowska, E. D. Sharing and Protection in a Single Address Space Operating

System. Tech. Rep. TR 93-04-02, University of Washington at Seattle, Seattle, WA, USA, Apr. 1993.

[127] Chaum, D. Design Concepts for Tamper Responding Systems. In *Advances in Cryptology (CRYPTO '83)* (Santa Barbara, CA, USA, Aug. 1983), D. Chaum, Ed., Lecture Notes in Computer Science, International Association for Cryptologic Research, Plenum Press, pp. 387–392.

[128] Cheheyl, M. H., Gasser, M., Huff, G. A., and Millen, J. K. Verifying Security. *ACM Computing Surveys 13*, 3 (Sept. 1981), 279–339.

[129] Cheswick, B. The Design of a Secure Internet Gateway. In *Proceedings of the Summer 1990 USENIX Conference* (Seattle, WA, USA, June 1990), USENIX, Ed., pp. 233–237.

[130] Chiles, J. R. *Inviting Disaster: Lessons from the Edge of Technology*. HarperCollins Publishers, New York, NY, USA, 2001.

[131] Chinchani, R., and Atkins, D. A Proactive Enhancement to the Distributed Firewall. In *Proceedings from the Third Annual IEEE SMC Information Assurance Workshop, United States Military Academy* (West Point, NY, USA, June 2002), IEEE Press, pp. 35–40.

[132] Chitturi, A. Implementing Mandatory Network Security in a Policy-flexible System. Master's thesis, University of Utah Department of Computer Science, Salt Lake City, UT, USA, June 1998.

[133] Chomicki, J., and Toman, D. Temporal Logic in Information Systems. In *Logics for Databases and Information Systems*, J. Chomicki and G. Saake, Eds. Kluwer Academic Publishers, 1998, ch. 3, pp. 31–70.

[134] Clark, A. J. Physical Protection of Cryptographic Devices. In *Advances in Cryptology (EUROCRYPT '87)* (Amsterdam, the

Netherlands, Apr. 1988), D. Chaum and W. L. Price, Eds., vol. 304 of *Lecture Notes in Computer Science*, International Association for Cryptologic Research, Springer Verlag, pp. 83–93.

[135] CLARK, D. D., AND WILSON, D. R. A Comparison of Commercial and Military Computer Security Policies. In *Proceedings of the 1987 IEEE Symposium on Security and Privacy (SOSP '87)* (Oakland, CA, USA, Apr. 1987), IEEE Computer Society, pp. 87–97.

[136] CLAVIER, C., CORON, J.-S., AND DABBOUS, N. Differential Power Analysis in the Presence of Hardware Countermeasures. In *Cryptographic Hardware and Embedded Systems – CHES 2000. Proceedings of the Second International Workshop* (Worcester, MA, USA, Aug. 2000), Ç. Koç and C. Paar, Eds., vol. 1965 of *Lecture Notes in Computer Science*, Springer Verlag, pp. 252–261.

[137] CLINT, M., AND HOARE, C. A. R. Program Proving: Jumps and Functions. *Acta Informatica 1* (1972), 214–224.

[138] CMU. Network Server Design. Tech. rep., Carnegie Mellon University, Pittsburgh, PA, USA, Aug. 1989.

[139] COHEN, E. Information Transmission in Computational Systems. *ACM Operating Systems Review 11*, 5 (1977), 133–139. Proceedings of the Sixth Symposium on Operating Systems Principles (6th SOSP '77).

[140] COHEN, E., AND JEFFERSON, D. Protection in the HYDRA Operating System. *ACM Operating Systems Review 9*, 5 (1975), 141–160. Proceedings of the Fifth ACM Symposium on Operating Systems Principles.

[141] COHN, A. J. A proof of correctness of the VIPER microprocessor: The first level. In Birtwistle and Subrahmanyam [79], pp. 27–71.

[142] COLBY, C., LEE, P., AND NECULA, G. C. A Proof-Carrying Code Architecture for Java. In *Proceedings of Computer Aided Verification, 12th International Conference (CAV 2000)* (Chicago, IL,

USA, July 2000), E. A. Emerson and A. P. Sistla, Eds., vol. 1855 of *Lecture Notes in Computer Science*, Springer Verlag, pp. 557–560.

[143] COMMISSION OF THE EUROPEAN COMMUNITIES. *Information Technology Security Evaluation Criteria (ITSEC), version 1.2*, June 1991.

[144] COMPAQ CORPORATION. *Software Product Description MLS+ Operating System, Version 4.0D SPD 46.21.04*, Apr. 1999.

[145] CONWAY, R. W., MAXWELL, W. L., AND MORGAN, H. L. On the Implementation of Security Measures in Information Systems. *Communications of the Association for Computing Machinery 15*, 4 (Apr. 1972), 211–220.

[146] COOK, J. V., CROCKER, S. D., PAGE, JR., T., POPEK, G., AND REIHER, P. Truffles: Secure File Sharing With Minimal System Administrator Intervention. In *Proceedings of the Second International Conference on Systems and Network Security and Management (SANS-II)* (Arlington, VA, USA, Apr. 1993), IEEE Computer Society Press.

[147] CORBATÓ, F. J., AND VYSSOTSKY, V. A. Introduction and Overview of the Multics System. In *Proceedings of the AFIPS Fall Joint Computer Conference (1965 FJCC)* (Washington D.C., USA, Nov. 1965), vol. 27 part 1, AFIPS, Spartan Books, pp. 185–196.

[148] CORNWELL, M. R., AND MOORE, A. P. Security Architecture for a Secure Military Message System. Tech. rep., Naval Research Laboratory, Washington D.C., USA, Apr. 1989. NRL Memorandum 9187.

[149] CORON, J.-S. Resistance against Differential Power Analysis for Elliptic Curve Cryptosystems. In *Cryptographic Hardware and Embedded Systems – CHES 1999. Proceedings of the First International Workshop* (Worcester, MA, USA, Aug. 1999), Ç. Koç and C. Paar, Eds., vol. 1717 of *Lecture Notes in Computer Science*,

International Association for Cryptologic Research, Springer Verlag, pp. 292–302.

[150] Coron, J.-S., and Goubin, L. On Boolean and Arithmetic Masking against Differential Power Analysis. In *Cryptographic Hardware and Embedded Systems – CHES 2000. Proceedings of the Second International Workshop* (Worcester, MA, USA, Aug. 2000), Ç. Koç and C. Paar, Eds., vol. 1965 of *Lecture Notes in Computer Science*, Springer Verlag, pp. 231–237.

[151] Cousot, P. Methods and Logics for Proving Programs. In *Handbook of Theoretical Computer Science*, J. van Leeuwen, Ed., vol. B. MIT Press, Cambridge, MA, USA, 1990, ch. 15, pp. 841–993.

[152] Cowan, C., Beattie, S., Pu, C., Wagle, P., and Gligor, V. SubDomain: Parsimonious Server Security. In *Proceedings of the 14th USENIX Systems Administration Conference (LISA 2000)* (San Francisco, CA, USA, Aug. 2000), USENIX, Ed.

[153] Cox, I. J., Miller, M. L., and Bloom, J. A. *Digital Watermarking*. The Morgan Kaufmann Series in Multimedia Information and Systems. Morgan Kaufmann Publishers, San Francisco, CA, USA, 2002.

[154] Crow, J., and Di Vito, B. L. Formalizing Space Shuttle Software Requirements: Four Case Studies. *ACM Transactions on Software Engineering and Methodology 7*, 3 (July 1998), 296–332.

[155] Cullen, H. F. *Introduction to General Topology*. Heath, Boston, MA, USA, 1968.

[156] Cummings, P. T., Fullam, D. A., Goldstein, M. J., Gosselin, M. J., Piccitto, J., Woodward, J. P. L., and Wynn, J. Compartmented Mode Workstation: Results Through Prototyping. In *Proceedings of the 1987 IEEE Symposium on Security and Privacy (SOSP '87)* (Oakland, CA, USA, Apr. 1987), pp. 2–12.

[157] Custer, H. *Inside Microsoft Windows NT*, 1st ed. Microsoft Press, Redmond, WA, USA, 1993.

[158] Custer, H. *Inside the Microsoft Windows NT File System*. Microsoft Press, Redmond, WA, USA, 1994.

[159] Daley, R. C., and Neumann, P. G. A General-Purpose File System for Secondary Storage. In *Proceedings of the AFIPS Fall Joint Computer Conference (1965 FJCC)* (Washington D.C., USA, Nov. 1965), vol. 27 part 1, AFIPS, Spartan Books, pp. 213–229.

[160] Davida, G. I., DeMillo, R. A., and Lipton, R. J. A System Architecture to Support a Verifiable Secure Multilevel Security System. In *Proceedings of the 1980 IEEE Symposium on Security and Privacy (SOSP '80)* (Oakland, CA, USA, Apr. 1980), IEEE Computer Society, pp. 137–145.

[161] Davis, M. The Early History of Automated Deduction. In *Handbook of Automated Reasoning*, J. A. Robinson and A. Voronkov, Eds., vol. I. MIT Press, Cambridge, MA, USA, 2001, ch. 1, pp. 3–15.

[162] Davis, R. E. *Truth, Deduction, and Computation: Logic and Semantics for Computer Science*. Computer Science Press, New York, NY, USA, 1989.

[163] Davis, R. G. *VAXCluster Principles*. Butterworth-Heinemann, Maynard, MA, USA, 1993.

[164] de Bakker, J. W. *Mathematical Theory of Program Correctness*. Prentice Hall, Englewood Cliffs, NJ, USA, 1980.

[165] de Cervantes Saavedra, M. *Don Quixote: Of the good fortune which the valiant Don Quixote had in the Terrible and Undreamed-of Adventure of the Windmills, with Other Occurrences Worthy to be Fitly Recorded*. W. W. Norton, New York, NY, USA, 1981. Translation by J. Ormsby, edited by J. R. Jones and K. Douglas.

[166] de Swart, H., Ed. *Automated Reasoning with Analytic Tableaux and Related Methods — International Conference TABLEAUX 1998* (Oisterwijk, The Netherlands, June 1998), vol. 1397 of *Lecture Notes in Computer Science*, Springer Verlag.

[167] de Vázquez Gómez, J. Multidomain Security. *Computers & Security Journal 13*, 2 (Apr. 1994), 161–184.

[168] Dekel, E., and Scotchmer, S. On the Evolution of Attitudes toward Risk in Winner-Take-All Games. *Journal of Economic Theory 87*, 1 (Jan. 1999), 125–143.

[169] Denning, D. E. *Secure Information Flow in Computer Systems.* PhD thesis, Computer Sciences Department, Purdue University, West Lafayette, IN, USA, 1975. CSD TR 145.

[170] Denning, D. E. A Lattice Model of Secure Information Flow. *Communications of the Association for Computing Machinery 19*, 5 (May 1976), 236–243.

[171] Denning, D. E. *Cryptography and Data Security.* Addison-Wesley, Reading, MA, USA, 1983.

[172] Denning, D. E. Cryptographic Checksums for Multilevel Database Security. In *Proceedings of the 1984 IEEE Symposium on Security and Privacy (SOSP '84)* (Oakland, CA, USA, Apr. 1984), IEEE Computer Society, pp. 52–61.

[173] Denning, D. E., Denning, P. J., Garland, S. J., Harrison, M. A., and Ruzzo, W. L. Proving Protection Systems Safe. Tech. rep., Computer Sciences Department, Purdue University, West Lafayette, IN, USA, Feb. 1978. Results are summarized in [171].

[174] Denning, D. E., Denning, P. J., and Graham, G. S. Seletively Confined Subsystems. In *Proceedings of the International Workshop on Protection in Operating Systems* (Rocquencourt, France, 1974), pp. 55–61.

[175] Denning, D. E., Lunt, T., Schell, R. R., Shockley, W. R., and Heckman, M. The SeaView Security Model. In *Proceedings of the 1988 IEEE Symposium on Security and Privacy (SOSP '88)* (Oakland, CA, USA, Apr. 1988), IEEE Computer Society, pp. 218–233.

[176] Dennis, J. B., and VanHorn, E. C. Programming Semantics for Multiprogrammed Computations. *Communications of the Association for Computing Machinery 9*, 3 (Mar. 1966), 143–155.

[177] Dershowitz, N., and Plaisted, D. A. Rewriting. In *Handbook of Automated Reasoning*, J. A. Robinson and A. Voronkov, Eds., vol. I. MIT Press, Cambridge, MA, USA, 2001, ch. 9, pp. 535–610.

[178] Dierks, T., and Allen, C. RFC 2246: The TLS Protocol Version 1.0, Jan. 1999.

[179] DiGiorgio, A. The Smart Ship Is Not the Answer. *Proceedings of the United States Naval Institute 124*, 6 (June 1998), 61–64.

[180] Digital Equipment Corporation. *decsystem-1080/1090 SYSTEM DESCRIPTION*. Marlborough, MA, USA, Jan. 1977. Document EK-1080U-SD-003.

[181] Digital Equipment Corporation. *TOPS-10 Operator's Guide*. Marlborough, MA, USA, Apr. 1986. Document AA-H283B-TB, covers TOPS-10 V7.03 and the GALAXY V5.1, RSX-20F V15-15, and BACKUP V5 subsystems.

[182] Director of Central Intelligence (DCI). *DCID 1/7: Control of Dissemination of Intelligence Information*. Langley, VA, USA, May 1981.

[183] Dubois, D., and Prade, H., Eds. *Handbook of Defeasible Reasoning and Uncertainty Management Systems: Belief Change*, vol. 3. Kluwer Academic Publishers, Dordrecht, the Netherlands, 1998.

[184] Dumas, L. J. *Lethal Arrogance: Human Fallibility and Dangerous Technologies*. St. Martin's Press, New York, NY, USA, 1999.

[185] Dyer, J. G., Lindemann, M., Perez, R., , Sailer, R., van Doorn, L., Smith, S. W., and Weingart, S. Building the IBM 4758 Secure Coprocessor. *IEEE Computer 34*, 10 (Oct. 2001), 57–66.

[186] Eddon, G., and Eddon, H. *Inside Distributed COM*. Microsoft Press, Redmond, WA, USA, 1998.

[187] Edwards, N., Berger, J., and Choo, T. H. A Secure Linux Platform. In *Proceedings of the 5th Annual Linux Showcase* (Oakland, CA, USA, Nov. 2001).

[188] Eilenberg, S. *Automata, Languages, and Machines*, vol. A. Academic Press, San Diego, CA, USA, 1974.

[189] Eilenberg, S. *Automata, Languages, and Machines*, vol. B. Academic Press, San Diego, CA, USA, 1976.

[190] Emerson, E. A. Temporal and Modal Logic. In *Handbook of Theoretical Computer Science*, J. van Leeuwen, Ed., vol. B. MIT Press, Cambridge, MA, USA, 1990, ch. 16, pp. 995–1072.

[191] Enderton, H. B. *Elements of Set Theory*. Academic Press, San Diego, CA, USA, 1977.

[192] Engeler, E., Ed. *Procedures and Parameters: An axiomatic Approach* (Berlin, Germany, June 1971), vol. 188 of *Lecture Notes in Computer Science*, Springer Verlag.

[193] Epstein, J. Architecture and Concepts of the ARGuE Guard. In *Proceedings 15th Annual Computer Security Applications Conference (ACSAC'99)* (Phoenix, AZ, USA, Dec. 1999), IEEE Computer Society Press, pp. 45–54.

[194] Epstein, J., McHugh, J., Pascale, R., Orman, H., Benson, G., Martin, C., Marmor-Squires, A., Danner, B., and Branstad, M. A Prototype B3 Trusted X Window System. In *Proceedings*

of the 7th Annual Computer Security Applications Conference, San Antonio, TX, USA (San Antonio, TX, USA, Dec. 1991), IEEE Computer Society Press, pp. 211–218.

[195] FABRY, R. S. Capability-Based Addressing. *Communications of the Association for Computing Machinery 17*, 7 (July 1974), 403–412.

[196] FAHN, P. N., AND PEARSON, P. K. IPA: A New Class of Power Attacks. In *Cryptographic Hardware and Embedded Systems – CHES 1999. Proceedings of the First International Workshop* (Worcester, MA, USA, Aug. 1999), Ç. Koç and C. Paar, Eds., vol. 1717 of *Lecture Notes in Computer Science*, Springer Verlag, pp. 173–186.

[197] FALL, K., AND FLOYD, S. Simulation-based Comparisons of Tahoe, Reno, and SACK TCP. *ACM Computer Communication Review 26*, 3 (July 1996), 5–21.

[198] FANO, R. M., AND CORBATÓ, F. J. Time-Sharing on Computers. *Scientific American 215*, 3 (Sept. 1966), 128–140.

[199] FANTECHI, A., GNESI, S., AND SEMINI, L. Formal Description and Validation for an Integrity Policy Supporting Multiple Levels of Criticality. In *Proceedings of the 7th International IFIP Conference on Dependable Computing for Critical Applications (DCCA-7)* (San Jose, CA, USA, Jan. 1999), C. B. Weinstock and J. Rushby, Eds., IEEE Press, pp. 129–146.

[200] FEHR, E., AND GÄCHTER, S. Altruistic punishment in humans. *Nature 415*, 6868 (Jan. 2002), 137–140.

[201] FEIERTAG, R. J. A Technique for Proving Specifications are Multilevel Secure. Tech. Rep. CSL-109, Computer Science Laboratory, SRI International, Menlo Park, CA, USA, Jan. 1980.

[202] FEIERTAG, R. J., LEVITT, K. N., AND ROBINSON, L. Proving Multilevel Security of a System Design. *ACM Operating Systems*

Review (1977), 57–65. Proceedings of the Sixth ACM Symposium on Operating Systems Principles.

[203] FEIERTAG, R. J., AND NEUMANN, P. G. The Foundations of a Provably Secure Operating System. In *Proceedings of the National Computer Conference* (New York, NY, USA, Nov. 1979), vol. 48, AFIPS, AFIPS Press, pp. 329–334.

[204] FEINLER, E. J., JACOBSEN, O. J., STAHL, M. K., AND WARD, C. A., Eds. *DDN Protocol Handbook*, vol. 1. SRI International DDN Network Information Center, Menlo Park, CA, USA, Dec. 1985. Prepared for the United States Defense Communications Agency.

[205] FENTON, J. S. *Information Protection Systems*. PhD thesis, Cambridge University, Cambridge, UK, 1973.

[206] FERRAIOLO, D. F., CUGINI, J. A., AND KUHN, D. R. RBAC: Features and Motivations. In *Proceedings 11th Annual Computer Security Applications Conference (ACSAC'95)* (New Orleans, LA, USA, Dec. 2001), IEEE Computer Society Press, pp. 185–192.

[207] FERRAIOLO, D. F., AND KUHN, D. R. Role-Based Access Control. In *Proceedings 15th NIST-NCSC National Computer Security Conference* (Baltimore, MD, USA, Oct. 1992), pp. 554–563.

[208] FERRAIOLO, D. F., KUHN, D. R., AND CHANDRAMOULI, R. *Role-based Access Controls*. The Artech House Computer Security Series. Artech House, Boston, MA, USA, 2003.

[209] FERRAIOLO, D. F., SANDHU, R. S., GAVRILA, S., KUHN, D. R., AND CHANDRAMOULI, R. Proposed NIST Standard for Role-Based Access Control. *ACM Transactions on Information and System Security 4*, 3 (Aug. 2001), 224–274.

[210] FIELDING, R., GETTYS, J., MOGUL, J., FRYSTYK, H., MASINTER, L., LEACH, P., AND BERNERS-LEE, T. RFC 2616: Hypertext Transfer Protocol – HTTP/1.1, June 1999.

[211] Fine, T. Defining Noninterference in the Temporal Logic of Actions. In *Proceedings of the 1996 IEEE Symposium on Security and Privacy (SOSP '96)* (Oakland, CA, USA, May 1996), IEEE Computer Society, pp. 12–23.

[212] Fine, T., and Minear, S. E. Assuring Distributed Trusted Mach. In *Proceedings of the 1993 IEEE Symposium on Security and Privacy (SOSP '93)* (Oakland, CA, USA, May 1993), IEEE Computer Society, pp. 206–218.

[213] Floyd, R. W. Assigning Meanings to Programs. In *Proceedings of a Symposium on Applied Mathematics* (1967), American Mathematical Society, pp. 19–32.

[214] Focardi, R., and Gorrieri, R. A Taxonomy of Trace-Based Security Properies for CCS. In *Proceedings of the 7th IEEE Computer Security Foundations Workshop (CSFW-7)* (Franconia, NH, USA, June 1994), IEEE Computer Society, pp. 126–136.

[215] Foley, M., and Hoare, C. A. R. Proof of a Recursive Program: QUICKSORT. *Computer Journal 14*, 4 (1971), 391–395.

[216] ΟΜΗΡΟΥ. ΙΛΙΑΣ. Artemis & Winkler, Zürich, Switzerland, 1994.

[217] Ford, B., Back, G., Benson, G., Lepreau, J., Lin, A., and Shivers, O. . *ACM Operating Systems Review 31*, 5 (Oct. 1997), 38–51. Proceedings of the 15th Symposium on Operating Systems Principles.

[218] Ford, B., Hibler, M., Lepreau, J., McGrath, R., and Tullmann, P. Interface and Execution Models in the Fluke Kernel. In *Proceedings of the 3rd Symposium on Operating Systems Design and Implementation (OSDI '99)* (New Orleans, LA, USA, Feb. 1999), USENIX, Ed., ACM Press, pp. 101–116. Published as an ACM Operating Systems Review Special Issue.

[219] Ford, B., Lepreau, J., Clawson, S., and van Maren, K. The Flux OS Toolkit: Reusable Components for OS Implementation. In *The Sixth Workshop on Hot Topics in Operating Systems* (Cape Cod, MA, USA, 1997), IEEE Computer Society Press, pp. 14–19.

[220] Ford Aerospace and Communications Corporation. Provably Secure Operating System (PSOS) Final Report. Tech. Rep. Contract MDA 904-80-C-0470, Ford Aerospace and Communications Corporation, June 1980.

[221] Fraim, L. SCOMP: A Solution to the Multilevel Security Problem. *IEEE Computer 16*, 7 (July 1983), 26–34.

[222] Frantz, W. S. KeyKOS – A Secure, High-Performance Environment for S/370. In *Proceedings of SHARE 70 I* (Chicago, IL, USA, Feb. 1988), SHARE, Inc., pp. 465–471.

[223] Frantz, W. S., Hardy, N., Jonekait, J., and Landau, C. R. GNOSIS - A Prototype Operating System for the 1990's. In *Proceedings of SHARE 52 I* (Chicago, IL, USA, Mar. 1979), SHARE, Inc., pp. 3–17.

[224] Fraser, T. LOMAC: Low Water-Mark Integrity Protection for COTS Environments. In *Proceedings of the 2000 IEEE Symposium on Security and Privacy (SOSP '00)* (Oakland, CA, USA, May 2000), IEEE Computer Society, pp. 230–245.

[225] Fraser, T. LOMAC: MAC You Can Live With. In *Proceedings of the FREENIX Track, 2001 USENIX Annual Technical Conference* (San Francisco, CA, USA, Aug. 2001), USENIX, Ed., pp. 1–13.

[226] Fraser, T., Badger, L., and Feldman, M. Hardening COTS Software with Generic Software Wrappers. In *Proceedings of the 1999 IEEE Symposium on Security and Privacy (SOSP '99)* (Oakland, CA, USA, May 1999), IEEE Computer Society, pp. 2–16.

[227] Fraser, T. J. An Object-Oriented Framework For Security Policy Representation. Master's thesis, University of Illinois at Urbana-Champaign Department of Computer Science, Urbana, IL, USA, Jan. 1997.

[228] Frederick, R. *Fundamentals of Statistical and Thermal Physics.* McGraw-Hill, New York, NY, USA, 1974.

[229] Friedman, D., and Nagle, D. F. Building Scalable Firewalls with Intelligent Network Interface Cards. Tech. Rep. CMU-CS-00-173, Carnegie Mellon University School of Computer Science, Pittsburgh, PA, USA, Dec. 2000.

[230] Gabbay, D. M., Hodkinson, I., and Reynolds, M. A., Eds. *Temporal Logic: Mathematical Foundations and Computational Aspects, Volume I*, vol. 28 of *Oxford Logic Guides*. Oxford University Press, Oxford, UK, 1993.

[231] Gabbay, D. M., Hodkinson, I., and Reynolds, M. A., Eds. *Temporal Logic: Mathematical Foundations and Computational Aspects, Volume II*, vol. 40 of *Oxford Logic Guides*. Oxford University Press, Oxford, UK, 1993.

[232] Gabbay, D. M., Hogger, C. J., Robinson, J. A., and Galton, A. P., Eds. *Epistemic and Temporal Reasoning*, vol. IV of *Handbook of Logic in Artificial Intelligence and Logic Programming*. Oxford University Press, Oxford, UK, 1993.

[233] Ganger, G. R. Blurring the Line Between OSes and Storage Devices. Tech. Rep. CMU-CS-01-166, Carnegie Mellon University School of Computer Science, Pittsburgh, PA, USA, Dec. 2001.

[234] Ganger, G. R., and Nagle, D. F. Enabling Dynamic Security Management of Networked Systems via Device-Embedded Security. Tech. Rep. CMU-CS-00-174, Carnegie Mellon University School of Computer Science, Pittsburgh, PA, USA, Dec. 2000.

[235] Ganger, G. R., and Nagle, D. F. Better Security via Smarter Devices. In *The Eighth Workshop on Hot Topics in Operating Systems* (Elmau, Germany, May 2001), IEEE Computer Society Press, pp. 100–105.

[236] Ganter, B., and Wille, R. *Formal Concept Analysis – Mathematical Foundations*. Springer Verlag, Heidelberg, Germany, 1998. Originally released in German as "Formale Begriffsanalyse – Mathematische Grundlagen".

[237] Ganzinger, H., Ed. *Proceedings of the 16th International Conference on Automated Deduction (CADE-16)* (Trento, Italy, July 1999), vol. 1632 of *Lecture Notes in Computer Science*, Springer Verlag.

[238] Gärdenfors, P. Belief Revision and Nonmonotonic Logic: Two Sides of the Same Coin? In *Proceedings of the European Workshop on Logics in AI (JELIA'90)* (Amsterdam, the Netherlands, Sept. 1990), J. van Eijck, Ed., vol. 478 of *Lecture Notes in Computer Science*, Springer Verlag, pp. 52–54.

[239] Garey, M. R., and Johnson, D. S. *Computers and Intractability: A Guide to the Theory of NP-Completeness*. W.H. Freeman & Company, San Francisco, CA, USA, 1978.

[240] Gasser, M. *Building a Secure Computer System*. Van Nostrand Reinhold, New York, NY, USA, 1988.

[241] Genesereth, M., and Nilsson, N. J. *Logical Foundations of Artificial Intelligence*. Morgan Kaufmann Publishers, Los Altos, CA, USA, 1988.

[242] Gentzen, G. Untersuchungen über das logische Schliessen I. *Mathematische Zeitschrift 39* (1934), 176–210.

[243] Gentzen, G. Untersuchungen über das logische Schliessen II. *Mathematische Zeitschrift 39* (1934), 405–431.

[244] Gibbons, A. *Algorithmic Graph Theory*. Cambridge University Press, Cambridge, UK, 1985.

[245] Girard, J. Y. *Proof Theory and Logical Complexity*. Bibliopolis, Naples, Italy, 1987.

[246] Glasgow, J., and MacEwen, G. Obligation as the Basis of Integrity Specification. In *Proceedings of the Computer Security Foundations Workshop II* (Franconia, NH, USA, 1989), IEEE Computer Society Press, pp. 64–70.

[247] Glasgow, J., MacEwen, G., and Panangaden, P. Reasoning about Knowledge and Permission in Secure Distributed Systems. In *Proceedings of the Computer Security Foundations Workshop* (Franconia, NH, USA, 1988), IEEE Computer Society Press, pp. 122–128.

[248] Glasgow, J., MacEwen, G., and Panangaden, P. A Logic For Reasoning about Security. *ACM Transactions on Computer Systems 10*, 3 (Aug. 1992), 226–264.

[249] Gligor, V., Chandersekaran, C., Chapman, R., Dotterer, L., Hecht, M., Jiang, W., Johri, A., Luckenbaugh, G., and Vasudevan, N. Design and Implementation of Secure Xenix. *IEEE Transactions on Software Engineering 13*, 2 (Feb. 1987), 208–220.

[250] Gligor, V. D. A Note on the Denial-of-Service Problem. In *Proceedings of the 1983 IEEE Symposium on Security and Privacy (SOSP '83)* (Oakland, CA, USA, Apr. 1983), IEEE Computer Society, pp. 139–149.

[251] Gligor, V. D. 20 Years of Operating Systems Security. In *Proceedings of the 1999 IEEE Symposium on Security and Privacy (SOSP '99)* (Oakland, CA, USA, May 1999), IEEE Computer Society, pp. 108–110.

[252] Gödel, K. Die Vollständigkeit der Axiome des logischen Funktionenkalküls. *Monatshefte für Mathematik und Physik 37* (1930), 349–360.

[253] GÖDEL, K. Über formal unentscheidbare Sätze der Principia Mathematica. *Monatshefte für Mathematik und Physik 38* (1931), 175–198.

[254] GÖDEL, K. Ein Spezialfall des Entscheidungsproblems der theoretischen Logik. *Ergebnisse eines mathematischen Kolloquiums 2* (1932), 27–28.

[255] GOGUEN, J. A., AND MESEGUER, J. Security Policy and Security Models. In *Proceedings of the 1982 IEEE Symposium on Security and Privacy (SOSP '82)* (Oakland, CA, USA, Apr. 1982), IEEE Computer Society, pp. 11–20.

[256] GOGUEN, J. A., AND MESEGUER, J. Unwinding and Inference Control. In *Proceedings of the 1984 IEEE Symposium on Security and Privacy (SOSP '84)* (Oakland, CA, USA, Apr. 1984), IEEE Computer Society, pp. 75–86.

[257] GOHEEN, S. M., AND FISKE, R. S. OS/360 Computer Security Penetration Exercise. Tech. rep., The MITRE Corporation, Bedford, MA, USA, Oct. 1976. Number WP-4467.

[258] GOLD, B. D., LINDE, R. R., PEELER, R. J., SCHAEFER, M., SCHEID, J. F., AND WARD, P. D. A Security Retrofit of VM/370. In *Proceedings of the National Computer Conference* (New York, NY, USA, Nov. 1979), vol. 48, AFIPS, AFIPS Press, pp. 335–344.

[259] GOLDENBERG, R. E., AND SARAVANAN, S. *Open VMS AXP Internals and Data Structures: Version 1.5.* Digital Press, Maynard, MA, USA, 1994.

[260] GOLDSTINE, H. H., AND VON NEUMANN, J. Planning and Coding of Problems for an Electronic Computing Instrument. In *Collected Works of J. von Neumann*, A. Taub, Ed., vol. 5. Pergamon Press, New York, NY, USA, 1965, pp. 80–151. Report for the U.S. Ordinance Department.

[261] Golub, D. B., Dean, R., Forin, A., and Rashid, R. UNIX as an Application Program. In *Proceedings of the Summer 1990 USENIX Conference* (Seattle, WA, USA, June 1990), USENIX, Ed., USENIX, pp. 87–95.

[262] Good, D. I., Akers, R., and Smith, L. Report on Gypsy 2.05. Tech. Rep. 48, Computational Logic, Inc., Austin, TX, USA, 1986. Revised from [263].

[263] Good, D. I., Cohen, R. M., Hoch, C. G., Hunter, L. W., and Hare, D. F. Report on the Language Gypsy: Version 2.0. Tech. Rep. ICSCA-10, University of Texas at Austin Institute for Computing Science and Computer Applications, Austin, TX, USA, Sept. 1978.

[264] Good, D. I., and Young, W. D. Mathematical Methods for Digital Systems Development. In *Proceedings of VDM '91 – Formal Software Development, 4th International Symposium of VDM Europe* (Noordwijkerhout, The Netherlands, Oct. 1991), S. Prehn and W. J. Toetenel, Eds., vol. 552 of *Lecture Notes in Computer Science*, Springer Verlag, pp. 406–430.

[265] Goodheart, B., and Cox, J. *The Magic Garden Explained: The Internals of Unix System V Release 4*. Prentice Hall, Englewood Cliffs, NJ, USA, 1994.

[266] Goodwin, J., Mitchell, G., and Tasker, P. S. Concept of Operations for Message Handling at CINCPAC. Tech. rep., The MITRE Corporation, Bedford, MA, USA, Oct. 1976. Number MTR-3323.

[267] Gordon, L. A., and Loeb, M. P. The Economics of Information Security Investment. *ACM Transactions on Information and System Security 5*, 4 (Nov. 2002), 438–457.

[268] Gordon, M. J. C., and Melham, T. F. *Introduction to HOL*. Cambridge University Press, Cambridge, UK, 1993.

[269] Goré, R., Leitsch, A., and Nipkow, T., Eds. *Proceedings of Automated Reasoning: First International Joint Conference (IJCAR 2001)* (Siena, Italy, June 2001), vol. 1397 of *Lecture Notes in Computer Science*, Springer Verlag.

[270] Goubin, L., and Patarin, J. DES and Differential Power Analysis (The "Duplication" Method). In *Cryptographic Hardware and Embedded Systems – CHES 1999. Proceedings of the First International Workshop* (Worcester, MA, USA, Aug. 1999), Ç. Koç and C. Paar, Eds., vol. 1717 of *Lecture Notes in Computer Science*, Springer Verlag, pp. 158–172.

[271] Graham, G. S., and Denning, P. J. Protection — Principles and Practice. In *Proceedings of the AFIPS Spring Joint Computer Conference (1972 SJCC)* (Montvale, NJ, USA, 1972), vol. 40, AFIPS, AFIPS Press, pp. 417–427.

[272] Graham, M. The SPIREL System for the Rice University Computer,. Tech. rep., Rice University, Houston, TX, USA, Apr. 1964.

[273] Grätzer, G. A. *General Lattice Theory*, 2nd ed. Birkhäuser, Providence, RI, USA, 1998.

[274] Gray, J. W. Probabilistic Interference. In *Proceedings of the 1990 IEEE Symposium on Security and Privacy (SOSP '90)* (Oakland, CA, USA, Apr. 1990), IEEE Computer Society, pp. 170–179.

[275] Gray, J. W. Towards a Mathematical Foundation for Information Flow Security. In *Proceedings of the 1991 IEEE Symposium on Security and Privacy (SOSP '91)* (Oakland, CA, USA, Apr. 1991), IEEE Computer Society, pp. 21–34.

[276] Gray, J. W., and Syverson, P. F. A Logical Approach to Multilevel Security of Probabilistic Systems. In *Proceedings of the 1992 IEEE Symposium on Security and Privacy (SOSP '92)* (Oakland, CA, USA, Apr. 1992), IEEE Computer Society, pp. 164–176.

[277] GREENWALD, S. J. A New Security Policy for Distributed Resource Management and Access Control. In *Proceedings of the Workshop on New Security Paradigms* (Lake Arrowhead, CA, USA, Sept. 1996), ACM Press, pp. 74–86.

[278] GRIMSHAW, A. S., WULF, W. A., FRENCH, J. C., WEAVER, A. C., AND REYNOLDS, JR., P. F. A Synopsis of the Legion Project. Tech. Rep. CS-94-20, University of Virginia Department of Computer Science, Charlottesville, VA, USA, June 1994.

[279] GUDGIN, M., HADLEY, M., MOREAU, J.-J., AND NIELSEN, H. F. Simple Object Access Protocol (SOAP) 1.2, Dec. 2001. W3C Working Draft, consists of three parts.

[280] GUESSARIAN, I., AND NIAR-DINEDANE, W. Fairness and Regularity for SCCS Processes. *Informatique Théorique et Applications 23* (1989), 59–86.

[281] HÄHNLE, R. Tableaux and Related Methods. In *Handbook of Automated Reasoning*, J. A. Robinson and A. Voronkov, Eds., vol. I. MIT Press, Cambridge, MA, USA, 2001, ch. 3, pp. 100–178.

[282] HÄHNLE, R., KERBER, M., AND WEIDENBACH, C. Common Syntax of the DFG-Schwerpunktprogramm Deduktion Version 1.3. Tech. rep., Universität Karlsruhe, Fakultät für Informatik, Karlsruhe, Germany, Oct. 1996.

[283] HAIGH, J. T. A Comparison of Formal Security Models. In *Proceedings of the 7th DoD / NBS National Computer Security Conference* (Gaithersburg, MD, USA, Sept. 1984), pp. 88–119.

[284] HALL, D. L. *Mathematical Techniques in Multisensor Data Fusion*. The Artech House Radar Library. Artech House, Boston, MA, USA, 1992.

[285] HALL, D. L., AND LLINAS, J., Eds. *Handbook of Multisensor Data Fusion*. The Electrical Engineering and Applied Signal Processing Series. CRC Press, Boca Raton, FL, USA, 2001.

[286] Hallyn, S., and Kearns, P. Domain and Type Enforcement for Linux. In *Proceedings of the 4th Annual Linux Showcase* (Atlanta, GA, USA, Oct. 2000).

[287] Handschuh, H., Paillier, P., and Stern, J. Probing Attacks on Tamper-Resistant Devices. In *Cryptographic Hardware and Embedded Systems – CHES 1999. Proceedings of the First International Workshop* (Worcester, MA, USA, Aug. 1999), Ç. Koç and C. Paar, Eds., vol. 1717 of *Lecture Notes in Computer Science*, International Association for Cryptologic Research, Springer Verlag, pp. 303–315.

[288] Hardin, G. The Tragedy of the Commons. *Science 162*, 3859 (Dec. 1968), 1243–1248.

[289] Hardy, N. The KeyKOS Architecture. *ACM Operating Systems Review 19*, 4 (1985), 8–25.

[290] Harkins, D., and Carrel, D. RFC 2409: The Internet Key Exchange (IKE), Nov. 1998.

[291] Harrison, A. The Problem of Privacy in the Computer Age: An Annotated Bibliography. Tech. rep., The RAND Corporation, Santa Monica, CA, USA, Dec. 1967. Number RM-5495-PR/RC.

[292] Harrison, M. A., and Ruzzo, W. L. Monotonic Protection Systems. In *Foundations of Secure Computation*, R. DeMillo, D. Dobkin, A. Jones, and R. Lipton, Eds. Academic Press, New York, NY, USA, 1978, pp. 337–365.

[293] Harrison, M. A., Ruzzo, W. L., and Ullman, J. D. Protection in Operating Systems. *Communications of the Association for Computing Machinery 19*, 8 (Aug. 1976), 461–471.

[294] Hasan, M. A. Power Analysis Attacks and Algorithmic Approaches to their Countermeasures for Koblitz Curve Cryptosystems. In *Cryptographic Hardware and Embedded Systems*

– *CHES 2000. Proceedings of the Second International Workshop* (Worcester, MA, USA, Aug. 2000), Ç. Koç and C. Paar, Eds., vol. 1965 of *Lecture Notes in Computer Science*, Springer Verlag, pp. 93–108.

[295] Hauert, C., De Monte, S., Hofbauer, J., and Sigmund, K. Volunteering as Red Queen Mechanism for Cooperation in Public Goods Games. *Science 296*, 5570 (May 2002), 1129–1132.

[296] Heitmeyer, C. L., and Cornwell, M. Specifications for Three Members of the Military Message System (MMS) Family. Tech. rep., Naval Research Laboratory, Washington D.C., USA, Mar. 1982. NRL Memorandum 5654.

[297] Heitmeyer, C. L., and Landwehr, C. Designing Secure Message Systems: The Military Message Systems (MMS) Project. In *Proceedings of the IFIP TC6.5 Working Conference on Computer-Based Message Services* (Nottingham, UK, May 1984), IFIP, Kluwer Academic Publishers, pp. 245–255.

[298] Herbrand, J. Recherches sur la theorie de la démonstration. *Travaux de la Societé des Sciences et des Lettres de Varsovie 33* (1930).

[299] Hilpinen, R. *Deontic Logic: Introductory and Systematic Readings*. D. Reidel Publishing Company, Dordrecht, The Netherlands, 1981.

[300] Hinchey, M. G., and Bowen, J. P. *Applications of Formal Methods*. Prentice Hall International Series in Computer Science. Prentice Hall, London, UK, 1995.

[301] Hintikka, J. *Knowledge and Belief*. Cornell University Press, Ithaca, NY, USA, 1962.

[302] Hoare, C. A. R. An Axiomatic Basis for Computer Programming. *Communications of the Association for Computing Machinery 12*, 10 (Oct. 1969), 576–580.

[303] Hoare, C. A. R. Proof of Correctness of Data Representations. *Acta Informatica 1* (1972), 271–281.

[304] Hoare, C. A. R. Towards a Theory of Parallel Programming. In *Operating System Techniques*, C. A. R. Hoare and R. H. Prescott, Eds. Academic Press, San Diego, CA, USA, 1972, pp. 61–71.

[305] Hoare, C. A. R. Monitors: An Operating System Structuring Concept. *Communications of the Association for Computing Machinery 17*, 10 (Oct. 1974), 549–557.

[306] Hoare, C. A. R. Parallel Programming: An Axiomatic Approach. *Computer Languages 1*, 2 (1975), 151–160.

[307] Hoare, C. A. R. Communicating Sequential Processes. *Communications of the Association for Computing Machinery 21*, 8 (Aug. 1978), 666–677.

[308] Hoare, C. A. R. A Calculus of Total Correctness for Communicating Processes. *Science of Computer Programming 1*, 1-2 (1981), 49–72.

[309] Hoare, C. A. R. How Did Software Get So Reliable Without Proof? In *Proceedings of FME '96: Industrial Benefit and Advances in Formal Methods, Third International Symposium of Formal Methods Europe* (Oxford, UK, Mar. 1996), M.-C. Gaudel and J. Woodcock, Eds., vol. 1051 of *Lecture Notes in Computer Science*, Springer Verlag, pp. 1–17.

[310] Hoffman, L. J. Computers and Privacy: A Survey. Tech. rep., Stanford Linear Accelerator Center, Stanford University, Stanford, CA, USA, Aug. 1968. Number SLAC-PUB-479.

[311] Hosmer, H. Integrating Security Policies. In *Proceedings of the Third RADC Database Security Workshop* (Castile, NY, USA, Oct. 1990), The MITRE Corporation, pp. 169–173.

[312] Hosmer, H. H. Shared Sensitivity Label. In *Proceedings of Database Security IV: Status and Prospects. Results of the IFIP WG*

11.3 Workshop on Database Security (Halifax, UK, Sept. 1990), S. Jajodia and C. E. Landwehr, Eds., pp. 251–264.

[313] Hosmer, H. H. Metapolicies I. In *Proceedings of the ACM SIGSAC Data Management Workshop* (San Antonio, TX, USA, Dec. 1991), pp. 18–43. Workshop was colocated with ACSAC '91, proceedings were published in ACM SIGSAC Review 1992.

[314] Hosmer, H. H. The Multipolicy Machine: A New Paradigm for Multilevel Secure Systems. In *Proceedings of Standard Security Label for GOSIP, an Invitational Workshop* (Gaithersburg, MD, USA, Apr. 1991).

[315] Hosmer, H. H. Metapolicies II. In *Proceedings 15th NIST-NCSC National Computer Security Conference* (Baltimore, MD, USA, Oct. 1992), pp. 369–378.

[316] Hosmer, H. H. The Multipolicy Paradigm. In *Proceedings 15th NIST-NCSC National Computer Security Conference* (Baltimore, MD, USA, Oct. 1992), pp. 409–422.

[317] Hughes, G., and Cresswell, M. J. *A New Introduction to Modal Logic*. Routledge, London, UK, 1996.

[318] Hughes, J., O'Keefe, M., Feist, C., Hawkinson, S., Perrault, J., and Corcoran, D. A Universal Access, Smart-Card-Based Secure File System. In *Proceedings of the Atlanta Linux Showcase* (Atlanta, GA, USA, Oct. 1999).

[319] Huisman, M. *Java Program Verification in Higher-Order Logic with PVS and Isabelle*. PhD thesis, University of Nijmwegen, Nijmwegen, The Netherlands, 2001.

[320] Hungerford, T. W. *Algebra*. Graduate Texts in Mathematics. Springer Verlag, Heidelberg, Germany, 1997.

[321] Hustadt, U., and Schmidt, R. A. MSPASS: Modal Reasoning by Translation and First-Order Resolution. In *Automated*

Reasoning with Analytic Tableaux and Related Methods — International Conference TABLEAUX 2000 (St. Andrews, UK, July 2000), R. Dyckhoff, Ed., vol. 1847 of *Lecture Notes in Computer Science*, Springer Verlag, pp. 67–71.

[322] IBM. *z/OS V1R2.0 Elements and Features Bookshelf.* Poughkeepsie, NY, USA, Dec. 2001. Electronic collection; part number for CD distribution is SK3T-4269-03.

[323] Iliffe, J. K. *Basic Machine Principles*, 2nd ed. Elsevier / MacDonald, New York, NY, USA, 1972.

[324] Iliffe, J. K., and Jodeit, J. G. A Dynamic Storage Allocation Scheme. *Computer Journal 5*, 3 (1962), 200–209.

[325] International Business Machines Corporation. *Resource Access Control Facility (RACF): General Information.* No. GC28-D722-17 in IBM Red Books. International Business Machines Corporation, 1976.

[326] International Organization for Standardization and International Electrotechnical Committee. Information Processing Systems - Open Systems Interconnection - The Basic Model. International Standard 7498-1. Tech. rep., ISO, 1994.

[327] International Organization for Standardization and International Electrotechnical Committee. *Information Technology - Open Systems Interconnection - Security Frameworks for Open Systems - Part 3: Access Control. International Standard 10181-3*, 1995.

[328] International Organization for Standardization and International Electrotechnical Committee. *Common Criteria for Information Technology Security Evaluation – International Standard 15408*, 1999. Version 2.1.

[329] International Organization for Standardization and International Electrotechnical Committee. *Information Tech-*

nology - Code of Practice for Information Security Management. International Standard 17799, 2000.

[330] Ioannidis, S., Keromytis, A. D., Bellovin, S. M., and Smith, J. M. Implementing a Distributed Firewall. In *Proceedings of the 7th ACM Conference on Computer and Communications Security (CCS 2000)* (Athens, Greece, Nov. 2000), ACM SIGSAC, ACM Press, pp. 190–199.

[331] Irvine, C. E., Anderson, J. P., Robb, D. A., and Hackerson, J. High Assurance Multilevel Services for Off-the-Shelf Workstation Applications. In *Proceedings 21st NIST-NCSC National Information Systems Security Conference* (Washington D.C., USA, Oct. 1998), pp. 421–431.

[332] Jackson, P., Reichgelt, H., and van Harmelen, F. *Logic-Based Knowledge Representation.* MIT Press, Cambridge, MA, USA, 1988.

[333] Jacky, J. *The Way of Z, Practical programming with formal methods.* Cambridge University Press, Cambridge, UK, 1997.

[334] Jacob, J. Categorising Non-interference. In *Proceedings of the Computer Security Foundations Workshop III* (Franconia, NH, USA, 1990), IEEE Computer Society Press, pp. 44–50.

[335] Jacobson, V., and Braden, R. RFC 1072: TCP Extensions for Long-Delay Paths, Oct. 1988.

[336] Jacobson, V., Braden, R., and Borman, D. RFC 1072: TCP Extensions for High Performance, May 1992.

[337] Jajodia, S., and Landwehr, C. E., Eds. *Proceedings of Database Security IV: Status and Prospects. Results of the IFIP WG 11.3 Workshop on Database Security* (Halifax, UK, Sept. 1991).

[338] Jajodia, S., Samarati, P., Sapino, M. L., and Subrahmanian, V. S. Flexible Support for Multiple Access Control Policies.

ACM Transactions on Database Systems 26, 2 (June 2001), 214–260.

[339] Jajodia, S., Samarati, P., and Subrahmanian, V. S. A Logical Language for Expressing Authorizations. In *Proceedings of the 1997 IEEE Symposium on Security and Privacy (SOSP '97)* (Oakland, CA, USA, May 1997), IEEE Computer Society, pp. 31–42.

[340] Jajodia, S., Samarati, P., Subrahmanian, V. S., and Bertino, E. A Unified Framework for Enforcing Multiple Access Control. In *Proceedings ACM SIGMOD International Conference on Management of Data* (Tucson, AZ, USA, May 1997), ACM Press, pp. 474–485.

[341] Jelen, G. F. Information Security: An Elusive Goal. Program on Information Resources Policy, Harvard University, June 1985.

[342] Jensen, K. *Coloured Petri Nets: Basic Concepts, Analysis Methods and Practical Use, Volume 1.* Monographs in Theoretical Computer Science. Springer Verlag, Heidelberg, Germany, 1997.

[343] Jensen, K. *Coloured Petri Nets: Basic Concepts, Analysis Methods and Practical Use, Volume 2.* Monographs in Theoretical Computer Science. Springer Verlag, Heidelberg, Germany, 1997.

[344] Jensen McCollum, C., Messing, J. R., and Notargiacomo, L. Beyond the Pale of MAC and DAC – Defining New Forms of Access Control. In *Proceedings of the 1990 IEEE Symposium on Security and Privacy (SOSP '90)* (Oakland, CA, USA, Apr. 1990), IEEE Computer Society, pp. 190–200.

[345] Johnson, D. M., and Thayer, F. J. Security and the Composition of Machines. In *Proceedings of the Computer Security Foundations Workshop* (Franconia, NH, USA, June 1988), IEEE Computer Society Press, pp. 72–89.

[346] Jones, A., and Ohlund, J. *Network Programming for Microsoft Windows.* Microsoft Press, Redmond, WA, USA, 1999.

[347] JONES, A. K. *Protection in Programmed Systems*. PhD thesis, Carnegie Mellon University, Pittsburgh, PA, USA, June 1973.

[348] JONES, A. K., CHANSLER, R. J., DURHAM, I., SCHWANS, K., AND VEGDAHL, S. R. StarOS, a Multiprocessor Operating System for the Support of Task Forces. *ACM Operating Systems Review 13*, 5 (1979), 117–121. Proceedings of the Seventh ACM Symposium on Operating Systems Principles.

[349] JONES, A. K., LIPTON, R. J., AND SNYDER, L. A Linear Time Algorithm for Deciding Security. In *Proceedings of the 17th Annual Symposium on Foundations of Computer Science (FOCS '76)* (Houston, TX, USA, Oct. 1976), IEEE Computer Society, pp. 33–41.

[350] JONES, A. K., AND WULF, W. A. EROS: A Principle-Driven Operating System from the Ground Up. *IEEE Software 5*, 4 (Oct./Dec. 1975), 321–336.

[351] KAHN, K. C., CORWIN, W. M., POLLACK, F. J., DENNIS, T. D., D'HOOGE, H., HUBKA, D. E., HUTCHINSON, L. A., MONTAGUE, J. T., AND GIFKINS, M. R. iMAX: A Multiprocessor Operating System for an Object-Based Computer. *ACM Operating Systems Review 15*, 5 (1981), 127–136. Proceedings of the Eighth ACM Symposium on Operating Systems Principles.

[352] KAMP, H. W. *Tense Logic and the Theory of Linear Order*. PhD thesis, University of California, Los Angeles, CA, USA, 1968.

[353] KAN, S. *Metrics and Models in Software Quality Engineering*. Addison-Wesley, Reading, MA, USA, 1995.

[354] KANG, M. H., AND MOSKOWITZ, I. S. A Pump for Rapid, Reliable, Secure Communications. In *Proceedings of the 1st ACM Conference on Computer and Communications Security* (Fairfax, VA, USA, Nov. 1993), ACM Press, pp. 118–129.

[355] Kang, M. H., Moskowitz, I. S., and Lee, D. C. A Network Pump. *IEEE Transactions on Software Engineering 22*, 5 (May 1996), 329–338. Revised from [356].

[356] Kang, M. H., Moskowitz, I. S., and Lee, D. C. A Network Version of the Pump. In *Proceedings of the 1996 IEEE Symposium on Security and Privacy (SOSP '96)* (Oakland, CA, USA, May 1996), IEEE Computer Society, pp. 144–154.

[357] Karger, P. A. A VMM Security Kernel for the VAX Architecture. In *Proceedings of the 1990 IEEE Symposium on Security and Privacy (SOSP '90)* (Oakland, CA, USA, May 1990), IEEE Computer Society, pp. 2–19.

[358] Karger, P. A., and Herbert, A. J. An Augmented Capability Architecture to Support Lattice Security and Traceability of Access. In *Proceedings of the 1984 IEEE Symposium on Security and Privacy (SOSP '84)* (Oakland, CA, USA, Apr. 1984), IEEE Computer Society, pp. 2–12.

[359] Karger, P. A., and Schell, R. R. Multics Security Evaluation: Vulnerability Analysis. Tech. rep., The MITRE Corporation, Bedford, MA, USA, Mar. 1974. Number ESD-TR-74-193. Performed for the Electronic Systems Division, Air Force Systems Command, United States Air Force.

[360] Karger, P. A., and Schell, R. R. Thirty Years Later: Lessons from the Multics Security Evaluation. In *Proceedings 18th Annual Computer Security Applications Conference (ACSAC'02)* (Las Vegas, NV, USA, Dec. 2002), IEEE Computer Society Press, pp. 205–216.

[361] Kelem, N. L., and Feiertag, R. J. A Separation Model for Virtual Machine Monitors. In *Proceedings of the 1991 IEEE Symposium on Security and Privacy (SOSP '91)* (Oakland, CA, USA, Apr. 1991), IEEE Computer Society, pp. 78–86.

[362] Kemmerer, R. A. A Practical Approach To Identifying Storage and Timing Channels. In *Proceedings of the 1982 IEEE Symposium on Security and Privacy (SOSP '82)* (Oakland, CA, USA, Apr. 1982), IEEE Computer Society, pp. 66–72.

[363] Kemmerer, R. A. Shared Resource Matrix Methodology: An Approach to Identifying Storage and Timing Channels. *ACM Transactions on Computer Systems 1*, 3 (Aug. 1983), 256–277.

[364] Kenah, L. J. *VAX/VMS Internals and Datastrucutures Version 4.4.* Butterworth-Heinemann, Maynard, MA, USA, 1988.

[365] Kenah, L. J., and Goldenberg, R. E. *VAX/VMS Internals and Datastrucutures Version 5.2.* Butterworth-Heinemann, Maynard, MA, USA, 1991.

[366] Kent, S. RFC 1108: U.S. Department of Defense Security Options for the Internet Protocol, Nov. 1991. Status: HISTORIC. Obsoletes RFC 1038.

[367] Kent, S., and Atkinson, R. RFC 2401: Security Architecture for the Internet Protocol, Nov. 1998.

[368] Kent, S., and Atkinson, R. RFC 2402: IP Authentication Header, Nov. 1998.

[369] Kent, S., and Atkinson, R. RFC 2406: IP Encapsulating Security Payload (ESP), Nov. 1998.

[370] Kern, C., and Greenstreet, M. R. Formal Verification in Hardware Design: A Survey. *ACM Transactions on Design Automation of Electronic Systems (TODAES) 4*, 2 (Apr. 1999), 123–193.

[371] Keromytis, A. D., and Wright, J. L. Transparent Network Security Policy Enforcement. In *Proceedings of the 2000 USENIX Annual Technical Conference* (San Diego, CA, USA, June 2000), USENIX Association, Ed., USENIX, pp. 215–226.

[372] Kirchner, C., and Kirchner, H., Eds. *Proceedings of the 15th International Conference on Automated Deduction (CADE-15)* (Lindau, Germany, July 1998), vol. 1421 of *Lecture Notes in Computer Science*, Springer Verlag.

[373] Knight, J., DeJong, C., Gibble, M., and Nakano, L. Why Are Formal Methods Not Used More Widely? In *Proceedings of the Fourth NASA Langley Formal Methods Workshop (LFM '97)* (Hampton, VA, USA, Sept. 1999), C. M. Holloway and K. J. Hayburst, Eds., vol. 3356 of *NASA Conference Publications*, NASA, pp. 1–12.

[374] Ko, C., Ruschitzka, M., and Levitt, K. Execution Monitoring of Security-Critical Programs in a Distributed Systems: A Specification-Based Approach. In *Proceedings of the 1997 IEEE Symposium on Security and Privacy (SOSP '97)* (Oakland, CA, USA, May 1997), IEEE Computer Society Press, pp. 175–187.

[375] Kocher, P., Jaffe, J., and Jun, B. Differential Power Analysis. In *Advances in Cryptology (CRYPTO '99)* (Santa Barbara, CA, USA, Aug. 1999), M. Wiener, Ed., vol. 1666 of *Lecture Notes in Computer Science*, International Association for Cryptologic Research, Springer Verlag, pp. 399–397.

[376] Kocher, P. C. Timing Attacks on Implementations of Diffie--Hellman, RSA, DSS, and Other Systems. In *Advances in Cryptology (CRYPTO '96)* (Santa Barbara, CA, USA, Aug. 1996), N. Koblitz, Ed., vol. 1109 of *Lecture Notes in Computer Science*, International Association for Cryptologic Research, Springer Verlag, pp. 104–113.

[377] Kozen, D., and Tiuryn, J. Logics of Programs. In *Handbook of Theoretical Computer Science*, J. van Leeuwen, Ed., vol. B. MIT Press, Cambridge, MA, USA, 1990, ch. 14, pp. 789–840.

[378] Kramer, S. Linus IV : An Experiment in Computer Security. In *Proceedings of the 1984 IEEE Symposium on Security and Privacy*

(SOSP '84) (Oakland, CA, USA, Apr. 1984), IEEE Computer Society, pp. 24–33.

[379] Kripke, S. Semantical Analysis of Modal Logic I: Normal Propositional Calculi. *Zeitschrift für mathematische Logik und Grundlagen der Mathematik 9* (1963), 67–96.

[380] Krishnamurthy, E. V., and McGuffin, A. On the Design and Administration of Secure Database Transactions. *ACM SIGSAC Review* (Spring/Summer 1992), 63–70.

[381] Kuball, S., May, J., and Hughes, G. Building a System Failure Rate Estimator by Identifying Component Failure Rates. In *Proceedings of the 10th International Symposium on Software Reliability Engineering (ISSRE '99)* (Boca Raton, FL, USA, Nov. 1999), IEEE Computer Society, pp. 32–41.

[382] Kubiatowicz, J. OceanStore: Data Security in an Insecure World, Sept. 2001. Presented as a Panel at the departmental Networking Day at the University of California at Berkeley.

[383] Kubiatowicz, J., Bindel, D., Chen, Y., Eaton, P., Geels, D., Gummadi, R., Rhea, S., Weatherspoon, H., Weimer, W., Wells, C., and Zhao, B. OceanStore: An Extremely Wide Area Storage System. Tech. Rep. UCB//CSD-00-1102, University of California at Berkeley, Berkeley, CA, USA, May 1999.

[384] Kühnhauser, W. E. A Paradigm For User-Defined Security Policies. In *Proceedings of the 14th Symposium on Reliable Distributed Systems* (Bad Neuenahr, Germany, May 1995), IEEE Computer Society Press, pp. 135–144.

[385] Kühnhauser, W. E. A Classification of Interdomain Actions. *ACM Operating Systems Review 32*, 4 (Oct. 1998), 47–61.

[386] Kühnhauser, W. E., and von Kopp Ostrowski, M. A Framework to Support Multiple Security Policies. In *Proceedings of*

the 7th Annual Canadian Computer Security Symposium (Ottawa, Canada, 1995).

[387] LACKEY, R. D. Penetration of Computer Systems, an Overview. *Honeywell Computer Journal 8*, 2 (Dec. 1968), 81–85.

[388] LAMACCHIA, B., LANGE, S., LYONS, M., MARTIN, R., AND PRICE, K. *.NET Framework Security*. Addison-Wesley, Reading, MA, USA, 2002.

[389] LAMPORT, L. Proving the Correctness of Multiprocess Programs. *IEEE Transactions on Software Engineering 3*, 2 (Mar. 1977), 125–143.

[390] LAMPORT, L. Time, Clocks, and the Ordering of Events in a Distributed System. *Communications of the Association for Computing Machinery 21*, 7 (July 1978), 558–565.

[391] LAMPORT, L. Logical Foundation. In *Distributed Systems: Methods and Tools for Specification. An Advanced Course.*, M. W. Alford, J. P. Ansart, G. Hommel, L. Lamport, B. Liskov, G. P. Mullery, F. B. Schneider, M. Paul, and H. J. Siegert, Eds., vol. 190 of *Lecture Notes in Computer Science*. Springer Verlag, Heidelberg, Germany, 1985, pp. 119–130.

[392] LAMPORT, L. The Temporal Logic of Actions. Tech. Rep. 79, Digital Equipment Corporation, Systems Research Center, Dec. 1991.

[393] LAMPORT, L. Specifying Concurrent Systems with TLA+. In *Calculational System Design*, M. Broy and R. Steinbrüggen, Eds. IOS Press, Amsterdam, The Netherlands, 1999.

[394] LAMPORT, L. *Specifying Systems: The TLA+ Language and Tools for Hardware and Software Engineers*. Addison-Wesley, Reading, MA, USA, 2002.

[395] Lamport, L., Shostak, R., and Pease, M. The Byzantine Generals Problem. *ACM Transactions on Programming Languages and Systems 4*, 3 (July 1982), 382–401.

[396] Lampson, B. W. Protection. In *Proceedings of the 5th Annual Princeton Conference on Information Sciences and Systems* (Princeton University, NJ, USA, 1971), pp. 437–443. Reprinted in [398].

[397] Lampson, B. W. A Note on the Confinement Problem. *Communications of the Association for Computing Machinery 16*, 10 (Oct. 1973), 613–615.

[398] Lampson, B. W. Protection. *ACM Operating Systems Review 8*, 1 (Jan. 1974), 18–24.

[399] Landau, C. R. Security in a Secure Capability-Based System. *ACM Operating Systems Review 23*, 4 (1989), 2–4.

[400] Landin, P. J. The Mechanical Evaluation of Expressions. *Computer Journal 6*, 4 (1964), 308–320.

[401] Landwehr, C. E., and Heitmeyer, C. L. Military Message Systems: Requirements and Security Model. Tech. rep., Naval Research Laboratory, Washington D.C., USA, Sept. 1982. NRL Memorandum 4925.

[402] Landwehr, C. E., Heitmeyer, C. L., and McLean, J. A Security Model for Military Message System. *ACM Transactions on Computer Systems 2*, 3 (Aug. 1984), 198–222.

[403] LaPadula, L. J. Formal modeling in a generalized framework for access control. In *Proceedings of the Computer Security Foundations Workshop III* (June 1990), IEEE Computer Society Press, pp. 100–109.

[404] Leffler, S. J., McKusick, M. K., Karels, M. J., and Quarterman, J. S. *The Design and Implementation of the 4.3BSD UNIX Operating System.* Addison-Wesley, Reading, MA, USA, 1989.

[405] LEITSCH, A. *The Resolution Calculus*. Texts in Theoretical Computer Science. Springer Verlag, Heidelberg, Germany, 1996.

[406] LEPREAU, J., FORD, B., AND HIBLER, M. The Persistent Relevance of the Local Operating System to Global Applications. In *Proceedings of the Seventh SIGOPS European Workshop* (Connemara, Ireland, Sept. 1996), ACM Press, pp. 60–67.

[407] LEVESQUE, H. J., AND LAKEMEYER, G. *The Logic of Knowledge Bases*. MIT Press, Cambridge, MA, USA, 2001.

[408] LEVIN, R., COHEN, E. S., CORWIN, W. M., POLLACK, F. J., AND WULF, W. A. Policy/mechanism separation in HYDRA. In *Proceedings of the Third ACM Symposium on Operating System Principles* (Austin, TX, USA, Nov. 1975), ACM SIGOPS, pp. 132–140.

[409] LEVITT, K. N., AND NEUMANN, P. G. Recent SRI Work in Verification. *ACM SIGSOFT Software Engineering Notes 6*, 3 (July 1981), 33–47.

[410] LICHTENSTEIN, O., PNUELI, A., AND ZUCK, L. The Glory of the Past. In *Proceedings of the Conference on Logics of Programs* (Pittsburgh, PA, USA, July 1985), R. Parikh, Ed., vol. 193 of *Lecture Notes in Computer Science*, Springer Verlag, pp. 196–218.

[411] LIEDTKE, J. Toward Real Microkernels. *Communications of the Association for Computing Machinery 39*, 9 (Sept. 1996), 70–77.

[412] LINDE, R. R. Operating System Penetration. In *Proceedings of the National Computer Conference* (Anaheim, CA, USA, Nov. 1975), vol. 44, AFIPS, AFIPS Press, pp. 361–368.

[413] LINDE, R. R., WEISSMAN, C., AND FOX, C. The ADEPT-50 timesharing system. In *Proceedings of the AFIPS Fall Joint Computer Conference (1969 FJCC)* (Las Vegas, NV, USA, Nov. 1969), vol. 35, AFIPS, AFIPS Press, pp. 39–50.

[414] Linden, T. A. Operating System Structures to Support Security and Reliable Software. *ACM Computing Surveys 8*, 4 (Dec. 1976), 409–445.

[415] Lindqvist, U. Observations on the Nature of Computer Security Intrusions. Tech. Rep. 253L, School of Electrical and Computer Engineering, Chalmers University of Technology, Göteborg, Sweden, Dec. 1996.

[416] Lions, J. *Lions' Commentary on UNIX 6th Edition, with Source Code*. Computer Classics Revisited. Peer-to-Peer Communications, San Jose, CA, USA, 1996. With forewords by Dennis M. Ritchie and Ken Thompson. Prefatory notes by Peter H. Salus and Michael Tilson; a Historical Note by Peter H. Salus. Originally circulated as two restricted-release volumes: "UNIX Operating System Source Code Level Six", and "A Commentary on the UNIX Operating System".

[417] Lipner. A Comment on the Confinement Problem. *ACM Operating Systems Review 9*, 5 (Dec. 1975), 192–196. Proceedings of the Fifth ACM Symposium on Operating Systems Principles.

[418] Lipner, S. B. Non-Discretionary Controls for Commercial Applications. In *Proceedings of the 1982 IEEE Symposium on Security and Privacy (SOSP '82)* (Oakland, CA, USA, May 1982), IEEE Computer Society, pp. 2–10.

[419] Lipton, R. J., and Snyder, L. Monotonic Protection Systems. In *Foundations of Secure Computation*, R. DeMillo, D. Dobkin, A. Jones, and R. Lipton, Eds. Academic Press, New York, NY, USA, 1978, pp. 367–385.

[420] Loeckx, J., and Sieber, K. *The Foundations of Program Verification*. John Wiley & Sons, Inc., New York, NY, USA, 1987.

[421] Loscocco, P. A Dynamic Network Labeling Scheme for a MLS LAN. In *Proceedings 12th NIST-NCSC National Computer Secu-*

rity Conference (Gaithersburg, MD, USA, Oct. 1989), pp. 277–285.

[422] LOSCOCCO, P., AND SMALLEY, S. Integrating Flexible Support for Security Policies into the Linux Operating System. Tech. rep., U.S. National Security Agency (NSA), Oct. 2000.

[423] LOSCOCCO, P., AND SMALLEY, S. Meeting Critical Security Objectives with Security-Enhanced Linux. In *Proceedings of the 2001 Ottawa Linux Symposium* (Ottawa, Canada, July 2001).

[424] LOSCOCCO, P. A., SMALLEY, S. D., MUCKELBAUER, P. A., TAYLOR, R. C., TURNER, S. J., AND FARRELL, J. F. The Inevitability of Failure: The Flawed Assumption of Security in Modern Computing Environments. In *Proceedings of the 21st National Information Systems Security Conference* (1998), National Computer Security Center, pp. 303–314.

[425] LOW, S. H., MAXEMCHUK, N. F., AND LAPONE, A. P. Document Identification for Copyright Protection Using Centroid Detection. *IEEE Transactions on Communications 46*, 3 (Mar. 1998), 372–383.

[426] LOWE, G. Quantifying Information Flow. In *Proceedings of the 15th IEEE Computer Security Foundations Workshop (CSFW-15)* (Cape Breton, NS, Canada, June 2002), IEEE Computer Society, pp. 18–31.

[427] LUBY, M. *Pseudorandomness and Cryptographic Applications*. Princeton Computer Science Notes. Princeton University Press, Princeton, NJ, USA, 1996.

[428] LUDWIG, S., AND KALKA, W. File System Encryption with Integrated User Management. *ACM Operating Systems Review 35*, 4 (Oct. 2001), 88–93.

[429] ŁUKASZEWICZ, W. Considerations on Default Logic. In *Proceedings of the International Workshop on Non-monotonic Reasoning* (1984), R. Reiter, Ed., pp. 165–193. New Paltz, NY, USA.

[430] Luo, J., and Bridges, S. M. Mining Fuzzy Association Rules and Fuzzy Frequency Episodes for Intrusion Detection. *International Journal of Intelligent Systems 15*, 8 (2000), 687–703.

[431] Lupu, E. C., and Sloman, M. S. A Policy Based Role Object Model. In *Proceedings of the First International Enterprise Distributed Object Computing Workshop (EDOC'97)* (Gold Coast, Australia, Oct. 1997), Chapman & Hall, pp. 430–443.

[432] Lupu, E. C., and Sloman, M. S. Conflict Analysis for Management Policies. In *Proceedings of the Fifth IFIP/IEEE International Symposium on Integrated Network Management (IM'97)* (San Diego, CA, USA, May 1997), Aurel A. Lazar and Roberto Saracco and Rolf Stadler, Ed., Chapman & Hall, pp. 430–443.

[433] Lupu, E. C., and Sloman, M. S. Reconciling Role Based Management and Role Based Access Control. In *Proceedings of the Second ACM Workshop on Role-Based Access Control* (Fairfax, VA, USA, Nov. 1997), ACM Press, pp. 135–141.

[434] Mac Lane, S. *Categories for the Working Mathematician*, 2nd ed. Graduate Texts in Mathematics. Springer Verlag, Heidelberg, Germany, 1998.

[435] MacKenzie, D. The Automation of Proof: an Historical and Sociological Exploration. *IEEE Annals of the History of Computing 17*, 3 (Fall 1995), 7–29.

[436] MacKenzie, D. *Mechanizing Proof.* MIT Press, Cambridge, MA, USA, 2001.

[437] Mally, E. *Grundgesetze des Sollens: Elemente der Logik des Willens*. Leuschner & Lubensky, Graz, Austria, 1926.

[438] Manna, Z. *Mathematical Theory of Computation*. McGraw-Hill, New York, NY, USA, 1974.

[439] MANNA, Z., AND PNUELI, A. Adequate Proof Principles for Invariance and Liveness Properties of Concurrent Programs. *Science of Computer Programming 4*, 3 (1984), 257–290.

[440] MANNA, Z., AND PNUELI, A. The Anchored Version of the Temporal Framework. In *Linear Time, Branching Time and Partial Order in Logics and Models for Concurrency* (Noordwijkerhout, The Netherlands, May 1988), J. de Bakker, W.-P. de Roever, and G. Rozenberg, Eds., vol. 345 of *Lecture Notes in Computer Science*, Springer Verlag, pp. 201–284. Proceedings of the REX Workshop 1988.

[441] MANTEL, H. Possibilistic Definitions of Security – An Assembly Kit. In *Proceedings of the Computer Security Foundations Workshop* (Cambridge, UK, July 2000), IEEE Computer Society Press, pp. 185–199.

[442] MANTEL, H. Unwinding Possibilistic Security Properties. In *Proceedings of the 6th European Symposium on Research in Computer Security (ESORICS 2000)* (Toulouse, France, Oct. 2000), F. Cuppens, Y. Deswarte, D. Gollmann, and M. Waidner, Eds., vol. 1895 of *Lecture Notes in Computer Science*, Springer Verlag, pp. 238–254.

[443] MANTEL, H. Information Flow Control and Applications – Bridging a Gap. In *FME 2001: Formal Methods for Increasing Software Productivity, International Symposium of Formal Methods Europe* (Berlin, Germany, 2001), J. N. Olivera and P. Zave, Eds., vol. 2021 of *Lecture Notes in Computer Science*, Springer Verlag, pp. 153–172.

[444] MARKHAM, T., AND PAYNE, C. Security at the Network Edge: A Distributed Firewall Architecture. In *Proceedings of the DARPA Information Survivability Conference (DISCEX II)* (Anaheim, CA, USA, June 2001), pp. 18–27.

[445] MARSH (ED.), R. T. Critical Foundations: Protecting America's Infrastructures. U.S. Government Printing Office, Oct. 1997.

Prepared by the President's Commission on Critical Infrastructure Protection. Unclassified.

[446] Massacci, F. Reasoning About Security: A Logic and a Decision Method for Role-Based Access Control. In *Proceedings of the First International Joint Conference on Qualitative and Quantitative Practical Reasoning (ECSQARU-FAPR'97)* (Bad Honnef, Germany, June 1997), D. M. Gabbay, R. Kruse, and O. Andreas Nonnengart H.-J, Eds., vol. 1244 of *Lecture Notes in Computer Science*, Springer Verlag, pp. 421–435.

[447] Mathis, M., Mahdavi, J., Floyd, S., and Romanow, A. RFC 2018: TCP Selective Acknowledgment Options, Oct. 1996.

[448] Maughan, D., Schertler, M., Schneider, M., and Turner, J. RFC 2408: Internet Security Association and Key Management Protocol (ISAKMP), Nov. 1998.

[449] Mauriello, E. TCFS: Transparent Cryptographic File System. *Linux Journal 40* (Aug. 1997).

[450] Mauro, J., and McDougall, R. *Solaris Internals: Core Kernel Architecture, Vol. 1.* Prentice Hall, Englewood Cliffs, NJ, USA, 2000.

[451] May, D., Muller, H. L., and Smart, N. P. Random Register Renaming to Foil DPA. In *Cryptographic Hardware and Embedded Systems – CHES 2001. Proceedings of the Third International Workshop* (Paris, France, May 2001), Ç. Koç, D. Naccache, and C. Paar, Eds., vol. 2162 of *Lecture Notes in Computer Science*, Springer Verlag, pp. 28–38.

[452] Mayer-Sommer, R. Smartly Analyzing the Simplicity and the Power of Simple Power Analysis on Smartcards. In *Cryptographic Hardware and Embedded Systems – CHES 2000. Proceedings of the Second International Workshop* (Worcester, MA, USA, Aug. 2000), Ç. Koç and C. Paar, Eds., vol. 1965 of *Lecture Notes in Computer Science*, Springer Verlag, pp. 78–92.

[453] McAllester, D., Ed. *Proceedings of the 15th International Conference on Automated Deduction (CADE-16)* (Pittsburgh, PA, USA, June 2000), vol. 1831 of *Lecture Notes in Computer Science*, Springer Verlag.

[454] McCarthy, J. *Formalization of Common Sense: Papers by John McCarthy*. Ablex Publishing, Norwood, NJ, USA, 1990. V. Lifschitz, Ed.

[455] McCarthy, J., and Hayes, P. J. Some Philosophical Problems from the Standpoint of Artificial Intelligence. In *Machine Intelligence 4*, B. Meltzer and D. Michie, Eds. Edinburgh University Press, 1969, pp. 463–502. Reprinted in [454].

[456] McCauley, E. J., and Drongowski, P. J. KSOS — The Design of a Secure Operating System. In *Proceedings of the National Computer Conference* (New York, NY, USA, June 1979), vol. 48, AFIPS, AFIPS Press, pp. 345–353.

[457] McCoy, K. *VMS File System Internals*. Butterworth-Heinemann, Maynard, MA, USA, 1990.

[458] McCullough, D. Specifications for Multi-Level Security and a Hook-Up Property. In *Proceedings of the 1987 IEEE Symposium on Security and Privacy (SOSP '87)* (Oakland, CA, USA, Apr. 1987), pp. 161–166.

[459] McCullough, D. Foundations of Ulysses: The Theory of Security. Tech. Rep. RADC-TR-87-222, Odyssey Research Associates, Inc., July 1988.

[460] McCullough, D. Noninterference and the Composability of Security Properties. In *Proceedings of the 1988 IEEE Symposium on Security and Privacy (SOSP '88)* (Oakland, CA, USA, Apr. 1988), pp. 177–187.

[461] McCullough, D. A Hookup Theorem for Multilevel Security. *IEEE Transactions on Software Engineering 16*, 6 (June 1990), 563–568. Special Section on Security and Privacy.

[462] McDermott, D., and Doyle, J. Nonmonotonic Logic 1. *Artificial Intelligence 13*, 1–2 (1980), 41–72.

[463] McHugh, J., and Good, D. I. An Information Flow Tool for Gypsy. In *Proceedings of the 1985 IEEE Symposium on Security and Privacy (SOSP '85)* (Oakland, CA, USA, Apr. 1985), IEEE Computer Society, pp. 46–48.

[464] McKusick, M. K., Bostic, K., Karels, M. J., and Quarterman, J. S. *The Design and Implementation of the 4.4 BSD Operating System*. Addison Wesley, Reading, MA, USA, 1996.

[465] McLean, J. A Comment on the 'Basic Security Theorem' of Bell and LaPadula. *Information Processing Letters 20*, 2 (Feb. 1985), 67–70.

[466] McLean, J. Security Models and Information Flow. In *Proceedings of the 1990 IEEE Symposium on Security and Privacy (SOSP '90)* (Oakland, CA, USA, Apr. 1990), IEEE Computer Society, pp. 180–187.

[467] McLean, J. The Specification and Modeling of Computer Security. *IEEE Computer 23*, 1 (Jan. 1990), 9–16.

[468] McLean, J. Proving Noninterference and Functional Correctness Using Traces. *Journal of Computer Security 1*, 1 (1992), 37–57.

[469] McLean, J. A General Theory of Composition for Trace Sets Closed Under Selective Interleaving Functions. In *Proceedings of the 1994 IEEE Symposium on Security and Privacy (SOSP '94)* (Silver Spring, MD, USA, May 1994), IEEE Computer Society, pp. 79–93.

[470] McLean, J. Security Models. In *Encyclopaedia of Software Engineering*, J. J. Marciniak, Ed., 1st ed. John Wiley & Sons, Inc., New York, NY, USA, 1994, pp. 1136–1145.

[471] McLean, J. A General Theory of Composition for a Class of "Possibilistic" Properties. *IEEE Transactions on Software Engineering 22*, 1 (Jan. 1996), 53–67. Special Section on Security and Privacy.

[472] Meadows, C. Extending the Brewer-Nash Model to a Multilevel Context. In *Proceedings of the 1990 IEEE Symposium on Security and Privacy (SOSP '90)* (Oakland, CA, USA, Apr. 1990), IEEE Computer Society, pp. 95–102.

[473] Menezes, A. J., van Oorschot, P. C., and Vanstone, S. A. *Handbook of Applied Cryptography*. CRC Press, Boca Raton, FL, USA, 1996.

[474] Messerges, T. S. Using Second-Order Power Analysis to Attack DPA Resistant Software. In *Cryptographic Hardware and Embedded Systems – CHES 2000, Proceedings of the Second International Workshop* (Worcester, MA, USA, Aug. 2000), Ç. Koç and C. Paar, Eds., vol. 1965 of *Lecture Notes in Computer Science*, Springer Verlag, pp. 238–251.

[475] Messerges, T. S., Dabbish, E. A., and Sloan, R. H. Power Analysis Attacks of Modular Exponentiation in Smartcards. In *Cryptographic Hardware and Embedded Systems – CHES 1999. Proceedings of the First International Workshop* (Worcester, MA, USA, Aug. 1999), Ç. Koç and C. Paar, Eds., vol. 1717 of *Lecture Notes in Computer Science*, Springer Verlag, pp. 144–157.

[476] Microsoft Corporation. Portable Executable and Common Object File Format Specification Version 6.0. Tech. rep., Microsoft Corporation, One Microsoft Way, Redmond, WA, USA, Feb. 1999.

[477] Millen, J. K. Security Kernel Validation in Practice. *Communications of the Association for Computing Machinery 19*, 5 (May 1976), 243–250.

[478] Millen, J. K. Covert Channel Capacity. In *Proceedings of the 1987 IEEE Symposium on Security and Privacy (SOSP '87)* (Oakland, CA, USA, Apr. 1987), IEEE Computer Society, pp. 60–66.

[479] Millen, J. K. A Resource Allocation Model for Denial of Service. In *Proceedings of the 1992 IEEE Symposium on Security and Privacy (SOSP '92)* (Oakland, CA, USA, Apr. 1992), IEEE Computer Society, pp. 137–147.

[480] Minear, S. E. Providing Policy Control Over Object Operations in a Mach Based System. In *Proceedings of the 5th USENIX UNIX Security Symposium* (Salt Lake City, UT, USA, June 1995), USENIX, Ed., USENIX, pp. 141–155.

[481] Mockapetris, P. V. RFC 1034: Domain Names — Concepts and Facilities, Nov. 1987. Obsoletes RFC 973, RFC 882, RFC 883. See also STD 13. Updated by RFC 1101, RFC 1183, RFC1348, RFC 1876, RFC 1982, RFC 2065, RFC 2181, RFC 2308.

[482] Moffett, J. D., Jonscher, D., and McDermid, J. A. The Policy Obstacle Course: A Framework for Policies Embedded within Distributed Computer Systems. Tech. Rep. SCHEMA/York/93/1, University of York Department of Computer Science, York, UK, July 1993.

[483] Moffett, J. D., and Sloman, M. S. The Representation of Policies as System Objects. In *Proceedings of the Conference on Organizational Computing Systems* (Atlanta, GA, USA, Nov. 1991), pp. 171–184.

[484] Mogul, J. C., Rashid, R. F., and Accetta, M. J. The Packet Filter: an Efficient Mechanism for User-level Network Code. *ACM Operating Systems Review* (1987), 57–65. Proceedings of the 11th ACM Symposium on Operating Systems Principles.

[485] Monteith, E. Genoa TIE, Advanced Boundary Controller Experiment. In *Proceedings 17th Annual Computer Security Appli-*

cations Conference (ACSAC'01) (New Orleans, LA, USA, Dec. 2001), IEEE Computer Society Press, pp. 74–82.

[486] Morrisett, G., Walker, D., Crary, K., and Glew, N. From System F to Typed Assembly Language. In *Proceedings of the 25th Annual Symposium on Principles of Programming Languages* (San Diego, CA, USA, 1998), D. B. MacQueen and L. Cardelli, Eds., ACM Press, pp. 85–97.

[487] Mullender, S. J., and Tanenbaum, A. S. The Design of a Capability-Based Distributed Operating System. *Computer Journal 29*, 4 (Aug. 1986), 289–299.

[488] Munson, J. C., and Wimer, S. Watcher: The Missing Piece of the Security Puzzle. In *Proceedings 17th Annual Computer Security Applications Conference (ACSAC'01)* (New Orleans, LA, USA, Dec. 2001), IEEE Computer Society Press, pp. 230–239.

[489] Murray, N. V., Ed. *Automated Reasoning with Analytic Tableaux and Related Methods — International Conference TABLEAUX 1999* (Saratoga Springs, NY, USA, June 1999), vol. 1617 of *Lecture Notes in Computer Science*, Springer Verlag.

[490] Myers, G. *Advances in Computer Architecture*. John Wiley & Sons, Inc., New York, NY, USA, 1978.

[491] Myers, G., and Buckingham, B. R. S. A Hardware Implementation of Capability-Based Addressing. *ACM Operating Systems Review 14*, 4 (Oct. 1980), 13–25.

[492] Myers, P. A. Subversion: The Neglected Aspect of Computer Security. Master's thesis, Naval Postgraduate School, Monterey, CA, USA, 1980.

[493] Nagar, R. *Windows NT File System Internals*. O'Reilly & Associates, Sebastopol, CA, USA, 1997.

[494] Nash, J. *Essays on Game Theory*. Edward Elgar Publishing Limited, Cheltenham, UK, 1996.

[495] National Computer Security Center. *Trusted Network Interpretation Environments Guideline*. Fort George G. Meade, MD, USA, July 1987. NCSC-TG-011.

[496] National Computer Security Center. *Trusted Network Interpretation of the TCSEC*. Fort George G. Meade, MD, USA, July 1987. NCSC-TG-005.

[497] National Computer Security Center. Final Evaluation Report: Boeing Corporation MLS LAN. Tech. Rep. CSC-EPL-94/006, U.S. National Security Agency, Ft. George G. Meade, MD, USA, July 1994.

[498] National Computer Security Center. Final Evaluation Report: Gemini Computers Incorporated Gemini Trusted Network Processor. Tech. Rep. NCSC-FER-94/008, U.S. National Security Agency, Ft. George G. Meade, MD, USA, June 1994.

[499] National Information Assurance Partnership. *Common Criteria Evaluation and Validation Scheme Validation Report: Windows 2000 Professional, Server, and Advanced Server with SP3 and Q326886 Hotfix*. Gaithersburg, MD, USA, Jan. 2003. To appear.

[500] National Security Agency Information Systems Security Organization. *Controlled Access Protection Profile*. Fort George G. Meade, MD, USA, Oct. 1999. Version 1.D.

[501] National Security Agency Information Systems Security Organization. *Labeled Security Protection Profile*. Fort George G. Meade, MD, USA, Oct. 1999. Version 1.D.

[502] National Security Telecommunications and Information Systems Security Committee (NSTISSC) Secretariat (I42). *National Information Systems Security (INFOSEC) Glossary*. Ft. George G. Meade, MD, USA, Sept. 2000. Unclassified.

[503] Nebbett, G. *Windows NT/2000 Native API Reference*. Macmillan Technical Publishing, Indianapolis, IN, USA, 2000.

[504] Nebel, B. Syntax-Based Approaches to Belief Revision. In *Belief Revision*, P. Gärdenfors, Ed., vol. 29 of *Cambridge Tracts in Theoretical Computer Science*. Cambridge University Press, Cambridge, UK, 1992, pp. 52–88.

[505] Necula, G. C. Proof-carrying Code. In *Proceedings of the 24th Annual Symposium on Principles of Programming Languages* (Paris, France, 1997), P. Lee, Ed., ACM Press, pp. 106–119.

[506] Necula, G. C. A Scalable Architecture for Proof-Carrying Code. In *Proceedings of Functional and Logic Programming, 5th International Symposium (FLOPS 2001)* (Tokyo, Japan, Mar. 2001), H. Kuchen and K. Ueda, Eds., vol. 2024 of *Lecture Notes in Computer Science*, Springer Verlag, pp. 21–39.

[507] Necula, G. C., and Lee, P. Safe Kernel Extensions Without Run-Time Checking. In *Proceedings of the 2nd Symposium on Operating System Design and Implementation (OSDI '96)* (Seattle, WA, USA, Oct. 1996), USENIX, Ed., USENIX, pp. 229–243.

[508] Necula, G. C., and Lee, P. The Design and Implementation of a Certifying Compiler. *ACM SIGPLAN Notices 33*, 5 (June 1998), 333–344. Proceedings of the ACM Conference on Programming Language Design and Implementation.

[509] Needham, R. M., and Walker, R. D. H. The Cambridge CAP Computer and Its Protection System. *ACM Operating Systems Review 11*, 5 (1977), 1–10. Proceedings of the Sixth ACM Symposium on Operating Systems Principles.

[510] Neumann, P. G., Boyer, R. S., Feiertag, R. J., Levitt, K. N., and Robinson, L. A Provably Secure Operating System: The System, its Applications, and Proof. Tech. Rep. CSL-116, Computer Science Laboratory, SRI International, May 1980. The first major report defining the system was dated June, 1975; the final report was February, 1977. This 1980 second edition was prepared in preparation for the Ford/Honeywell implementation study contract.

[511] NIST INFORMATION TECHNOLOGY LABORATORY. *FIPS PUB 140-2: Security Requirements for Cryptographic Modules.* Gaithersburg, MD, USA, Oct. 2001.

[512] NYANCHAMA, M., AND OSBORN, S. L. Access Rights Administration in Role-Based Security Systems. In *Proceedings of Database Security VIII: Status and Prospects. Results of the IFIP WG 11.3 Workshop on Database Security* (Bad Salzdetfurth, Germany, Aug. 1994), J. Biskup, M. Morgenstern, and C. E. Landwehr, Eds., pp. 37–56.

[513] OFFICE OF THE CHIEF OF NAVAL OPERATIONS N6: FLEET AND ALLIED REQUIREMENTS DIVISION AND IT-21. Joint, Allied, and Coalition Interoperability. Flag briefing at AFCEA TECHNET 2001 by VADM Mayo, June 2001.

[514] OFFICE OF THE CHIEF OF NAVAL OPERATIONS N60: FLEET AND ALLIED REQUIREMENTS DIVISION AND IT-21. Information Technology for the 21st Century. Briefing at N6 Conference, June 1998.

[515] OFFICE OF THE PRESIDENT OF THE UNITED STATES OF AMERICA. EO 12356: National Security Information. Executive Order, May 1990. Original EO 12356 is dated April 2, 1982, date given is of last amendment.

[516] OFFICE OF THE SECRETARY OF DEFENSE. *Department of Defense Joint Technical Architecture; Version 4.0.* Washington D.C., USA, Apr. 2001.

[517] OFFICE OF THE U.S. ASSISTANT SECRETARY OF DEFENSE (COMMAND, CONTROL, COMMUNICATIONS & INTELLIGENCE). *Technical Architecture Framework for Information Management; Version 3.0.* Washington D.C., USA, Mar. 1998. Unclassified; Consists of seven volumes.

[518] Office of the U.S. Secretary of Defense. *C⁴ISR Architecture Framework; Version 2.0.* Washington D.C., USA, Dec. 1997. Unclassified; Consists of seven volumes.

[519] Office of the U.S. Secretary of Defense. *Department of Defense Directive 3600.1: Information Operations.* Washington D.C., USA, Oct. 2001. Unclassified.

[520] Office of the U.S. Under Secretary of Defense for Policy. *Department of Defense Directive 5210.48: DoD Polygraph Program.* Washington D.C., USA, Dec. 1984. Unclassified.

[521] Office of the U.S. Under Secretary of Defense for Policy. *Department of Defense Directive 5205.8: Access to Classified Cryptographic Information.* Washington D.C., USA, Feb. 1991. Unclassified.

[522] Office of the U.S. Under Secretary of Defense for Policy. *Department of Defense Directive 5220.22-M: National Industrial Security Program Operating Manual.* Washington D.C., USA, Jan. 1995. Unclassified.

[523] Ohlbach, H. J. Optimized Translation of Multi-Modal Logic into Predicate Logic. In *Proceedings of the 4th Conference in Logic Programming and Automated Reasoning (LPAR '93)* (St. Petersburg, Russia, July 1993), A. Voronkov, Ed., vol. 698 of *Lecture Notes in Computer Science,* Springer Verlag, pp. 253–264.

[524] Olawsky, D., Fine, T., Schneider, E., and Spencer, R. Developing and Using a "Policy Neutral" Access Control Policy. In *Proceedings of the Workshop on New Security Paradigms* (Great Langdale, UK, Sept. 1997), ACM Press, pp. 60–67.

[525] Orman, H. RFC 2412: The OAKLEY Key Determination Protocol, Nov. 1998.

[526] Osborn, S., Sandhu, R. S., and Munawer, Q. Configuring Rule-based Access Control to Enforce Mandatory and Discretionary

Access Control Policies. *ACM Transactions on Information and System Security 3*, 2 (Feb. 2000), 85–106.

[527] Oswald, E., and Aigner, M. Randomized Addition-Subtraction Chains as a Countermeasure against Power Attacks. In *Cryptographic Hardware and Embedded Systems – CHES 2001. Proceedings of the Third International Workshop* (Paris, France, May 2001), Ç. Koç, D. Naccache, and C. Paar, Eds., vol. 2162 of *Lecture Notes in Computer Science*, Springer Verlag, pp. 39–50.

[528] Ott, A. *Mandatory Rule Set Based Access Control in Linux*. PhD thesis, University of Hamburg, Hamburg, Germany, 2002.

[529] Owre, S., Rushby, J. M., and Shankar, N. PVS: A Prototype Verification System. In *Proceedings of the 11th International Conference on Automated Deduction (CADE-11)* (Saratoga Springs, NY, USA, June 1992), D. Kapur, Ed., vol. 607 of *Lecture Notes in Computer Science*, Springer Verlag.

[530] P. J. Denning. An Undergraduate Course on Operating System Principles. Tech. rep., National Academy of Engineering, Washington D.C., USA, June 1971. Report of the Cosine Commission on Education.

[531] Padlipsky, M. A., Biba, K. J., and Neely, R. B. KSOS — Computer Network Applications. In *Proceedings of the National Computer Conference* (New York, NY, USA, June 1979), vol. 48, AFIPS, AFIPS Press, pp. 365–371.

[532] Page, J., Heaney, J., Adkins, M., and Dolsen, G. Evaluation of Security Model Rule Bases. In *Proceedings 12th NIST-NCSC National Computer Security Conference* (Gaithersburg, MD, USA, Oct. 1989), pp. 98–111.

[533] Page, Jr., T. W., Guy, R. G., Popek, G. J., and Heidemann, J. S. Architecture of the Ficus Scalable Replicated File System. Tech. Rep. UCLA-CSD 910005, University of California at Los Angeles, Los Angeles, CA, USA, Mar. 1991.

[534] Palmer, E. R. An Introduction to Citadel — A Secure Crypto Coprocessor for Workstations. Tech. Rep. RC 18373, IBM T.J. Watson Research Center, Yorktown Heights, NY, USA, Sept. 1992.

[535] Palmer, E. R. Introduction to Citadel: A Secure Crypto Coprocessor for Workstations. In *10th International Information Security Conference (IFIP SEC '94)* (Curacao, Dutch Antilles, May 1994), Technical Committee TC 11 of the International Federation for Information Processing IFIP, Ed.

[536] Palmer, E. R. Personal Communication. Discussion on the cryptographic and random number generation hardware primitives implemented in the IBM 4758 model 2., Apr. 2000.

[537] Papadimitriou, C. H. *Computational Complexity*. Addison-Wesley, Reading, MA, USA, 1994.

[538] Park, D. Concurrency and Automata on Infinite Sequences. In *Theoretical Computer Science* (Heidelberg, Germany, 1981), P. Deussen, Ed., vol. 104 of *Lecture Notes in Computer Science*, Springer Verlag, pp. 167–183.

[539] Parnas, D. L. A Technique for Software Module Specification with Examples. *Communications of the Association for Computing Machinery 15*, 5 (May 1972), 330–336.

[540] Payne, C. Using Composition and Refinement to Support Security Architecture Trade-off Analysis. In *Proceedings 22nd NIST-NCSC National Information Systems Security Conference* (Crystal City, VA, USA, Oct. 1999), pp. 191–197.

[541] Payne, C., and Markham, T. Architecture and Applications for a Distributed Firewall. In *Proceedings 17th Annual Computer Security Applications Conference (ACSAC'01)* (New Orleans, LA, USA, Dec. 2001), IEEE Computer Society Press, pp. 329–336.

[542] Pearsall, J., and Trumble, W., Eds. *The Oxford English Reference Dictionary*, 2nd ed. Oxford University Press, Oxford, UK, 1996.

[543] Pendharkar, N. S., and Gopinath, K. Formal Verification of an O.S. Submodule. In *Proceedings of the 18th Conference on the Foundations of Software Technology and Theoretical Computer Science* (Madras, India, Dec. 1998), V. Arvind and K. Ramanujin, Eds., vol. 1530 of *Lecture Notes in Computer Science*, Springer Verlag, pp. 197–208.

[544] Perez, R. Personal Communication. Discussion on the tamper-detection mechanisms of the IBM 4758 model 2., Feb. 2000.

[545] Perrine, T., Codd, J., and B, H. An Overview of the Kernelized Secure Operating System (KSOS). In *Proceedings of the 7th DoD /NBS National Computer Security Conference* (Gaithersburg, MD, USA, Sept. 1984), pp. 146–160.

[546] Perrow, C. *Normal Accidents.* Princeton University Press, Princeton, NJ, USA, 1999.

[547] Pfitzmann, B., Riordan, J., Stüble, C., Waidner, M., and Weber, A. The PERSEUS System Architecture. Tech. rep., International Business Machines Corporation, Zürich, Switzerland, Apr. 2001. IBM Research Report RZ 3335.

[548] Pierce, B. C. *Basic Category Theory for Computer Scientists.* MIT Press, Cambridge, MA, USA, 1991.

[549] Pinsky, S. Absorbing Covers and Intransitive Non-Interference. In *Proceedings of the 1995 IEEE Symposium on Security and Privacy (SOSP '95)* (Oakland, CA, USA, May 1995), IEEE Computer Society, pp. 102–113.

[550] Piper, D. RFC 2407: The Internet IP Security Domain of Interpretation for ISAKMP, Nov. 1998.

[551] Plump, D. Term Graph Rewriting. Tech. Rep. CSI-R9822, Computing Science Institute, University of Nijmegen, Nijmegen, The Netherlands, Sept. 1998.

[552] Pnueli, A. The Temporal Logic of Programs. In *Proceedings of the 18th Annual IEEE Symposium on Foundations of Computer Science (FOCS 18)* (Providence, RI, USA, 1977), IEEE Computer Society Press, pp. 46–57.

[553] Pnueli, A. The Temporal Semantics of Concurrent Programs. In *Semantics of Concurrent Computations* (Evian, France, July 1979), G. Kahn, Ed., vol. 70 of *Lecture Notes in Computer Science*, Springer Verlag, pp. 1–20.

[554] Pnueli, A. Applications of Temporal Logic to the Specification and Verification of Reactive Systems: a Survey of Current Trends. In *Current Trends in Concurrency: Overviews and Tutorials* (Sept. 1986), J. W. de Bakker, W.-P. de Roever, and G. Rozenberg, Eds., vol. 224 of *Lecture Notes in Computer Science*, Springer Verlag, pp. 510–584.

[555] Pnueli, A., and Rosner, R. On the Synthesis of a Reactive Module. In *Proceedings of the 16th Annual Symposium on Principles of Programming Languages* (Austin, TX, USA, 1989), ACM Press, pp. 179–290.

[556] Pnueli, A., and Rosner, R. On the Synthesis of an Asynchronous Reactive Module. In *Proceedings of the 16th International Colloquium on Automata, Languages, and Programming* (Stresa, Italy, July 1989), G. Ausiello et al., Eds., vol. 372 of *Lecture Notes in Computer Science*, Springer Verlag, pp. 652–671.

[557] Popek, G. J., and Farber, D. A Model for Verification of Data Security in Operating Systems. *Communications of the Association for Computing Machinery 21*, 9 (Sept. 1978), 737–749.

[558] Popek, G. J., Kampe, M., Kline, C. S., Stoughton, A., Urban, M., and Walton, E. J. UCLA Secure Unix. In *Proceedings of the National Computer Conference* (New York, NY, USA, June 1979), vol. 48, AFIPS, AFIPS Press, pp. 355–364.

[559] Popek, G. J., and Kline, C. S. A Verifiable Protection System. In *Proceedings of the International Conference on Reliable Software* (Los Angeles, CA, USA, 1975), ACM Press, pp. 294–304.

[560] Popper, K. R. *Logik der Forschung*. Julius Springer, Vienna, Austria, 1934.

[561] Postel, J. RFC 791: Internet Protocol: DARPA Internet Program Protocol Specification, Sept. 1981. Status: STANDARD. Replaces RFC 760, IENs 128, 123, 111, 80, 54, 44, 41, 28 and 26; equivalent to MIL-STD 1777.

[562] Pottinger, G. Proof Requirements in the Orange Book: Origins, Implemementation, and Implications. Tech. rep., Cornell University Mathematical Sciences Institute, Ithaca, NY, USA, Feb. 1994. Performed for the Naval Research Laboratory, United States Navy.

[563] Presotto, D., Pike, R., Thompson, K., and Trickey, H. Plan 9, a Distributed System. In *Proceedings of the USENIX Workshop on Micro-Kernels and Other Kernel Architectures* (Seattle, WA, USA, Apr. 1992), USENIX, Ed., USENIX, pp. 31–37.

[564] Priese, L., Rehrmann, R., and Willecke-Klemme, U. An Introduction to the Regular Theory of Fairness. *Theoretical Computer Science 54*, 2 /3 (Oct. 1987), 139–163.

[565] Prior, A. N. *Papers on Time and Tense*. Oxford University Press, Oxford, UK, 1968.

[566] Rademer, E., and Wolthusen, S. Transparent Access To Encrypted Data Using Operating System Network Stack Extensions. In *Communications and Multimedia Security Issues of the New Century: Proceedings of the IFIP TC6/TC11 Fifth Joint Working Conference on Communications and Multimedia Security (CMS'01)* (Darmstadt, Germany, May 2001), R. Steinmetz, J. Dittman, and M. Steinebach, Eds., IFIP, Kluwer Academic Publishers, pp. 213–226.

[567] Rajunas, S. A., Hardy, N., Bomberger, A. C., Frantz, W. S., and Landau, C. R. Security in KeyKOS. In *Proceedings of the 1986 IEEE Symposium on Security and Privacy (SOSP '86)* (Oakland, CA, USA, Apr. 1986), IEEE Computer Society, pp. 78–85.

[568] Rashid, R. F. From RIG to Accent to Mach: The Evolution of a Network-Operating System. In *Proceedings of the ACM/IEEE Computer Society Fall Joint Computer Conference* (Dallas, TX, USA, Nov. 1986), ACM/IEEE Computer Society, ACM Press, pp. 1128–1137.

[569] Rashid, R. F., and Robertson, G. Accent: A Communication--Oriented Network-Operating System Kernel. *ACM Operating Systems Review 15*, 5 (Dec. 1981), 64–75. Proceedings of the 8th Symposium on Operating Systems Principles.

[570] Rebensburg, K., Youman, C. E., and Atluri, V., Eds. *Proceedings of the Fifth ACM Workshop on Role-Based Access Control (RBAC 2000)* (Berlin, Germany, July 2000).

[571] Redell, D. R., and Fabry, R. S. Selective Revocation and Capabilities. In *Proceedings of the International Workshop on Protection in Operating Systems* (Roquencourt, Le Chesnay, France, Aug. 1974), Institute de Recherche d'Informatique, Roquencourt, pp. 197–209.

[572] Reiter, R. A Logic for Default Reasoning. *Artificial Intelligence 13*, 1–2 (1980), 81–132.

[573] Rescorla, E. RFC 2818: HTTP Over TLS, May 2000.

[574] Rich, E., and Knight, K. *Artificial Intelligence.* McGraw-Hill, New York, NY, USA, 1990.

[575] Richter, J. *Applied Microsoft .NET Framework Programming.* Microsoft Press, Redmond, WA, USA, 2002.

[576] RITCHIE, D. M., AND THOMPSON, K. The UNIX Time-Sharing System. *Communications of the Association for Computing Machinery 17*, 7 (July 1974), 365–375.

[577] ROBINSON, J. A. A Machine-Oriented Logic Based on the Resolution Principle. *Journal of the ACM 12*, 1 (Jan. 1965), 23–41. Reprinted in [656], pp. 397–415.

[578] ROBINSON, J. A., AND VORONKOV, A., Eds. *Handbook of Automated Reasoning*, vol. I. MIT Press, Cambridge, MA, USA, 2001.

[579] ROBINSON, J. A., AND VORONKOV, A., Eds. *Handbook of Automated Reasoning*, vol. II. MIT Press, Cambridge, MA, USA, 2001.

[580] ROBINSON, L. *The HDM Handbook, Volume I: The Foundations of HDM*. Menlo Park, CA, USA, June 1979. SRI Project 4828.

[581] ROBINSON, L., LEVITT, K. N., NEUMANN, P. G., AND SAXENA, A. R. A Formal Methodology for the Design of Operating Systems. In *Current Trends in Programming Methodology*, R. T. Yeh, Ed. Prentice Hall, Englewood Cliffs, NJ, USA, 1977.

[582] ROSCOE, A. W., AND GOLDSMITH, M. H. What is Intransitive Noninterference? In *Proceedings of the Computer Security Foundations Workshop* (Mordano, Italy, June 1999), IEEE Computer Society Press, pp. 228–238.

[583] RUSHBY, J. A Trusted Computing Base for Embedded Systems. In *Proceedings of the 7th DoD / NBS National Computer Security Conference* (Gaithersburg, MD, USA, Sept. 1984), pp. 294–311.

[584] RUSHBY, J. Noninterference, Transitivity, and Channel-Control Security Policies. Tech. Rep. CSL-92-02, Computer Science Laboratory, SRI International, Menlo Park, CA, USA, Dec. 1992.

[585] RUSHBY, J., AND VON HENKE, F. Formal Verification of Algorithms for Critical Systems. *IEEE Transactions on Software Engineering 19*, 1 (Jan. 1993), 13–23.

[586] Russell, S., and Norvig, P. *Artificial Intelligence: A Modern Approach*. Prentice Hall, Englewood Cliffs, NJ, USA, 1995.

[587] Russinovich, M., and Cogswell, B. Examining the Windows NT Filesystem. *Dr. Dobb's Journal of Software Tools 22*, 2 (Feb. 1997), 42–50.

[588] Ryan, P., and Schneider, S. *Modelling and Analysis of Security Protocols*. Addison-Wesley, Reading, MA, USA, 2001.

[589] Ryan, P. Y. A. A CSP Formulation of Non-interference. In *Proceedings of the IEEE Computer Security Foundations Workshop IV* (Franconia, NH, USA, June 1991), IEEE Computer Society. Paper was not ready at press time for proceedings, appeared in [590].

[590] Ryan, P. Y. A. A CSP Formulation of Non-interference and Unwinding. *IEEE Cipher* (Mar. 1991), 19–30.

[591] Ryan, P. Y. A. Mathematical Models of Computer Security. In *Foundations of Security Analysis and Design*, R. Focardi and R. Gorrieri, Eds., vol. 2171 of *Lecture Notes in Computer Science*. Springer Verlag, Bertinoro, Italy, Sept. 2001, pp. 1–62. Lecture notes from the 1st International School on Foundations of Security Analysis and Design (FOSAD'01).

[592] Ryan, P. Y. A., and Schneider, S. A Process Algebra and Non-Interference. In *Proceedings of the 12th IEEE Computer Security Foundations Workshop (CSFW-12)* (Mordano, Italy, June 1999), IEEE Computer Society, pp. 214–227.

[593] Sagan, S. D. *The Limits of Safety*. Princeton University Press, Princeton, NJ, USA, 1995.

[594] Saltzer, J. H., and Schroeder, M. D. The Protection of Information in Computer Systems. *Proceedings of the IEEE 63*, 9 (Sept. 1975), 1278–1308.

[595] SANDHU, R., AND BERTINO, E., Eds. *Proceedings of the 7th ACM Symposium on Access Control Models and Technologies (SACMAT 2002)* (Monterey, CA, USA, June 2002).

[596] SANDHU, R., AND JAEGER, T., Eds. *Proceedings of the 6th ACM Symposium on Access Control Models and Technologies (SACMAT '01)* (Chantilly, VA, USA, May 2001).

[597] SANDHU, R. S. Transaction Control Expressions for Separation of Duties. In *Proceedings 4th Aerospace Computer Security Applications Conference (ACSAC'88)* (Orlando, FL, USA, Dec. 1988), IEEE Computer Society Press, pp. 282–286.

[598] SANDHU, R. S. Role Hierarchies and Constraints for Lattice-Based Access Controls. In *Proceedings of the 4th European Symposium on Research in Computer Security (ESORICS 96)* (Rome, Italy, Sept. 1996), E. Bertino, H. Kurth, G. Martella, and E. Montolivo, Eds., vol. 1146 of *Lecture Notes in Computer Science*, Springer Verlag, pp. 65–79.

[599] SANDHU, R. S., AND CHEN, F. The Multilevel Relational (MLR) Data Model. *ACM Transactions on Information and System Security 1*, 1 (Nov. 1998), 93–132.

[600] SANDHU, R. S., COYNE, E. J., FEINSTEIN, H. L., AND YOUMAN, C. E. Role-Based Access Control Models. *IEEE Computer 29*, 2 (Feb. 1996), 38–47.

[601] SANDHU, R. S., AND GANTA, S. On the Expressive Power of the Unary Transformation Model. In *Proceedings of the 3rd European Symposium on Research in Computer Security (ESORICS 94)* (Brighton, UK, 1994), D. Gollmann, Ed., vol. 875 of *Lecture Notes in Computer Science*, Springer Verlag, pp. 301–318.

[602] SANDHU, R. S., AND SURI, G. S. Non-Monotonic Transformation of Access Rights. In *Proceedings of the 1992 IEEE Symposium on Security and Privacy (SOSP '92)* (Oakland, CA, USA, May 1902), IEEE Computer Society, pp. 148–161.

[603] Saydjari, O. S., Beckman, J. M., and Leaman, J. R. LOCKing Computers Securely. In *Proceedings 10th NIST-NCSC National Computer Security Conference* (Gaithersburg, MD, USA, Oct. 1987), pp. 129–141.

[604] Saydjari, O. S., Beckman, J. M., and Leaman, J. R. LOCK Trek: Navigating Uncharted Space. In *Proceedings of the 1989 IEEE Symposium on Security and Privacy (SOSP '89)* (Oakland, CA, USA, Apr. 1989), IEEE Computer Society, pp. 167–175.

[605] Saydjari, O. S., Turner, S. J., Peele, D. E., Farrell, J. F., Loscocco, P. A., Kutz, W., and Bock, G. L. SYNERGY: A Distributed, Microkernel-based Security Architecture. Tech. Rep. R231, U.S. National Security Agency INFOSEC Directorate, Ft. George G. Meade, MD, USA, Nov. 1993.

[606] SCC. Software Requirements Specification for Distributed Trusted Mach. Secure Computing Corporation DTMach CDRL A005, Oct. 1992.

[607] SCC. Final Report for the Distributed Trusted Mach Program. Secure Computing Corporation (RL-TR-93-235). Performed for RADC, Griffiss Air Force Base, NY, USA., Dec. 1993.

[608] SCC. Formal Top Level Specification for Distributed Trusted Mach. Secure Computing Corporation DTMach CDRL A012, Apr. 1993.

[609] SCC. System / Segment Design Document for Distributed Trusted Mach. Secure Computing Corporation DTMach CDRL A006; Version 2, Mar. 1993.

[610] SCC. DTOS Covert Channel Analysis Report. Tech. rep., Secure Computing Corporation, Roseville, MN, USA, May 1995. Contract no. MDA904-93-C-4209, CDRL sequence no. A007; Distribution limited to U.S. Government Agencies Only. This document contains NSA information. Request for the document must be referred to the Director, NSA.

[611] SCC. DTOS Software System Development Test and Evaluation Plan. Tech. rep., Secure Computing Corporation, Roseville, MN, USA, Apr. 1995. Contract no. MDA904-93-C-4209, CDRL sequence no. A012; Distribution limited to U.S. Government Agencies Only. This document contains NSA information. Request for the document must be referred to the Director, NSA.

[612] SCC. DTOS Formal Securiry Policy Model (FSPM). Tech. rep., Secure Computing Corporation, Roseville, MN, USA, Sept. 1996. Contract no. MDA904-93-C-4209, CDRL sequence no. A004; Distribution limited to U.S. Government Agencies Only. This document contains NSA information. Request for the document must be referred to the Director, NSA.

[613] SCC. DTOS Formal Top-Level Specification (FTLS). Tech. rep., Secure Computing Corporation, Roseville, MN, USA, Dec. 1996. Contract no. MDA904-93-C-4209, CDRL sequence no. A005; Distribution limited to U.S. Government Agencies Only. This document contains NSA information. Request for the document must be referred to the Director, NSA.

[614] SCC. DTOS Kernel and Security Server Software Design Document. Tech. rep., Secure Computing Corporation, Roseville, MN, USA, Dec. 1996. Contract no. MDA904-93-C-4209, CDRL sequence no. A002; Distribution limited to U.S. Government Agencies Only. This document contains NSA information. Request for the document must be referred to the Director, NSA.

[615] SCC. DTOS Software Requirements Specification. Secure Computing Corporation DTOS CDRL A001. This is an updated version of the DTMach CDRL [606], June 1996.

[616] SCC. *DTOS User Manual*. Roseville, MN, USA, Sept. 1996. Distribution unlimited.

[617] SCC. DTOS Composability Study. Tech. rep., Secure Computing Corporation, Roseville, MN, USA, June 1997. Contract no.

MDA904-93-C-4209, CDRL sequence no. A020; Distribution limited to U.S. Government Agencies Only. This document contains NSA information. Request for the document must be referred to the Director, NSA.

[618] SCC. DTOS Composability Study. Tech. rep., Secure Computing Corporation, Roseville, MN, USA, June 1997. Contract no. MDA904-93-C-4209, CDRL sequence no. A020; Distribution limited to U.S. Government Agencies Only. This document contains NSA information. Request for the document must be referred to the Director, NSA.

[619] SCC. DTOS Formal Security Policy Model (FSPM) (Non-Z Version). Tech. rep., Secure Computing Corporation, Roseville, MN, USA, June 1997. SCC internal distribution only.

[620] SCC. DTOS General System Security and Assurability Assessment Report. Tech. rep., Secure Computing Corporation, Roseville, MN, USA, June 1997. Contract no. MDA904-93-C-4209, CDRL sequence no. A011; Distribution limited to U.S. Government Agencies Only. This document contains NSA information. Request for the document must be referred to the Director, NSA.

[621] SCC. DTOS Generalized Security Policy Specification. Tech. rep., Secure Computing Corporation, Roseville, MN, USA, June 1997. Contract no. MDA904-93-C-4209, CDRL sequence no. A019; Distribution limited to U.S. Government Agencies Only. This document contains NSA information. Request for the document must be referred to the Director, NSA.

[622] SCC. DTOS Generalized Security Policy Specification. Tech. rep., Secure Computing Corporation, Roseville, MN, USA, June 1997. Contract no. MDA904-93-C-4209, CDRL sequence no. A019; Distribution limited to U.S. Government Agencies Only. This document contains NSA information. Request for the document must be referred to the Director, NSA.

[623] SCC. DTOS Lessons Learned Report. Tech. rep., Secure Computing Corporation, Roseville, MN, USA, June 1997. Contract no. MDA904-93-C-4209, CDRL sequence no. A008; Distribution limited to U.S. Government Agencies Only. This document contains NSA information. Request for the document must be referred to the Director, NSA.

[624] SCC. Assurance in the Fluke Microkernel: Final Report. Tech. rep., Secure Computing Corporation, Roseville, MN, USA, Apr. 1999. Contract no. MDA904-97-C-3047, CDRL sequence no. A002; Distribution limited to U.S. Government Agencies Only. This document contains NSA information. Request for the document must be referred to the Director, NSA.

[625] SCC. Assurance in the Fluke Microkernel: Formal Security Policy Model. Tech. rep., Secure Computing Corporation, Roseville, MN, USA, Feb. 1999. Contract no. MDA904-97-C-3047, CDRL sequence no. A003; Distribution limited to U.S. Government Agencies Only. This document contains NSA information. Request for the document must be referred to the Director, NSA.

[626] SCC. Assurance in the Fluke Microkernel: Formal Top Level Specification. Tech. rep., Secure Computing Corporation, Roseville, MN, USA, Feb. 1999. Contract no. MDA904-97-C-3047, CDRL sequence no. A004; Distribution limited to U.S. Government Agencies Only. This document contains NSA information. Request for the document must be referred to the Director, NSA.

[627] SCC. Assurance in the Fluke Microkernel: System Composition Study Report. Tech. rep., Secure Computing Corporation, Roseville, MN, USA, Mar. 1999. Contract no. MDA904-97-C-3047, CDRL sequence no. A005; Distribution limited to U.S. Government Agencies Only. This document contains NSA information. Request for the document must be referred to the Director, NSA.

[628] Schaefer, M. The Multilevel Policy Model for the Trusted ONTOS Prototype. Invited Lecture at the Naval Postgraduate School Center for Information Systems Security Studies and Research, July 1998.

[629] Schaffer, M. A., and Walsh, G. LOCK/ix: On Implementing Unix On The LOCK TCB. In *Proceedings 11th NIST-NCSC National Computer Security Conference* (Washington D.C., USA, Oct. 1988), pp. 319–329.

[630] Schell, R. R. Notes on an Approach for Design of Secure Military ADP Systems. In *Proceedings of the ACM Annual Conference* (Aug. 1972), ACM Press, pp. 665–666.

[631] Schell, R. R. Information Security: Science, Pseudoscience, and Flying Pigs. In *Proceedings 17th Annual Computer Security Applications Conference (ACSAC'01)* (New Orleans, LA, USA, Dec. 2001), IEEE Computer Society Press, pp. 205–216.

[632] Schell, R. R. Personal Communication. Conversation at ACSAC 2001., Dec. 2001.

[633] Schell, R. R., Tao, T. F., and Heckman, M. Designing the GEMSOS Security Kernel for Security and Performance. In *Proceedings of the 8th DoD / NBS National Computer Security Conference* (Gaithersburg, MD, USA, Oct. 1985), pp. 108–119.

[634] Schiller, W. L. The Design and Specification of a Security Kernel for the PDP-11/45. Tech. rep., The MITRE Corporation, Bedford, MA, USA, Mar. 1975. Number ESD-TR-75-69. Performed for the Electronic Systems Division, Air Force Systems Command, United States Air Force.

[635] Schiller, W. L. Design and Abstract Specification of a Multics Security Kernel. Tech. rep., The MITRE Corporation, Bedford, MA, USA, Aug. 1977. Number ESD-TR-77-259. Performed for the Electronic Systems Division, Air Force Systems Command, United States Air Force.

[636] Schimmel, J. A Historical Look at Firewall Technologies. *;login: the USENIX Association newsletter 22*, 1 (Feb. 1997), 21–23.

[637] Schindler, W. A Timing Attack against RSA with the Chinese Remainder Theorem. In *Cryptographic Hardware and Embedded Systems – CHES 2000. Proceedings of the Second International Workshop* (Worcester, MA, USA, Aug. 2000), Ç. Koç and C. Paar, Eds., vol. 1965 of *Lecture Notes in Computer Science*, Springer Verlag, pp. 109–117.

[638] Schindler, W. A Combined Timing and Power Attack. In *Proceedings of Public Key Cryptography, 5th International Workshop on Practice and Theory in Public Key Cryptosystems (PKC 2002)* (Paris, France, Feb. 2002), D. Naccache and P. Paillier, Eds., vol. 2274 of *Lecture Notes in Computer Science*, Springer Verlag, pp. 263–279.

[639] Schmidt, R. A. Decidability by Resolution for Propositional Modal Logics. *Journal of Automated Reasoning 22* (1999), 379–396.

[640] Schnackenberg, D. Applying the Orange Book to an MLS LAN. In *Proceedings 10th NIST-NCSC National Computer Security Conference* (Gaithersburg, MD, USA, Oct. 1987), pp. 51–55.

[641] Schneck, R. R., and Necula, G. C. A Gradual Approach to a More Trustworthy, Yet Scalable, Proof-Carrying Code. In *Proceedings of the 18th International Conference on Automated Deduction (CADE-18)* (Copenhagen, Denmark, July 2002), A. Vorkonkov, Ed., vol. 2392 of *Lecture Notes in Computer Science*, Springer Verlag, pp. 47–62.

[642] Schneider, F. B. Decomposing Properties into Safety and Liveness Using Predicate Logic. Tech. Rep. 87-874, Cornell University Department of Computer Science, Ithaca, NY, USA, Oct. 1987.

[643] Schneider, F. B. Enforceable Security Policies. *ACM Transactions on Information and System Security 3*, 1 (Feb. 2000), 30–50.

[644] Schneier, B. Computer Security: It's the Economics, Stupid. In *Workshop on Economics and Information Security at the University of California, Berkeley* (Berkeley, CA, USA, May 2002). Electronic proceedings only.

[645] Schreiber, S. B. *Undocumented Windows 2000 Secrets*. Addison-Wesley, Reading, MA, USA, 2001.

[646] Schütte, K. *Proof Theory*. Springer Verlag, Heidelberg, Germany, 1977.

[647] Science Applications International Corporation. *Windows 2000 Security Target*. Columbia, MD, USA, Oct. 2002. Version 2.0; prepared for Microsoft Corporation.

[648] Shamir, A. Protecting Smart Cards from Passive Power Analysis with Detached Power Supplies. In *Cryptographic Hardware and Embedded Systems – CHES 2000. Proceedings of the Second International Workshop* (Worcester, MA, USA, Aug. 2000), Ç. Koç and C. Paar, Eds., vol. 1965 of *Lecture Notes in Computer Science*, Springer Verlag, pp. 71–77.

[649] Shapiro, J. S. *EROS: A Capability System*. PhD thesis, University of Pennsylvania, Philadelphia, PA, USA, 1999.

[650] Shapiro, J. S., and Hardy, N. EROS: A Principle-Driven Operating System from the Ground Up. *IEEE Software 19*, 1 (Jan./ Feb. 2002), 26–33.

[651] Shapiro, J. S., Smith, J. M., and Farber, D. J. EROS: A Fast Capability System. *ACM Operating Systems Review 34*, 5 (Dec. 1999), 170–185. Proceedings of the 17th Symposium on Operating Systems Principles.

[652] Shapiro, J. S., and Weber, S. Verifying the EROS Confinement Mechanism. In *Proceedings of the 2000 IEEE Symposium on Security and Privacy (SOSP '00)* (Oakland, CA, USA, Apr. 2000), pp. 166–176.

[653] Sharp (ed.), R. Commercial IP Security Option (CIPSO 2.2), July 1992. IETF CIPSO Working Group Work in Progress, Expired January 1993.

[654] Shockley, W. R., and Downey, J. P. Is the Reference Monitor Concept Fatally Flawed? The Case for the Negative. In *Proceedings of the 1997 IEEE Symposium on Security and Privacy (SOSP '97)* (Oakland, CA, USA, May 1997), IEEE Computer Society, pp. 6–7. Position paper for the panel discussion: Is the Trusted Computing Base Concept Fundamentally Flawed?

[655] Shostak, R. E., Schwartz, R., and Mellier-Smith, P. M. STP: A Mechanized Logic for Specification and Verification. In *Proceedings of the 6th International Conference on Automated Deduction (CADE-6)* (Courant Institute, NY, USA, June 1982), D. Loveland, Ed., vol. 138 of *Lecture Notes in Computer Science*, Springer Verlag, pp. 93–101.

[656] Siekmann, J., and Wrightson, G., Eds. *Automation of Reasoning: 1957–1966*, vol. 1 of *Classical Papers on Computational Logic*. Springer Verlag, Heidelberg, Germany, 1983.

[657] Siekmann, J., and Wrightson, G., Eds. *Automation of Reasoning: 1967–1970*, vol. 2 of *Classical Papers on Computational Logic*. Springer Verlag, Heidelberg, Germany, 1983.

[658] Silverberg, B., Robinson, L., and Levitt, K. N. *The HDM Handbook, Volume II: The Languages and Tools of HDM*. Menlo Park, CA, USA, June 1979. SRI Project 4828.

[659] Silverman, J. M. Proving an Operating System Kernel Secure. Tech. Rep. 81SRC31, Honeywell Systems & Research Center, Minneapolis, MN, USA, Apr. 1981.

[660] Silverman, J. M. Reflections on the Verification of the Security of an Operating System Kernel. *ACM Operating Systems Review 17*, 5 (Dec. 1983), 143–154.

[661] Simmons, G. J. Results Concerning the Bandwidth of Subliminal Channels. *IEEE Journal on Selected Areas in Communications 16*, 4 (May 1988), 463–473.

[662] Simmons, G. J. The History of Subliminal Channels. *IEEE Journal on Selected Areas in Communications 16*, 4 (May 1988), 452–462.

[663] Skinner, G. C., and K., T. "Stacking" Vnodes: A Progress Report. In *Proceedings of the Summer 1993 USENIX Conference* (Seattle, WA, USA, June 1993), USENIX, Ed., USENIX, pp. 161–174.

[664] Smalley, S. Configuring the SELinux Policy. Tech. rep., NAI Labs, Santa Clara, CA, USA, June 2002. NSA Contract no. MDA904-01-C-0926. NAI Labs Report Number 02-007.

[665] Smalley, S., Vance, C., and Salamon, W. Implementing SELinux as a Linux Security Module. Tech. rep., NAI Labs, Santa Clara, CA, USA, May 2002. NSA Contract no. MDA904-01-C-0926. NAI Labs Report Number 01-043.

[666] Smith, M. A. S. *Formal Verification of TCP and T/TCP*. PhD thesis, Massachusetts Institute of Technology, Cambridge, MA, USA, 2001.

[667] Smith, R. E. Constructing a High Assurance Mail Guard. In *Proceedings 17th NIST-NCSC National Computer Security Conference* (Gaithersburg, MD, USA, Oct. 1994), pp. 247–253.

[668] Smith, R. E. Cost Profile of a Highly Assured, Secure Generating System. *ACM Transactions on Information and System Security 4*, 1 (Feb. 2001), 72–101.

[669] SMITH, S. W., AND AUSTEL, V. Trusting Trusted Hardware: Towards a Formal Model for Programmable Secure Coprocessors. In *Proceedings of the 3rd USENIX Workshop on Electronic Commerce* (Oakland, CA, USA, Aug. 1998), USENIX, Ed., USENIX, pp. 83–98.

[670] SMITH, S. W., PALMER, E. R., AND WEINGART, S. Using a High-Performance, Programmable Secure Coprocessor. In *Proceedings of Financial Cryptography: Second International Conference (FC '98)* (Anguilla, British West Indies, Feb. 1998), vol. 1465 of *Lecture Notes in Computer Science*, pp. 73–89.

[671] SMITH, S. W., PEREZ, R., WEINGART, S. H., AND AUSTEL, V. Validating a High-Performance, Programmable Secure Coprocessor. In *Proceedings 22nd NIST-NCSC National Information Systems Security Conference* (Crystal City, VA, USA, Oct. 1999), pp. 163–173.

[672] SMITH, S. W., AND WEINGART, S. Building a High-Performance, Programmable Secure Coprocessor. *Computer Networks 31*, 8 (Aug. 1999), 831–860.

[673] SNAPP, S. R., BRENTANO, J., DIAS, G. V., GOAN, T. L., HEBERLEIN, L. T., HO, C.-L., LEVITT, K. N., MUKHERJEE, B., SMAHA, S. E., GRANCE, T., TEAL, D. M., AND MANSUR, D. DIDS (Distributed Intrusion Detection System) – Motivation, Architecture, and an Early Prototype. In *Proceedings 14th NIST-NCSC National Computer Security Conference* (Washington D.C., USA, Oct. 1991), pp. 167–176.

[674] SNYDER, L. Formal Methods of Capability-Based Protection Systems. *IEEE Transactions on Computers C-30*, 3 (Mar. 1981), 172–181.

[675] SOLOMON, D. A., AND CUSTER, H. *Inside Microsoft Windows NT*, 2nd ed. Microsoft Press, Redmond, WA, USA, 1998.

[676] Solomon, D. A., and Russinovich, M. E. *Inside Microsoft Windows 2000*, 3rd ed. Microsoft Press, Redmond, WA, USA, 2000.

[677] Sowa, J. W. *Knowledge Representation: Logical, Philosophical, and Computational Foundations*. Brooks Cole Publishing, Pacific Grove, CA, USA, 2000.

[678] Spencer, R., Smalley, S., Loscocco, P., Hibler, M., Andersen, D., and Lepreau, J. The Flask Security Architecture: System Support for Diverse Security Policies. In *Proceedings of the 8th USENIX Security Symposium* (Washington D.C., USA, Aug. 1999), USENIX, Ed., pp. 123–139.

[679] Spivey, J. M. *The Z Notation: A Reference Manual*. Prentice Hall International Series in Computer Science. Prentice Hall, London, UK, 1992.

[680] Sripada, R., and Keefe, T. Version Management in the STAR MLS Database System. In *Proceedings of Database Security XII: Status and Prospects, IFIP TC11 WG 11.3 Twelfth International Conference on Database Security* (Chakaladiki, Greece, Aug. 1998), S. Jajodia, Ed., pp. 159–171.

[681] St. Johns, M. RFC 1038: Draft Revised IP Security Option, Jan. 1988. Status: HISTORIC.

[682] Steinberg, A. N., Bowman, C. L., and White, Jr., F. E. Revisions to the JDL Data Fusion Model. In *Proceedings of the 3rd NATO/IRIS Conference* (Quebec City, Canada, Oct. 1998), NATO Research and Technology Organization. Proceedings are classified, summarized in [285].

[683] Steinberg, A. N., and Washburn, R. B. Multi-level fusion for Warbreaker intelligence correlation. In *Proceedings of the 8th National Symposium on Sensor Fusion* (1995).

[684] Sterne, D. F., Branstad, M. A., Hubbard, B. S., Mayer, B. A., and Wolcott, D. M. An Analysis of Application-Specific Security Policies. In *Proceedings 14th NIST-NCSC National Computer*

Security Conference (Washington D.C., USA, Oct. 1991), pp. 25–36.

[685] Stevenson, J. M., and Julin, D. P. Mach-US: UNIX On Generic OS Object Servers. In *Proceedings of the 1995 USENIX Technical Conference* (New Orleans, LA, USA, Jan. 1995), USENIX Association, Ed., USENIX, pp. 119–130.

[686] Stoneburner, G. R., and Snow, D. A. The Boeing MLS LAN: Headed Towards an INFOSEC Security Solution. In *Proceedings 12th NIST-NCSC National Computer Security Conference* (Gaithersburg, MD, USA, Oct. 1989), pp. 254–266.

[687] Strachey, C. Time Sharing in Large Fast Computers. In *Proceedings of the International Conference on Information Processing* (London, UK, June 1959), Butterworth's, pp. B336–B341.

[688] Strunk, J. D., Goodson, G. R., Scheinholtz, M. L., Soules, C. A. N., and Ganger, G. R. Self-Securing Storage: Protecting Data in Compromised Systems. In *Proceedings of the 4th Symposium on Operating Systems Design and Implementation (OSDI '00)* (San Diego, CA, USA, Oct. 2000), USENIX, pp. 165–180.

[689] Sutherland, D. A Model of Information. In *Proceedings of the 9th DoD / NBS National Computer Security Conference* (Gaithersburg, MD, USA, Oct. 1986), pp. 175–183.

[690] Swift, M. M., Hopkins, A., Brundrett, P., van Dyke, C., Garg, P., Chang, S., Goertzel, M., and Jensenworth, G. Improving the Granularity of Access Control for Windows 2000. *ACM Transactions on Information and System Security 5*, 4 (Nov. 2002), 398–437.

[691] Syverson, P. F., and Gray III, J. W. The Epistemic Representation of Information Flow Security in Probabilistic Systems. In *Proceedings of the Computer Security Foundations Workshop VIII* (June 1995), IEEE Computer Society Press, pp. 152–166.

[692] TANENBAUM, A. S. *Distributed Operating Systems*. Prentice Hall, Englewood Cliffs, NJ, USA, 1995.

[693] TANENBAUM, A. S., AND MULLENDER, S. An Overview of the Amoeba Distributed Operating System. *ACM Operating Systems Review 15*, 3 (July 1981), 51–64.

[694] TANENBAUM, A. S., MULLENDER, S. J., AND VAN RENESSE, R. Using Sparse Capabilities in a Distributed Operating System. In *Proceedings of the 6th International Conference on Distributed Computing Systems* (Los Angeles, CA, USA, May 1986), IEEE Computer Society, pp. 558–563.

[695] TANENBAUM, A. S., VAN RENESSE, R., VAN STAVEREN, H., SHARP, G. J., MULLENDER, S. J., JANSEN, J., AND VAN ROSSUM, G. Experiences with the Amoeba Distributed Operating System. *Communications of the Association for Computing Machinery 33*, 12 (Dec. 1990), 46–63.

[696] TANGNEY, J. D. Minicomputer Architectures for Effective Security Kernel Implementations. Tech. rep., The MITRE Corporation, Bedford, MA, USA, Oct. 1978. Number ESD-TR-78-170. Performed for the Electronic Systems Division, Air Force Systems Command, United States Air Force.

[697] TANGNEY, J. D., AMES, JR., S. R., AND BURKE, E. L. Security Evaluation Criteria for MMP Message Service Selection. Tech. rep., The MITRE Corporation, Bedford, MA, USA, June 1977. Number MTR-3433.

[698] TAYLOR, T. Comparison Paper between the Bell and LaPadula Model and the SRI Model. In *Proceedings of the 1984 IEEE Symposium on Security and Privacy (SOSP '84)* (Oakland, CA, USA, Apr. 1984), IEEE Computer Society, pp. 195–203.

[699] TAYLOR, T. FTLS-Based Security Testing for LOCK. In *Proceedings 12th NIST-NCSC National Computer Security Conference* (Gaithersburg, MD, USA, Oct. 1989), pp. 136–145.

[700] THAI, T., AND LAM, H. Q. *.NET Framework Essentials*. O'Reilly & Associates, Sebastopol, CA, USA, 2002.

[701] THAYER, R., DORASWAMY, N., AND GLENN, R. RFC 2411: IP Security Document Roadmap, Nov. 1998. Status: INFORMATIONAL.

[702] THOMAS, W. Automata on Infinite Objects. In *Handbook of Theoretical Computer Science*, J. van Leeuwen, Ed., vol. B. MIT Press, Cambridge, MA, USA, 1990, ch. 4, pp. 789–840.

[703] THOMSEN, D. J. Role-Based Application Design and Enforcement. In *Proceedings of Database Security IV: Status and Prospects. Results of the IFIP WG 11.3 Workshop on Database Security* (Halifax, UK, Sept. 1990), S. Jajodia and C. E. Landwehr, Eds., pp. 151–168.

[704] TING, T. C., DEMURJIAN, S. A., AND HU, M.-Y. Requirements, Capabilities, and Functionalities of User-Role Based Security for an Object-Oriented Design Model. In *Proceedings of Database Security V: Status and Prospects. Results of the IFIP WG 11.3 Workshop on Database Security* (Shepherdstown, WV, USA, Nov. 1992), C. E. Landwehr and S. Jajodia, Eds., pp. 275–296.

[705] TIS. MK++ Kernel Executive Summary. Tech. Rep. DDOC-0034-95A, Trusted Information Systems, Glenwood, MD, USA, Nov. 1995.

[706] TIS. MK++ Kernel Interfaces. Tech. Rep. IDOC-0001-95A, Trusted Information Systems, Glenwood, MD, USA, Nov. 1995.

[707] TIS. Trusted Mach System Architecture. Tech. Rep. EDOC-0001-97A, Trusted Information Systems, Glenwood, MD, USA, Oct. 1995.

[708] TIS. Trusted Mach B3 or F-B3 Security Target. Tech. Rep. EDOC-0005-96A, Trusted Information Systems, Glenwood, MD, USA, Oct. 1996.

[709] TIS. Trusted Mach Development Practices and Procedures. Tech. Rep. EDOC-0002-96A, Trusted Information Systems, Glenwood, MD, USA, Oct. 1996.

[710] TIS. Trusted Mach Mathematical Model. Tech. Rep. EDOC-0017-96B, Trusted Information Systems, Glenwood, MD, USA, Oct. 1996.

[711] TIS. Trusted Mach Philosophy of Protection. Tech. Rep. EDOC-0003-96A, Trusted Information Systems, Glenwood, MD, USA, Oct. 1996.

[712] Torvalds, L. Linux Kernel Implementation. In *Proceedings of the AUUG94 Conference: Open Systems* (Kensington, NSW, Australia, 1994), AUUG Inc., pp. 9–14.

[713] Totel, E., Blanquart, J.-P., Deswarte, Y., and Powell, D. Supporting Multiple Levels of Criticality. In *Proceedings of the Twenty-Eighth Annual International Symposium on Fault-Tolerant Computing* (Munich, Germany, June 1998), IEEE Press, pp. 70–79.

[714] Tretick, B. Certification and Accreditation Approach for the WWMCCS Guard. In *Proceedings 16th NIST-NCSC National Computer Security Conference* (Baltimore, MD, USA, Oct. 1993), pp. 245–252.

[715] Tretick, B. Operational Requirements for Multilevel Security. In *Proceedings 9th Annual Computer Security Applications Conference (ACSAC'93)* (Orlando, FL, USA, Dec. 1993), IEEE Computer Society Press, pp. 30–35.

[716] Troelstra, A. S., and Schwichtenberg, H. *Basic Proof Theory*, 2nd ed., vol. 43 of *Cambridge Tracts In Theoretical Computer Science*. Cambridge University Press, Cambridge, UK, 2000.

[717] Trusted Standards Interoperability Group. *TSIX (RE) Trusted Security Information Exchange for Restricted Environments Version 1.1*, 1995.

[718] Tsai, C.-R., and Gligor, V. D. A Bandwidth Computation Model for Covert Storage Channels and its Applications. In *Proceedings of the 1988 IEEE Symposium on Security and Privacy (SOSP '88)* (Oakland, CA, USA, May 1988), IEEE Computer Society, pp. 108–121.

[719] Tsai, C.-R., Gligor, V. D., and Chandersekaran, C. S. On the Identification of Covert Storage Shannels in Secure Systems. *IEEE Transactions on Software Engineering 16*, 6 (June 1990), 569–580.

[720] Turing, A. M. On computable numbers with an application to the Entscheidungsproblem. *Proceedings of the London Mathematical Society Series 2 42* (1936/1937), 230–265.

[721] Turing, A. M. On Checking a Large Routine. In *Report of a Conference on High-speed Automatic Calculating Machines* (Cambridge, UK, 1949), Cambridge University Mathematics Laboratory, pp. 67–69.

[722] Turn, R., Ed. *Advances in Computer System Security*, vol. 2 of *Communications Security Library*. Artech House, Boston, MA, USA, 1984.

[723] Turn, R., Ed. *Advances in Computer System Security*, vol. 3 of *Communications Security Library*. Artech House, Boston, MA, USA, 1988.

[724] Tygar, J. D., and Yee, B. Dyad: A System for Using Physically Secure Coprocessors. Tech. Rep. CMU-CS-91-140R, Carnegie Mellon University School of Computer Science, Pittsburgh, PA, USA, May 1991. A revised version was published as [725].

[725] Tygar, J. D., and Yee, B. Dyad: a System Using Physically Secure Coprocessors. In *Technological Strategies for Protecting Intellectual Property in the Networked Multimedia Environment* (MIT, Program on Digital Open High-Resolution Systems, Jan. 1994), The Journal of the Interactive Multimedia Association

Intellectual Property Project, Coalition for Networked Information, Harvard University John F. Kennedy School of Government Interactive Multimedia Association, pp. 121–152.

[726] ULLMAN, J. D., AHO, A. V., AND HOPCROFT, J. E. *The Design and Analysis of Computer Algorithms.* Addison-Wesley, Reading, MA, USA, 1974.

[727] UNITED STATES DEFENSE INTELLIGENCE AGENCY. *DIA Manual 65-19: Standard Security Markings*, 1984. Unclassified.

[728] UNITED STATES DEFENSE INTELLIGENCE AGENCY. *DDS 2600-4537-87: Military Intelligence Integrated Data System Definition and Specification Document; Version 1*, 1987. Parts I, II, III, and IV are classified CONFIDENTIAL.

[729] UNITED STATES DEFENSE INTELLIGENCE AGENCY. *DDS 2600-5502-87: Security Requirements for System High and Compartmented Mode Workstations; Version 1*, 1987.

[730] UNITED STATES DEFENSE INTELLIGENCE AGENCY. *DDS 2600-6216-91: CMW Labeling Encoding Format*, 1991. Unclassified.

[731] UNITED STATES DEFENSE INTELLIGENCE AGENCY. *DDS 2600-6243-91: CMW Labeling Guidelines*, 1991. Unclassified.

[732] UNITED STATES DEFENSE INTELLIGENCE AGENCY. *DDS 2600-6243-92: Compartmented Mode Workstation Evaluation Criteria; Version 1*, 1992.

[733] UNITED STATES DEPARTMENT OF DEFENSE. *DoD 5200.28-STD: Department of Defense (DoD) Trusted Computer System Evaluation Criteria (TCSEC)*, 1985. Unclassified.

[734] UNITED STATES DEPARTMENT OF DEFENSE. *MIL-STD 2045-48501 Common Security Label (CSL)*, 1994. Unclassified.

[735] United States Department of Defense. *DoD O-5200.1-I: Department of Defense (DoD) Index of Security Classification Guides*, 1996. Classified.

[736] United States Department of Defense. *DoD 5200.1-R: Department of Defense (DoD) Information Security Program*, 1997. Unclassified.

[737] United States Department of Defense. *DoD 5200.1-H: Department of Defense (DoD) Handbook for Writing Security Classification Guidance*, 1999. Unclassified.

[738] United States Joint Chiefs of Staff. *Joint Doctrine for Information Operations*. Washington D.C., USA, Oct. 1998. Unclassified except for appendix A.

[739] Urquhart, A. I. F. The Complexity of Propositional Proofs. *Bulletin of Symbolic Logic 1*, 4 (Dec. 1995), 425–467.

[740] van Benthem, J. *The Logic of Time: a Model-Theoretic Investigation into the Varieties of Temporal Ontology and Temporal Discourse*, 2nd ed. No. 156 in Synthese Library. Reidel / Kluwer Academic Publishers, Dordrecht, the Netherlands, 1991.

[741] van der Hoek, W., and Verbrugge, L. C. Epistemic Logic: A Survey. Tech. Rep. 2001-13, University of Groningen, Groningen, The Netherlands, Dec. 2001.

[742] van Linder, B., van der Hoek, W., and Meyer, J.-J. Actions that Make you Change Your Mind. In *Knowledge and Belief in Philosophy and Artificial Intelligence*, A. Laux and H. Wansing, Eds. Akademie Verlag, Berlin, Germany, 1995, pp. 103–146.

[743] van Vleck, T. H. Personal Communication. E-Mail exchange in which Mr. van Vleck outlined the involvement of Dr. Schell during the early extension of Multics to support multilevel security in project GUARDIAN between 1972 and 1974 which

later led to the creation of the Multics Access Isolation Mechanism., Nov. 2000.

[744] Varian, H. R. System Reliability and Free Riding. In *Workshop on Economics and Information Security at the University of California, Berkeley* (Berkeley, CA, USA, May 2002). Electronic proceedings only.

[745] Venkatraman, B. R., and Newman-Wolfe, R. E. Capacity Estimation and Auditability of Network Covert Channels. In *Proceedings of the 1995 IEEE Symposium on Security and Privacy (SOSP '95)* (Oakland, CA, USA, May 1995), IEEE Computer Society, pp. 186–198.

[746] Vickers Benzel, T. C. Overview of the SCOMP Architecture and Security Mechanisms. Tech. Rep. MTR-9071, The MITRE Corporation, Bedford, MA, USA, Sept. 1983.

[747] Vickers Benzel, T. C. Analysis of a Kernel Verification. In *Proceedings of the 1984 IEEE Symposium on Security and Privacy (SOSP '84)* (Oakland, CA, USA, Apr. 1985), IEEE Computer Society, pp. 125–131.

[748] Vickers Benzel, T. C., and Tevilla, D. A. Trusted Software Verification: A Case Study. In *Proceedings of the 1985 IEEE Symposium on Security and Privacy (SOSP '85)* (Oakland, CA, USA, Apr. 1985), IEEE Computer Society, pp. 14–31.

[749] Voas, J. Protecting against What? The Achilles Heel of Information Assurance. *IEEE Software 16*, 1 (Jan./Feb. 1999), 28–29.

[750] von Neumann, J., and Morgenstern, O. *Theory of Games and Economic Behavior*. Princeton University Press, Princeton, NJ, USA, 1947.

[751] von Willich, M. A Technique with an Information-Theoretic Basis for Protecting Secret Data from Differential Power Attacks. In *Proceedings of Cryptography and Coding, 8th IMA International Conference* (Cirencester, UK, Dec. 2001), B. Honary, Ed.,

vol. 2260 of *Lecture Notes in Computer Science*, Springer Verlag, pp. 44–62.

[752] Voronkov, A., Ed. *Proceedings of the 18th International Conference on Automated Deduction (CADE-18)* (Copenhagen, Denmark, July 2002), vol. 2392 of *Lecture Notes in Computer Science*, Springer Verlag.

[753] Vyssotsky, V. A., Corbató, F. J., and Graham, R. M. Structure of the Multics Supervisor. In *Proceedings of the AFIPS Fall Joint Computer Conference (1965 FJCC)* (Washington D.C., USA, Nov. 1965), vol. 27 part 1, AFIPS, Spartan Books, pp. 203–212.

[754] Walter, K. G. Primitive Models for Computer Security. Tech. Rep. ESD-TR-74-117, Air Force Electronic Systems Division (AFSC), L. G. Hanscom Field, Bedford, MA, USA, Jan. 1974.

[755] Walter, K. G., Gilligan, J. M., Schaen, S. I., Ogden, W. F., Rounds, W. C., Shumway, D. G., Schaeffer, D. D., Biba, K. J., Bradshaw, F. T., and Ames, S. R. Structured Specification of a Security Kernel. In *Proceedings of the International Conference on Reliable Software* (Los Angeles, CA, USA, 1975), ACM Press, pp. 285–293.

[756] Ware, W. H. Security Controls for Computer Systems: Report of Defense Science Board Task Force on Computer Security. Tech. rep., The RAND Corporation, Santa Monica, CA, USA, Feb. 1970. Number R-609. Document was declassified in October 1975 and reissued as R-609/1 with a new introduction in 1979.

[757] Warrender, C., Forrest, S., and Pearlmutter, B. Detecting Intrusions Using System Calls: Alternative Data Models. In *Proceedings of the 1996 IEEE Symposium on Security and Privacy (SOSP '96)* (Oakland, CA, USA, May 1996), IEEE Computer Society Press, pp. 133–145.

[758] WEINGART, S. H. Physical Security for the μABYSS System. In *Proceedings of the 1987 IEEE Symposium on Security and Privacy (SOSP '87)* (Oakland, CA, USA, May 1987), IEEE Computer Society, pp. 52–58.

[759] WEINGART, S. H. Physical Security Devices for Computer Subsystems: A Survey of Attacks and Defenses. In *Cryptographic Hardware and Embedded Systems – CHES 2000. Proceedings of the Second International Workshop* (Worcester, MA, USA, Aug. 2000), Ç. Koç and C. Paar, Eds., vol. 1965 of *Lecture Notes in Computer Science*, Springer Verlag, pp. 302–317.

[760] WEINGART, S. H., WHITE, S., ARNOLD, W., AND DOUBLE, G. An Evaluation System for the Physical Security of Computing Systems. In *Proceedings 6th Annual Computer Security Applications Conference (ACSAC'90)* (Tucson, AZ, USA, Dec. 1990), IEEE Computer Society Press, pp. 232–243.

[761] WEISSMAN, C. Security controls in the ADEPT-50 time-sharing system. In *Proceedings of the AFIPS Fall Joint Computer Conference (1969 FJCC)* (Las Vegas, NV, USA, Nov. 1969), vol. 35, AFIPS, AFIPS Press, pp. 119–133.

[762] WEISSMAN, C. BLACKER: Security for the DDN. Examples of A1 Security Engineering Trades. In *Proceedings of the 1988 IEEE Symposium on Security and Privacy (SOSP '88)* (Oakland, CA, USA, Apr. 1988), IEEE Computer Society, pp. 26–26. The publication release of this paper was rescinded prior to the 1988 SOSP; it was published after publication release was re-obtained in the 1992 SOSP proceedings volume.

[763] WENSLEY, J. H., LAMPORT, L., GOLDBERG, J., GREEN, M. W., LEVITT, E. W., MELLIAR-SMITH, P. M., SHOSTAK, R. E., AND WEINSTOCK, C. B. SIFT: Design and Analysis of a Fault-Tolerant Computer for Aircraft Control. *Proceedings of the IEEE 66*, 10 (Oct. 1982), 1240–1255.

[764] White, B. S., Walker, M., Humphrey, M., and Grimshaw, A. S. LegionFS: A Secure and Scalable File System Supporting Cross-Domain High-Performance Applications. In *Proceedings of Supercomputing 2001* (Denver, CO, USA, Nov. 2001), ACM Press.

[765] White, S. R., and Comerford, L. ABYSS: A Trusted Architecture for Software Protection. In *Proceedings of the 1987 IEEE Symposium on Security and Privacy (SOSP '87)* (Oakland, CA, USA, May 1987), IEEE Computer Society, pp. 38–51.

[766] White, S. R., Weingart, S. H., Arnold, W. C., and Palmer, E. R. Introduction to the Citadel Architecture: Security in Physically Exposed Environments. Tech. Rep. RC 16672, IBM T.J. Watson Research Center, Yorktown Heights, NY, USA, May 1991.

[767] White, Jr., F. E. *Data Fusion Lexicon.* Joint Directors of Laboratories, Technical Panel for C^3, Data Fusion Subpanel, Naval Ocean Systems Center, San Diego, CA, USA, 1987.

[768] Wijesekera, D., and Jajodia, S. Policy Algebras for Access Control – The Propositional Case. In *Proceedings of the 8th ACM Conference on Computer and Communications Security* (Philadelphia, PA, USA, Nov. 2001), ACM Press, pp. 38–47.

[769] Willard, S. *General Topology.* Addison-Wesley, Reading, MA, USA, 1968.

[770] Wittbold, J. T., and Johnson, D. M. Information Flow in Nondeterministic Systems. In *Proceedings of the 1990 IEEE Symposium on Security and Privacy (SOSP '90)* (Oakland, CA, USA, May 1990), IEEE Computer Society, pp. 144–161.

[771] Woo, T. Y. C., and Lam, S. S. Authorization in Distributed Systems: A Formal Approach. In *Proceedings of the 1992 IEEE Symposium on Security and Privacy (SOSP '92)* (Oakland, CA, USA, Apr. 1992), IEEE Computer Society, pp. 33–50.

[772] Woodcock, J., and Davies, J. *Using Z*. Prentice Hall International Series in Computer Science. Prentice Hall, London, UK, 1996.

[773] Woodcock, J. C. P. The Rudiments of Algorithm Refinement. *The Computer Journal 35*, 5 (1992), 441–450.

[774] Woodward, J. P. L. Applications for Multilevel Secure Operating Systems. In *Proceedings of the National Computer Conference* (New York, NY, USA, Nov. 1979), vol. 48, AFIPS, AFIPS Press, pp. 319–328.

[775] Wray, J. C. An Analysis of Covert Timing Channels. In *Proceedings of the 1990 IEEE Symposium on Security and Privacy (SOSP '90)* (Oakland, CA, USA, Apr. 1990), IEEE Computer Society, pp. 180–187.

[776] Wright, C., Cowan, C., Smalley, S., Morris, J., and Kroah-Hartman, G. Linux Security Modules: General Security Support for the Linux Kernel. In *Proceedings of the 11th USENIX Security Symposium* (San Francisco, CA, USA, Aug. 2002), USENIX, Ed., pp. 17–31.

[777] Wulf, W. A., Cohen, E., Corwin, W., Jones, A., Levin, R., Pierson, C., and Pollack, F. HYDRA: The Kernel of a Multiprocessor System. *Communications of the Association for Computing Machinery 17*, 6 (June 1974), 337–345.

[778] Wulf, W. A., Wang, C., and Kienzle, D. A New Model of Security for Distributed Systems. In *Proceedings of the Workshop on New Security Paradigms* (Great Langdale, UK, Sept. 1997), ACM Press, pp. 34–44.

[779] Yee, B. *Using Secure Coprocessors*. PhD thesis, Carnegie Mellon University, Pittsburgh, PA, USA, May 1994. Computer Science Technical Report CMU-CS-94-149.

[780] Youman, C. E., Coyne, E. J., and Jaeger, T., Eds. *Proceedings of the Second ACM Workshop on Role-Based Access Control (RBAC '97)* (Fairfax, VA, USA, Nov. 1997).

[781] Youman, C. E., and Jaeger, T., Eds. *Proceedings of the Third ACM Workshop on Role-Based Access Control (RBAC '98)* (Fairfax, VA, USA, Oct. 1998).

[782] Youman, C. E., and Osborn, S., Eds. *Proceedings of the Fourth ACM Workshop on Role-Based Access Control (RBAC '99)* (Fairfax, VA, USA, Oct. 1999).

[783] Youman, C. E., Sandhu, R. S., and Coyne, E. J., Eds. *Proceedings of the First ACM Workshop on Role-Based Access Control (RBAC '95)* (Gaithersburg, MD, USA, Nov. 1995).

[784] Young, W. D., Boebert, W. E., and Kain, R. Y. Proving a Computer System Secure. *Scientific Honeyweller 6*, 2 (July 1985), 18–27. Reprinted in [723].

[785] Yu, C.-F., and Gligor, V. D. A Specification and Verification Method for Preventing Denial of Service. *IEEE Transactions on Software Engineering 16*, 6 (June 1990), 581–592. Special Section on Security and Privacy.

[786] Zadeh, L. A., and Kacprzyk, J., Eds. *Fuzzy Logic for the Management of Uncertainty*. John Wiley & Sons, New York, NY, USA, 1992.

[787] Zadok, E. Stackable File Systems as a Security Tool. Tech. Rep. CUCS-036-99, Columbia University Department of Computer Science, New York, NY, USA, Dec. 1999.

[788] Zadok, E. *FiST: A System for Stackable File System Code Generation*. PhD thesis, Columbia University, New York, NY, USA, May 2001.

[789] Zadok, E., Badulescu, I., and Shender, A. CryptFS: A Stackable VNode Level Encryption File System. Tech. Rep. CUCS-021-98, Columbia University Department of Computer Science, New York, NY, USA, June 1998.

[790] Zadok, E., Badulescu, I., and Shender, A. Extending File Systems Using Stackable Templates. In *Proceedings of the 1999 USENIX Annual Technical Conference* (Berkeley, CA, USA, June 1999), USENIX Association, Ed., USENIX, pp. 57–70.

[791] Zadok, E., and Nieh, J. FiST: A Language for Stackable File Systems. In *Proceedings of the 2000 USENIX Annual Technical Conference* (Berkeley, CA, USA, June 2000), USENIX Association, Ed., USENIX, pp. 55–70.

[792] Zakinthinos, A., and Lee, E. S. A General Theory of Security Properties. In *Proceedings of the 1997 IEEE Symposium on Security and Privacy* (Oakland, CA, USA, May 1997), IEEE Computer Society, pp. 94–102.

[793] Zhao, J., and Koch, E. Embedding Robust Labels Into Images For Copyright Protection. In *Intellectual Property Rights and New Technologies: Proceedings of the KnowRight '95 Conference* (Vienna, Austria, Aug. 1995), K. Brunnstein and P. P. Sint, Eds., vol. 82 of *Schriftenreihe der Österreichischen Computer Gesellschaft*, R. Oldenbourg Verlag, pp. 242–252.

[794] Zwicky, E. D., Cooper, S., and Chapman, D. B. *Building Internet Firewalls*, 2nd ed. O'Reilly & Associates, Newton, MA, USA, 2000.

INDEX

Colophon

The body of the book was typeset in 10pt URW Palladio based on the 1948 Linotype Palatino design by Hermann Zapf. Additional typefaces used are a 10pt URW Nimbus Sans based on the 1961 Neue Haas Grotesk designed by Max Miedinger 1951–53 and the 10pt 1994 URW Typewriter T Regular (trad.). The mathematical typefaces are based on URW Palladio with modifications by Young Ryu and the author. Additional symbols from the 1989 AMS Euler by Hermann Zapf were also used.